AF633813

Series in BioEngineering

The Series in Bioengineering serves as an information source for a professional audience in science and technology as well as for advanced students. It covers all applications of the physical sciences and technology to medicine and the life sciences. Its scope ranges from bioengineering, biomedical and clinical engineering to biophysics, biomechanics, biomaterials, and bioinformatics.

Indexed by WTI Frankfurt eG, zbMATH.

Almir Badnjević · Lemana Spahić

Biomedical Signals and Systems

Acquisition, Processing, and Clinical Insights

Foreword by Prof. Dr. Nenad Filipović

Springer

Almir Badnjević
Agency for Identification Documents, Registers, and Data Exchange of Bosnia and Herzegovina
Banja Luka, Bosnia and Herzegovina

Lemana Spahić
Research Institute Verlab
Sarajevo, Bosnia and Herzegovina

ISSN 2196-8861 ISSN 2196-887X (electronic)
Series in BioEngineering
ISBN 978-3-032-10981-1 ISBN 978-3-032-10982-8 (eBook)
https://doi.org/10.1007/978-3-032-10982-8

© The Editor(s) (if applicable) and The Author(s), under exclusive license to Springer Nature Switzerland AG 2026

This work is subject to copyright. All rights are solely and exclusively licensed by the Publisher, whether the whole or part of the material is concerned, specifically the rights of translation, reprinting, reuse of illustrations, recitation, broadcasting, reproduction on microfilms or in any other physical way, and transmission or information storage and retrieval, electronic adaptation, computer software, or by similar or dissimilar methodology now known or hereafter developed.
The use of general descriptive names, registered names, trademarks, service marks, etc. in this publication does not imply, even in the absence of a specific statement, that such names are exempt from the relevant protective laws and regulations and therefore free for general use.
The publisher, the authors and the editors are safe to assume that the advice and information in this book are believed to be true and accurate at the date of publication. Neither the publisher nor the authors or the editors give a warranty, expressed or implied, with respect to the material contained herein or for any errors or omissions that may have been made. The publisher remains neutral with regard to jurisdictional claims in published maps and institutional affiliations.

This Springer imprint is published by the registered company Springer Nature Switzerland AG
The registered company address is: Gewerbestrasse 11, 6330 Cham, Switzerland

If disposing of this product, please recycle the paper.

Foreword

Biomedical engineering has rapidly evolved into one of the most dynamic and impactful fields of science and technology. At its core lies the study of biomedical signals, expressions of physiological activity that enable us to understand, quantify, and intervene in human health. From the first electrocardiographic tracings to today's wearable devices and AI-driven analytics, biomedical signals have shaped both fundamental knowledge and clinical practice. Their importance cannot be overstated: they are the bridge between biological complexity and technological innovation.

This book, *Biomedical Signals and Systems—Acquisition, Processing, and Clinical Insights*, provides a comprehensive and systematic exploration of this vital domain. By combining theoretical principles with practical applications, it offers readers not only the tools for signal acquisition and processing but also the context to appreciate their clinical and societal significance. The structure of the text—from fundamental definitions to advanced applications in cardiology, neurology, respiratory medicine, and imaging—reflects the breadth of biomedical engineering as well as its capacity to transform healthcare.

Equally important, the book does not treat biomedical signals in isolation but situates them within the ethical and regulatory frameworks that govern modern medicine. In doing so, it recognizes that innovation must always be accompanied by responsibility. This perspective is particularly relevant today, as advances in digital health, machine learning, and personalized medicine expand both the possibilities and the challenges of biomedical signal analysis.

The value of this work lies in its clarity, depth, and interdisciplinarity. It speaks to students seeking a foundation, to researchers advancing the science, and to clinicians applying these methods in practice. By bridging engineering and medicine, this book highlights how knowledge of signals and systems can lead to safer, more effective, and more equitable healthcare.

It is with great anticipation that I commend *Biomedical Signals and Systems* to its readers. The insights it provides will not only aid in mastering current technologies but also inspire the next generation of innovations that define the future of biomedical engineering.

Prof. Dr. Nenad Filipović
Department of Applied Mechanics
and Automatic Control
University of Kragujevac
Kragujevac, Serbia

Preface

The field of biomedical engineering stands at the intersection of engineering, medicine, and the life sciences, seeking to transform fundamental physiological processes into actionable knowledge for the advancement of healthcare. Among its most critical domains is the study of biomedical signals, time-varying representations of the body's functions that provide insight into health, disease, and treatment response. This book, *Biomedical Signals and Systems—Acquisition, Processing, and Clinical Insights*, is designed to introduce readers to the principles, techniques, and applications of biomedical signal acquisition, processing, and analysis, bridging theory and practice for students, researchers, and healthcare professionals alike.

The motivation for this book arises from the growing integration of signal-based technologies in clinical practice, from electrocardiography and electroencephalography to respiratory monitoring, ultrasound imaging, and blood pressure analysis. With the advent of digital health, wearable devices, and artificial intelligence, biomedical signals are no longer confined to specialized laboratories or hospitals but are becoming essential components of everyday healthcare and personalized medicine. As such, a systematic understanding of biomedical signals is crucial for those who wish to engage with modern medical technologies and contribute to their development.

The book is structured into eight chapters. Chapter 1 introduces the fundamental concepts of biomedical signals, their acquisition, processing, and the ethical-legal dimensions of their use. Chapter 2 focuses on signal acquisition and processing techniques, while subsequent chapters address cardiac, respiratory, and neurological signals, as well as ultrasound and blood pressure analysis. The final Chap. 8 provides a review and practice problems to reinforce comprehension and encourage application of the material. Each chapter balances theoretical foundations with practical examples and clinical applications, reflecting the inherently interdisciplinary nature of biomedical engineering.

Rather than serving only as a university level textbook and professional reference, this book is intended as a bridge between disciplines, linking engineering principles with clinical needs and technological innovation. By engaging with its content, readers are encouraged not only to master the fundamentals of biomedical signal analysis but also to envision how these methods can shape the future of diagnostics, monitoring, and personalized care.

Banja Luka, Bosnia and Herzegovina — Almir Badnjević
Sarajevo, Bosnia and Herzegovina — Lemana Spahić

Acknowledgements The authors wish to express their sincere appreciation to Sara Deumić, M.Sc. for her dedicated work in the process of formatting and technical finalization of this book. Her attention to detail and commitment to quality greatly contributed to the presentation and coherence of the final manuscript.

We are deeply grateful to our students, whose curiosity, enthusiasm, and insightful questions continuously inspire and motivate our academic work. Their engagement has been instrumental in shaping many of the ideas and examples presented throughout this book.

Finally, we extend our heartfelt thanks to our colleagues and families for their patience, understanding, and constant support during the demanding process of writing and editing. Their encouragement and belief in our work provided the foundation on which this project was completed.

Competing Interests The authors have no competing interests to declare that are relevant to the content of this manuscript.

Contents

Chapter 1
Introduction to Biomedical Signals and Systems

This section introduces the foundations of biomedical signals and their importance in healthcare. In Sect. 1.1 it defines biomedical signals as time-varying manifestations of physiological activity across electrical, mechanical, acoustic, optical, and thermal domains, emphasizing their nonlinear, variable, and noise-sensitive nature. The principles of biomedical signal acquisition are then examined (Sect. 1.2), tracing the processes of sensing, transduction, conditioning, amplification, and digitization that enable biological activity to be transformed into analyzable data. Signal processing is presented as a cornerstone of biomedical engineering (Sect. 1.3), with attention to filtering, segmentation, feature extraction, and diagnostic applications. Ethical and legal issues (Sect. 1.4) are addressed, focusing on patient safety, privacy, regulatory compliance, and responsible use of biomedical technologies. Together, these chapters provide the conceptual and methodological basis for understanding and applying biomedical signal analysis.

1.1 What are Biomedical Signals

In the vast and intricate world of biomedical engineering, few concepts are as fundamental as biomedical signals. These signals serve as the language through which the body communicates its physiological processes, revealing insights into health, disease, and the intricate functions that sustain life [1]. Whether it is the rhythmic pulsing of the heart, the electrical symphony of neurons, or the subtle variations in blood pressure, biomedical signals form the basis of modern medical diagnostics and therapeutic interventions [2].

Imagine a physician listening to a patient's heartbeat through a stethoscope. What they hear is more than just sound—it is a representation of a biomedical signal, an auditory manifestation of the heart's mechanical activity. The human heart beats approximately 60–100 times per minute in a healthy adult, with each beat producing

© The Author(s), under exclusive license to Springer Nature Switzerland AG 2026
A. Badnjević and L. Spahić, *Biomedical Signals and Systems*, Series in BioEngineering,
https://doi.org/10.1007/978-3-032-10982-8_1

a distinct sound as blood flows through the chambers and valves. Abnormalities in these sounds, such as murmurs or irregular rhythms, can indicate underlying heart conditions [3].

However, beyond just sound, biomedical signals exist in many other forms. Consider the electrical impulses generated by the heart—these are not audible to the human ear, but they can be recorded using an electrocardiogram (ECG). This device translates the heart's electrical activity into a visual waveform that physicians can analyze. A typical ECG waveform follows a distinct pattern, with characteristic peaks and intervals corresponding to different phases of the cardiac cycle. Any deviation from the usual pattern may indicate arrhythmias, ischemia, or other cardiac disorders [4–6]. Similarly, the brain generates its electrical signals, which can be recorded through electroencephalography (EEG). The brain's activity consists of continuous waves, varying in amplitude and frequency. These waves reflect different states of consciousness, from deep sleep to full alertness. Abnormal EEG patterns can indicate conditions such as epilepsy, sleep disorders, or brain trauma [7, 8].

Biomedical signals extend far beyond the cardiovascular and nervous systems. They encompass every physiological process, from respiratory activity and muscle contractions to temperature fluctuations and metabolic processes. As technology advances, our ability to measure and interpret these signals continues to improve, leading to more accurate diagnoses, better patient monitoring, and increasingly personalized treatments [2, 9].

1.1.1 Understanding Biomedical Signals

At their core, biomedical signals are time-varying representations of physiological phenomena. These signals originate from biological systems and can take various forms, including electrical, mechanical, acoustic, thermal, or even optical [10]. Unlike the signals encountered in other engineering disciplines, biomedical signals are often complex, nonlinear, and highly individualized. They do not follow simple mathematical models and are influenced by a multitude of factors, including genetics, lifestyle, and momentary physiological changes [11]. This variability, while posing challenges in signal analysis, also provides a wealth of information about an individual's health status.

Biomedical signals exhibit several distinct properties that distinguish them from signals encountered in other engineering domains. These characteristics arise due to the inherent complexity of biological systems and their dynamic nature. Understanding these properties is essential for accurate signal acquisition, processing, and interpretation [12].

Biomedical signals are often nonlinear, meaning their behavior cannot be described by simple mathematical equations. This nonlinearity arises due to complex interactions within biological systems, including feedback loops, adaptive responses, and dependencies on multiple variables. For example, heart rate variability (HRV) is influenced by autonomic nervous system activity, which introduces nonlinear fluctuations in cardiac rhythms. Similarly, neuronal signals exhibit chaotic behavior that cannot be predicted using conventional linear models [7].

Biomedical signals exhibit significant variability, both between individuals (inter-individual variability) and within the same individual over time (intra-individual variability). This variability is influenced by factors such as age, genetic makeup, lifestyle, and environmental conditions. Moreover, these signals are non-stationary, meaning their statistical properties—such as mean, variance, and frequency content—change over time. Non-stationarity can result from physiological variations (e.g., fluctuations in heart rate due to stress or exercise), emotional states, or external influences (e.g., ambient temperature or the effects of medication). This characteristic makes biomedical signal processing challenging, requiring adaptive and time-dependent analysis techniques [13].

Many biomedical signals have relatively low amplitudes and are susceptible to external noise and artifacts. For instance, EEG signals, which measure brain activity, are in the microvolt (μV) range and can be easily contaminated by electrical interference from surrounding electronic devices, muscle movements (electromyographic artifacts), or eye blinks. Similarly, ECG signals can be affected by power line interference, electrode movement, and baseline drift caused by respiration. Effective noise reduction techniques, such as filtering and signal averaging, are essential for improving signal quality and ensuring accurate interpretation [14, 15].

Biomedical signals often consist of multiple overlapping components, each originating from different physiological processes. For example, an ECG waveform includes distinct features, such as the P wave (representing atrial depolarization), the QRS complex (representing ventricular depolarization), and the T wave (representing ventricular repolarization). Similarly, respiratory signals contain multiple frequency components associated with different phases of the breathing cycle. The multi-component nature of biomedical signals necessitates advanced signal processing methods, such as wavelet decomposition and independent component analysis, to effectively separate and analyze individual components.

Many biomedical signals exhibit periodic or quasi-periodic behavior, meaning they follow a repeating pattern over time. For example, ECG signals display characteristic rhythms that correspond to heartbeats, while respiratory waveforms follow cyclic patterns of inhalation and exhalation. However, these rhythms may be irregular due to underlying pathological conditions. Atrial fibrillation, for instance, causes irregular heart rhythms, making ECG signals less predictable [9]. Understanding periodicity and detecting abnormalities in rhythmic patterns is crucial for diagnosing and monitoring various medical conditions.

Biomedical signals are highly correlated with physiological states and can change significantly in response to internal and external factors. For example, blood pressure signals fluctuate in response to changes in body position, stress levels, and hydration status. Similarly, EEG signals vary depending on sleep stages, cognitive workload, and neurological disorders. By analyzing these correlations, clinicians can gain valuable insights into a patient's health status, detect early signs of disease, and optimize treatment strategies. Machine learning algorithms and artificial intelligence

(AI)-based approaches are increasingly being used to uncover hidden patterns in biomedical signals, enabling more precise and personalized healthcare interventions [16].

Biomedical signals are typically acquired through sensors and transducers, which convert physiological events into electrical signals that can be recorded, displayed, and analyzed. The process of capturing these signals is often non-invasive, as in the case of ECG, EEG, or pulse oximetry. However, some signals require minimally invasive or invasive methods, such as intracranial EEG, which involves placing electrodes inside the skull to directly measure brain activity [17]. A crucial aspect of biomedical signal analysis is distinguishing useful information from noise and artifacts. Noise can arise from various sources, including muscle activity, power-line interference, and movement artifacts. Filtering techniques and signal processing algorithms are employed to enhance the quality of biomedical signals, enabling clinicians and researchers to interpret the data accurately.

To illustrate the diversity of biomedical signals, let us examine a few common examples in greater detail [2, 18].

1. Electrical Signals

Electrical signals are among the most widely studied biomedical signals, as they reflect the activity of excitable cells such as neurons and muscle fibers. These signals are generated by the movement of ions across cellular membranes, creating voltage changes that can be recorded and analyzed.

- Electrocardiography: The heart generates electrical impulses that coordinate its rhythmic contractions. The ECG records these impulses through electrodes placed on the skin, providing valuable information about heart rate, rhythm, and conduction abnormalities.
- Electroencephalography: The brain's neurons communicate through electrical activity, forming characteristic wave patterns that can be recorded using electrodes placed on the scalp. EEG is widely used in neurology to diagnose epilepsy, monitor sleep disorders, and study brain function.
- Electromyography (EMG): Muscle activity generates electrical signals, which can be measured using EMG. This technique is crucial for diagnosing neuromuscular disorders, evaluating muscle function, and analyzing movement biomechanics (Figs. 1.1 and 1.2).

2. Mechanical Signals

The body's organs and tissues undergo continuous mechanical movements, many of which are crucial for life. These mechanical signals can be recorded to assess physiological function [19].

- Blood Pressure Waveforms: Blood pressure fluctuates as the heart pumps blood through the circulatory system. This signal provides insights into cardiovascular health and can be recorded using non-invasive cuffs or invasive catheter-based methods.

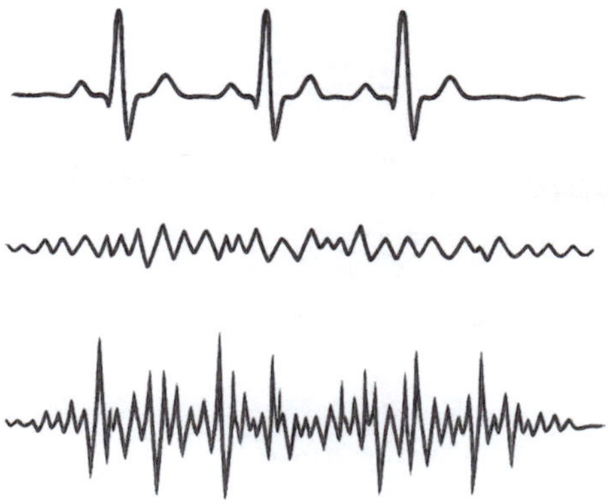

Fig. 1.1 ECG, EEG, and EMG waveforms

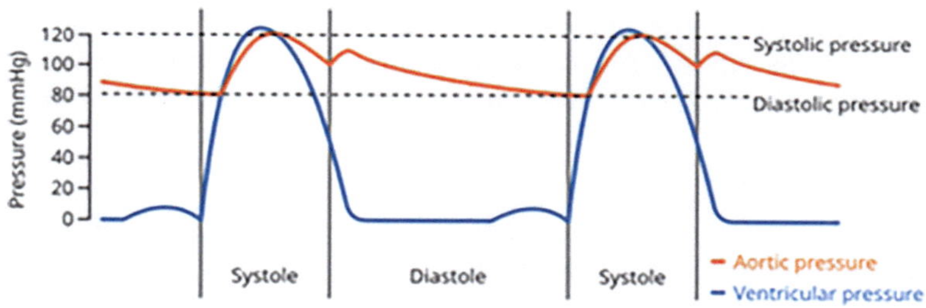

Fig. 1.2 Blood pressure waveforms (https://commons.wikimedia.org/wiki/File:Aortic_and_ventricular_pressure_over_two_cardiac_cycles.svg)

- Respiratory Mechanics: The expansion and contraction of the lungs generate mechanical movements that can be measured using spirometry or respiratory inductance plethysmography. These signals are crucial for diagnosing conditions such as chronic obstructive pulmonary disease (COPD) and sleep apnea.
- Ballistocardiography (BCG): The subtle movements of the body caused by the ejection of blood from the heart can be recorded using specialized sensors. BCG is a non-invasive technique for monitoring cardiac function.

3. Acoustic Signals

Acoustic signals in biomedicine arise from physiological activities that produce sound. These sounds can be recorded and analyzed to assess organ function.

- Heart Sounds (Phonocardiography): The heart produces distinct sounds during the opening and closing of valves. Phonocardiography (PCG) records these sounds, allowing for the detection of murmurs and valvular disorders.

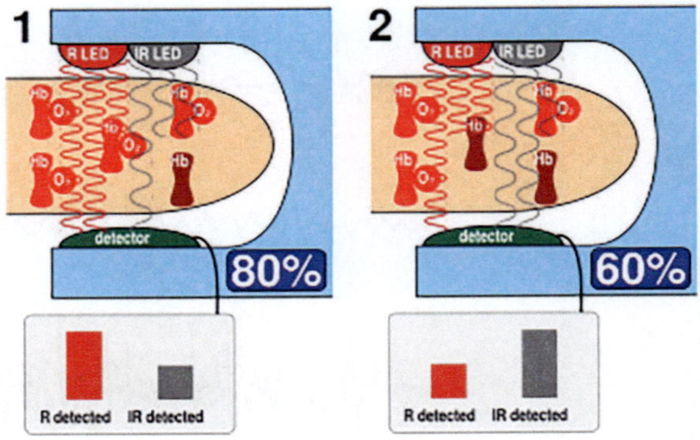

Fig. 1.3 An illustration of a pulse oximeter (https://commons.wikimedia.org/wiki/File:Pulse_oximetry_principle.svg)

- Lung Sounds (Auscultation): Respiratory sounds, such as wheezing, crackles, and rales, provide information about lung function. These sounds are commonly analyzed in the diagnosis of conditions such as asthma and pneumonia.
- Vocal Cord Vibrations: Voice production involves the vibration of the vocal cords, generating an acoustic signal that can be recorded and analyzed for the detection of speech disorders.
- Optical and Thermal Signals

Advancements in biomedical technology have enabled the use of optical and thermal signals for non-invasive diagnostics.

- Pulse Oximetry: This technique uses the light absorption properties of hemoglobin to estimate blood oxygen saturation. A small clip-on device, typically placed on the fingertip, emits red and infrared light to measure oxygen levels.
- Infrared Thermography: The human body emits infrared radiation, which varies in intensity according to its temperature. Infrared thermography detects heat patterns, helping identify inflammation, vascular disorders, or infections.
- Near-Infrared Spectroscopy (NIRS): This optical technique measures tissue oxygenation and blood flow, commonly used in brain imaging and studies of muscle metabolism (Fig. 1.3).

1.2 Overview of Biomedical Signal Acquisition

In the field of biomedical engineering, the ability to acquire and interpret physiological signals has revolutionized healthcare, enabling doctors to diagnose conditions, monitor patients, and even predict health risks before symptoms appear. The acquisition of biomedical signals is a process that turns biological activity into readable and analyzable data, allowing the unseen functions of the body to become visible through technology [20]. Every heartbeat, every neural impulse, every breath, and every minute change in temperature or pressure can be captured, converted, and processed to reveal critical information about human health.

At the core of this process lies a fundamental sequence of steps. The first is sensing, where a device comes into contact with the biological system and detects the physiological event of interest. Whether it is an electrical signal from the heart, a pressure change in the lungs, or the sound of blood flowing through arteries, the process always begins with detection. The second step is transduction, where the detected physiological activity is converted into an electrical or optical signal. The third step is signal conditioning, a crucial process where weak and often noisy signals undergo amplification and filtering to enhance their quality. Finally, the signal must be digitized so that it can be stored, analyzed, and interpreted using computational techniques [21–23]. Without these steps, modern biomedical signal analysis would not be possible, and the vast landscape of medical diagnostics and patient monitoring would remain confined to subjective observation rather than objective measurement (Figs. 1.4 and 1.5).

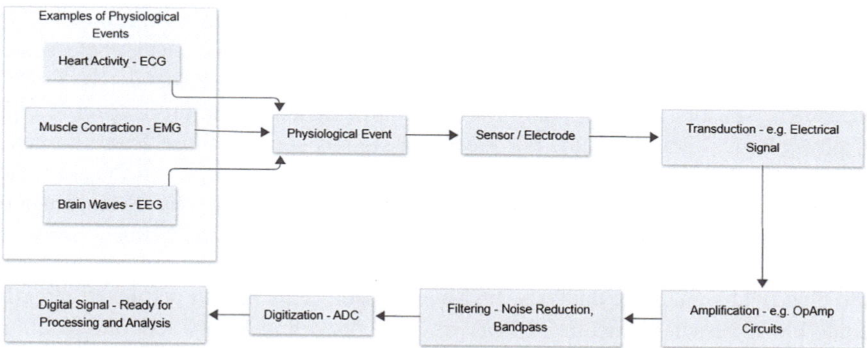

Fig. 1.4 The process of biomedical signal acquisition

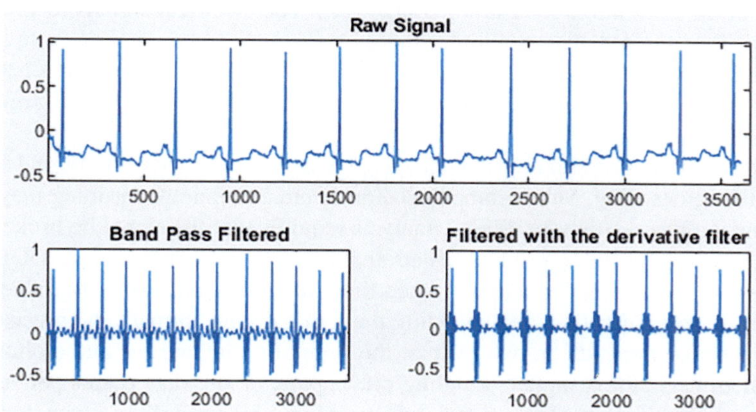

Fig. 1.5 A comparison of a raw ECG signal before and after amplification and filtering

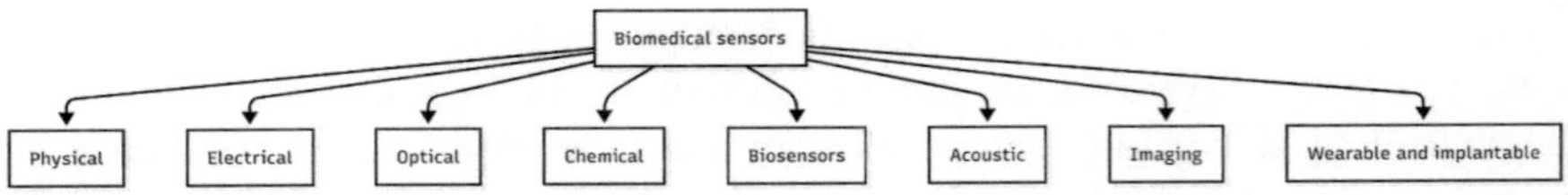

Fig. 1.6 Different types of biomedical sensors

To understand the process of biomedical signal acquisition, consider an ECG. When electrodes are placed on the skin, they detect the minute electrical impulses generated by the heart's activity. These impulses are far too weak to be analyzed directly, so they pass through an amplifier that strengthens the signal. Unwanted noise, such as interference from muscle activity or external electrical sources, is filtered out, ensuring that the ECG waveform remains clear and accurate [24]. The signal is then digitized, stored, and displayed on a monitor for a physician to analyze. This same fundamental process applies to a wide range of biomedical signals, from brain waves measured by EEG to respiratory sounds recorded through digital stethoscopes [25].

The first point of contact in biomedical signal acquisition is the sensor or electrode, a critical component that translates biological events into signals that can be further processed. Sensors come in many forms, each designed to capture a specific type of physiological activity [26]. For electrical signals such as those from the heart or brain, conductive electrodes are used to detect voltage changes on the skin. For optical measurements, such as those used in pulse oximetry, light sensors analyze how blood absorbs and reflects light to determine oxygen saturation. Acoustic sensors, such as digital stethoscopes, capture sound waves from the body and convert them into electrical signals for analysis. Invasive pressure sensors can be placed directly inside blood vessels to measure blood pressure, providing highly accurate readings in critical care situations (Fig. 1.6).

Once a physiological signal is detected, it often requires modification before it can be adequately analyzed. Biomedical signals are typically weak, requiring amplification to bring them into a measurable range [27]. The electrical signals recorded by an EEG, for example, have amplitudes in the microvolt range, making them highly susceptible to interference from external electrical sources and even the patient's muscle movements. Amplification boosts the signal strength while filtering removes unwanted noise [24, 28].

After amplification and filtering, the signal must be converted into a digital format for further processing. Most biomedical signals are continuous, meaning they exist in an analog form. However, digital analysis requires that the signal be broken into discrete points through a process called analog-to-digital conversion (ADC). This step involves sampling the signal at fixed time intervals, quantizing the values into numerical representations, and encoding them in a digital format. The precision of this conversion depends on two factors: the sampling rate and the bit resolution of the system [21, 29]. A higher sampling rate captures more data points per second, preserving finer details of the signal, while a higher bit resolution allows for more precise numerical representation (Fig. 1.7).

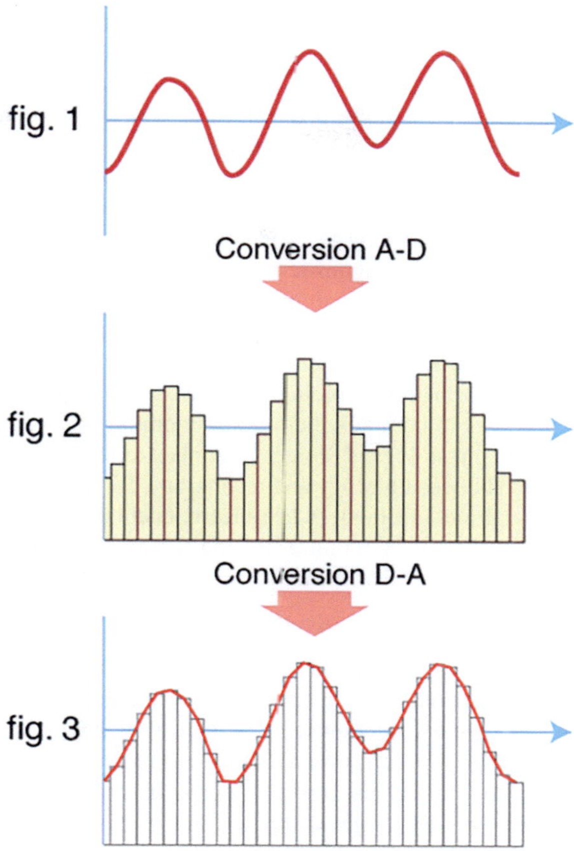

Fig. 1.7 A visual representation of analog-to-digital conversion (https://commons.wikimedia.org/wiki/File:Conversion_AD_DA.png)

Different biomedical signals require different acquisition techniques depending on their source, strength, and frequency characteristics. For example, ECG systems record heart activity using electrodes placed on the skin, while EMG systems measure muscle activity by detecting electrical impulses from contracting muscles. EEG systems capture brain activity through electrodes positioned on the scalp, requiring advanced filtering techniques to distinguish neural signals from background noise [26, 30]. Invasive methods, such as intracranial EEG, involve placing electrodes directly on the brain's surface to obtain more precise recordings.

Beyond electrical signals, other biomedical signals rely on different sensing techniques. Respiratory signals, for example, can be captured using airflow sensors that detect the movement of air in and out of the lungs. Photoplethysmography (PPG) utilizes optical sensors to measure changes in blood volume in tissues, providing valuable insights into heart rate and blood oxygen levels. Blood pressure measurements use either non-invasive cuff-based sensors or invasive catheter-based pressure transducers for continuous monitoring [31]. Each of these acquisition methods has its advantages and limitations, requiring careful selection based on the clinical application.

As technology advances, biomedical signal acquisition continues to evolve, integrating new methods for real-time, non-invasive, and wireless monitoring [30, 32, 33]. Wearable devices now enable the continuous acquisition of physiological data, allowing for remote healthcare monitoring of patients with chronic diseases. Miniaturized sensors embedded in smartwatches, clothing, and even implantable devices provide long-term health tracking, reducing the need for hospital visits while improving diagnostic accuracy. The integration of artificial intelligence further enhances signal acquisition by automating noise reduction, detecting anomalies, and predicting health risks before they become clinically apparent.

The acquisition of biomedical signals is at the heart of modern medicine, enabling the transformation of biological processes into quantifiable data. From simple ECG recordings to complex multimodal monitoring systems, the ability to measure and analyze physiological signals continues to shape the future of healthcare, bridging the gap between biology and technology. Biomedical signal acquisition systems are essential tools in modern healthcare, enabling the collection of vital physiological data for diagnosis, monitoring, and research. Below is an overview of some of the key devices and systems used in acquiring biomxcx (Table 1.1).

Sensing and Signal Detection

The first step in biomedical signal acquisition is sensing, where a device interacts with the human body to detect a physiological event. Sensors play a pivotal role in this process by converting biological activities into electrical or optical signals that can be further processed. These sensors are designed to detect a wide range of physiological signals, from the electrical impulses generated by the heart and brain to the pressure changes within the arteries or the sound of blood flowing through the body.

In the case of electrical signals, such as those from the heart, brain, and muscles, conductive electrodes are typically employed. These electrodes detect voltage changes on the skin's surface and convert them into measurable electrical signals. In ECG systems, for example, electrodes placed on the skin detect the heart's electrical impulses and relay them to an amplifier for further processing. The amplified signals can then be analyzed to assess the heart's rhythm, identify arrhythmias, and diagnose various cardiac conditions [34]. For optical measurements, such as those used in pulse oximetry, light sensors analyze how blood absorbs and reflects light. This technique enables the non-invasive measurement of blood oxygen levels, which is critical for assessing respiratory function, particularly in patients with respiratory diseases or those under anesthesia. The light emitted by the sensor passes through the skin and underlying tissues, and the amount of light reflected is used to estimate the oxygen saturation in the blood [12, 35].

Another important category of sensors is acoustic sensors, which capture sound waves from the body. Digital stethoscopes are a prime example of this technology, enabling clinicians to listen to heartbeats, lung sounds, and other bodily noises.

Table 1.1 Overview of biomedical signals and components of their acquisition system

Device/System	Purpose	Components	Key functionality	Applications
Electrocardiogram	Measures the electrical activity of the heart	Electrodes, amplifier, filter, digitizer	Detects heart electrical impulses, used to diagnose heart diseases like arrhythmias and ischemia	Cardiology, heart disease diagnostics, and ICU monitoring
Electroencephalogram	Measures electrical activity in the brain	Electrodes, amplifier, filter, ADC	Captures brain waves to diagnose neurological conditions like epilepsy, sleep disorders, and brain injuries	Neurology, sleep studies, brain-computer interfaces
Electromyogram	Measures electrical activity in muscles	Surface or needle electrodes, amplifier, filter	Records muscle electrical activity, used to diagnose neuromuscular disorders and muscle function	Neurology, physical therapy, sports medicine
Pulse oximeter	Measures blood oxygen saturation and pulse rate	Light emitter, photodetector, signal processor	Uses light to measure blood volume changes and determine oxygen saturation levels in blood	Respiratory medicine, anesthesia, and COPD monitoring
Blood pressure monitor	Measures blood pressure	Cuff, pressure transducer, amplifier, digitizer	Measures pressure in arteries, assisting in diagnosing hypertension and cardiovascular health	Cardiology, primary care, and ICU monitoring

(continued)

Table 1.1 (continued)

Device/System	Purpose	Components	Key functionality	Applications
Respiratory flow sensor	Measures airflow and lung volumes	Pressure sensors, flow transducers, and signal processors	Detects airflow in and out of the lungs, used to diagnose lung diseases like asthma and COPD	Pulmonology, sleep medicine, emergency care
Invasive pressure sensors	Measures internal physiological pressures	Pressure sensor, catheter, signal processor	Provides accurate internal pressure measurements, proper for critically ill patients	Intensive care, surgery, cardiac monitoring
Magnetoencephalography	Measures magnetic fields from brain activity	Superconducting quantum interference devices, magnetometers	Provides detailed brain activity mapping, especially for motor control and sensory processing	Neuroscience, epilepsy localization, and brain surgery planning
Cardiac event monitor	Monitors heart activity over a long period	Electrodes, transmitter, storage device, data analysis software	Long-term heart rhythm monitoring is helpful for detecting arrhythmias and other heart conditions	Cardiology, long-term cardiac monitoring, post-heart attack rehabilitation
Wearable health devices	Continuously monitors heart rate, ECG, SpO_2, etc	Integrated sensors (PPG, ECG electrodes), wireless communication	Real-time health data collection for personalized healthcare, and early detection of health issues	Health tracking, chronic disease management, preventive healthcare, and telemedicine
Ultrasound	Uses sound waves to create images of internal organs and blood flow	Transducer, signal processor, imaging software	Visualizes soft tissues and detects anomalies such as tumors, cardiac conditions, and fetal development	Obstetrics, cardiology, general diagnostics, emergency care

These sounds are converted into electrical signals for further analysis, helping physicians detect conditions such as heart murmurs, abnormal lung sounds, and arterial blockages [36].

Transduction: Converting Biological Signals into Electrical Signals

Once a sensor has detected a physiological event, the next step is transduction, where the biological signal is converted into an electrical signal that can be processed, amplified, and analyzed. In many cases, the signal detected by the sensor is weak, requiring amplification to make it suitable for further analysis. For example, the electrical signals recorded by an EEG are typically in the microvolt range, meaning they are too faint to be analyzed directly. The EEG electrodes detect the small voltage changes caused by brain activity, but these signals must be amplified to a level that allows them to be observed and interpreted [37].

Transduction can take various forms depending on the type of signal being acquired. In electrical signals, such as those from the heart, the signal is typically converted into an analog electrical signal, which is then amplified to increase its strength. Similarly, in optical systems such as pulse oximeters, the sensor detects changes in light intensity caused by the passage of light through the blood vessels. These light changes are then converted into electrical signals that can be processed and analyzed to determine blood oxygen saturation levels [38].

Signal Conditioning: Enhancing Signal Quality

One of the most critical steps in biomedical signal acquisition is signal conditioning, which involves amplifying and filtering the raw signal to improve its quality. In real-world environments, signals are often contaminated by noise from various sources, such as muscle movements, electrical interference, and environmental factors. If left unaddressed, this noise can obscure the accurate physiological signal, making it difficult to analyze or interpret [8].

Signal conditioning involves a series of processes designed to clean up the signal. Amplification is used to boost the strength of weak signals, while filtering removes unwanted noise. In the case of an ECG, for example, the electrical impulses generated by the heart are typically weak. They can be easily obscured by muscle activity or external electrical sources. By amplifying the signal and applying filters, clinicians can obtain a clean, clear ECG waveform that accurately reflects the heart's electrical activity [39]. Other forms of signal conditioning are also employed to enhance the signal quality. For example, in EEG systems, filtering techniques are employed to remove artifacts caused by muscle activity or eye movements. These filters enable the isolation of the brain's electrical activity from noise, providing a more accurate representation of brain function.

Digitization: Converting Analog Signals to Digital Format

Once a signal has been conditioned and cleaned, it must be converted into a digital format for further analysis. Biomedical signals are often continuous (analog) in nature, meaning that they exist as continuous waveforms. However, to process and analyze the data using computational techniques, the signal must be broken down into discrete data points through a process known as ADC.

ADC involves sampling the analog signal at fixed time intervals, quantizing the values into numerical representations, and encoding them into a digital format. The conversion process allows the signal to be stored, analyzed, and manipulated using digital signal processing (DSP) techniques. The precision of the conversion depends on factors such as the sampling rate and bit resolution. A higher sampling rate captures more data points per second, preserving finer details of the signal, while a higher bit resolution ensures that the numerical representation of the signal is as accurate as possible [40, 41].

For example, an ECG waveform is typically sampled at a rate of several hundred to thousands of samples per second. Each sample represents a snapshot of the heart's electrical activity at a specific moment in time. These discrete data points are then analyzed to assess the heart's rhythm, identify abnormalities, and diagnose conditions such as arrhythmias or ischemia.

Sampling, Quantization, and Filtering

Sampling is one of the cornerstones of signal processing, transforming continuous signals into a digital format. This transformation is essential for modern biomedical systems, which rely on digital signal processing for accurate analysis and decision-making [28].

Understanding Sampling Rate

The sampling rate (or sampling frequency) determines how often a signal is sampled, or in other words, how frequently measurements are taken from the continuous-time signal. The rate at which the signal is sampled should ideally exceed the Nyquist rate (twice the highest frequency component of the signal) to avoid aliasing.

Biomedical signals typically exhibit a wide range of frequencies. For instance, ECG signals generally range from 0.05 Hz to 100 Hz. In contrast, EEG signals can span frequencies from 0.5 Hz to 100 Hz or more. Therefore, understanding the frequency characteristics of each type of signal is crucial for selecting the appropriate sampling rate.

The Nyquist Rate in Biomedical Signals

To avoid aliasing, it is necessary to choose a sampling rate that's at least twice the highest frequency of interest in the signal. This rule is derived from the Nyquist-Shannon Sampling Theorem. Mathematically, the sampling theorem can be expressed as Eq. 1.1:

$$f_s \geq 2f_{\max} \tag{1.1}$$

where:

- f_s is the sampling rate (samples per second, or Hz),
- f_{max} is the maximum frequency component in the signal.

This means that if the signal has frequency components up to fmax, the sampling rate must be at least 2fmax to avoid aliasing, which is the phenomenon where higher frequency components are incorrectly represented as lower frequencies.

For example, if we are dealing with an ECG signal that contains frequencies up to 50 Hz, the Nyquist rate would be 100 Hz [29]. Therefore, a sampling rate of at least 100 Hz should be chosen. However, to ensure adequate resolution in practice, a typical sampling rate for ECG is 250–1000 Hz. This principle also applies to other biomedical signals. For instance:

- EEG signals, with frequency components up to 100 Hz or more, would require a sampling rate of at least 200 Hz, but in practice, EEG systems typically use sampling rates of 500–2000 Hz, depending on the specific applications.
- Audio signals (e.g., for phonocardiograms) may have higher frequency components so that the sampling rates may range from 8–44.1 kHz.

Aliasing in Sampling

If the signal is sampled at a rate lower than its Nyquist rate, aliasing occurs. Aliasing causes higher frequency components to "fold" into the lower-frequency range, distorting the signal in ways that make it impossible to recover the original information [23, 43]. This is a critical concern in biomedical signal processing, as accurately representing physiological signals is essential for diagnosis and treatment. One notable example of aliasing in biomedical signals can be found in EEG recordings. Suppose an EEG signal contains a high-frequency artifact, such as muscle contractions or electrical interference. If the sampling rate is too low, this high-frequency noise may overlap with brainwave frequencies, leading to incorrect signal interpretation.

Anti-Aliasing Filters

Before sampling, an anti-aliasing filter is typically employed. This low-pass filter is used to remove high-frequency components from the signal that cannot be captured due to the chosen sampling rate. The cutoff frequency of the anti-aliasing filter is typically set to just below the Nyquist frequency, ensuring that all frequency components higher than this cutoff are attenuated [19].

For example, suppose you are sampling an ECG signal at 1000 Hz (Nyquist rate = 500 Hz). In that case, the anti-aliasing filter typically has a cutoff frequency of around 450 Hz, thereby filtering out any frequency components above this threshold.

1.2.1 Quantization of Biomedical Signals

After the signal is sampled, it still exists in an analog form—each sample represents an analog value. Quantization is the process of converting these continuous amplitude values into discrete levels, which is necessary for digital representation. This process introduces quantization error, a form of distortion; however, with proper quantization, the error can be minimized to an acceptable level.

The Quantization Process

Quantization involves dividing the continuous range of signal amplitudes into a series of discrete levels. The signal is then rounded to the nearest discrete level, depending on the bit depth of the quantizer. The number of discrete levels L is determined by the number of bits used for each sample. If the signal has a range from Vmin to Vmax, the quantization levels are defined as Eq. 1.2:

$$Q_k = V_{\min} + \left(\frac{k}{L-1}\right) \cdot (V_{\max} - V_{\min}) \tag{1.2}$$

where:

- k is the index of the quantization level
- L is the number of levels, which is given by Eq. 1.3:

$$L = 2^n \tag{1.3}$$

Here:

- n is the number of bits used for quantization.

For example, with an 8-bit quantizer (n = 8), there would be 256 discrete levels (L = 28 = 256) for the signal's amplitude.

Quantization Error and Its Implications

The main drawback of quantization is quantization error—the difference between the actual signal value and the quantized value. This error is typically bounded by Eq. 1.4:

$$Q_e = \frac{1}{2} \cdot \frac{V_{\max} - V_{\min}}{2^n} \tag{1.4}$$

where:

- Qe is the maximum quantization error.

The quantization noise is a result of this error and manifests as a random signal that can degrade the quality of the original signal.

In biomedical signals, quantization error is particularly problematic when precision is crucial, such as in ECG or EEG signals, where even slight deviations can affect diagnosis. For instance, if a signal is quantized using an 8-bit depth, a person's heart rate might be misrepresented by a few beats per minute due to the quantization error.

Improving Quantization with Higher Bit Depth

One way to reduce quantization error is to increase the bit depth, thus increasing the number of discrete levels. A 16-bit quantization scheme, for instance, allows for 65,536 discrete levels (as opposed to 256 levels in the 8-bit scheme), reducing the quantization error by a factor of approximately 256.

For biomedical applications, where high accuracy is crucial, increasing the bit depth can significantly improve signal quality; however, this also leads to larger file sizes and higher computational requirements [21].

Filtering in Biomedical Signal Processing

After sampling and quantization, the digital signal may still contain unwanted noise or components that obscure the information of interest. Filtering is the process of modifying or suppressing specific frequencies from the signal to isolate the relevant components [22].

Types of Filters

In biomedical signal processing, various types of filters are used to address different forms of noise or unwanted signal components. These include:

- Low-pass filters: These filters allow low-frequency components to pass while attenuating high-frequency components. Low-pass filters are handy for removing high-frequency noise or smoothing a signal.
- High-pass filters: These filters allow high-frequency components to pass while blocking low-frequency components. They are typically used to remove baseline wander or slow drift in signals.
- Band-pass filters: These filters pass a specific range of frequencies while attenuating frequencies outside of this range. They are often used when a particular band of frequencies is of interest, such as the frequency range for EEG or ECG signals.
- Band-stop filters: These filters block a specific range of frequencies while allowing frequencies outside of this range to pass. They can be used to remove specific noise, such as power-line interference.

Filter Design and Characteristics

When designing filters, it is essential to consider several key factors, including the cutoff frequency, the filter order, and the transition band (the range between the passband and the stopband). A filter's cutoff frequency is the frequency at which the filter starts to attenuate the signal. The filter order determines the steepness of the transition from the passband to the stopband.

The most common filters used in biomedical signal processing are Butterworth filters, Chebyshev filters, and Elliptic filters, which all have different characteristics in terms of flatness in the passband and the steepness of the roll-off.

For example, a Butterworth low-pass filter has the following transfer function given by Eq. 1.5:

$$H(s) = \frac{1}{1 + (s/\omega_c)^{2n}} \tag{1.5}$$

where:

- ω_c is the cutoff frequency,
- n is the filter order,
- s is the Laplace transform variable.

Butterworth filters are often used in biomedical applications due to their smooth response in the passband and relatively flat frequency characteristics.

Real-World Application of Filtering in Biomedical Signals

In the context of EEG and ECG signal processing, filtering is crucial for removing unwanted noise [31]. For instance:

- In ECG signals, high-frequency noise may arise from muscle contractions or electrical interference, which can be removed using a low-pass filter. Similarly, low-frequency baseline wander can be removed using a high-pass filter.
- In EEG signals, artifacts like muscle activity or eye blinks can contaminate the signal. These artifacts can often be filtered out using band-pass filters that focus on the frequency range of brain waves, typically between 0.5 Hz and 50 Hz for most EEG applications.

By selecting and designing the appropriate filters, biomedical signal acquisition systems can significantly enhance signal quality, thereby making subsequent analysis more reliable and accurate.

1.2.2 Clinical Applications

Biomedical signal acquisition systems are widely used in clinical practice to monitor patients, diagnose conditions, and assess the effectiveness of treatments. For example, ECG systems are used to monitor the heart's electrical activity, providing valuable information about the heart's rhythm and function. Clinicians can use ECG readings to identify arrhythmias, detect signs of myocardial infarction (heart attack), and assess the overall health of the heart [24, 26].

EEG systems are employed to monitor brain activity, especially in patients with neurological conditions such as epilepsy, sleep disorders, and traumatic brain injuries. By analyzing the brain's electrical signals, doctors can detect abnormalities, identify potential seizure activity, and develop targeted treatment plans. Similarly, pulse oximeters are commonly used to assess blood oxygen levels in patients with respiratory conditions [42]. By measuring the amount of oxygen in the blood, clinicians can determine whether a patient is receiving sufficient oxygen and whether supplemental oxygen is necessary.

The acquisition of biomedical signals is not limited to traditional clinical settings. Wearable devices and mobile health technologies are increasingly being used for continuous monitoring of physiological signals outside the hospital. These devices, which include smartwatches, fitness trackers, and wearable ECG patches, allow patients to monitor their health in real-time, providing valuable data for both individuals and healthcare providers. These devices can detect early signs of health issues, allowing for prompt intervention and reducing the need for in-person visits.

As technology continues to evolve, new methods for signal acquisition are being developed. Non-invasive techniques, such as PPG, provide new methods for monitoring vital signs, including heart rate and blood oxygen levels, using optical sensors. These advances are making it easier to track health status remotely, leading to more personalized and proactive healthcare.

1.2.3 The Future of Biomedical Signal Acquisition

The future of biomedical signal acquisition is bright, with ongoing advancements in sensor technologies, signal processing algorithms, and data analysis techniques. Wearable devices and implantable sensors are becoming smaller, more accurate, and more affordable, enabling continuous health monitoring in a wide range of clinical and non-clinical settings [16, 22]. Additionally, the integration of artificial intelligence (AI) into biomedical signal acquisition systems is enhancing the ability to process and interpret large volumes of data, enabling the early detection of health issues before they become clinically apparent.

In particular, AI-driven signal processing algorithms are being developed to automatically identify and filter out noise, detect anomalies in physiological signals, and predict health risks. These algorithms can help clinicians make faster and more accurate diagnoses, improving patient outcomes and reducing the burden on healthcare systems.

The integration of wireless technologies is also transforming the landscape of biomedical signal acquisition. Remote patient monitoring systems, which enable patients to be monitored from the comfort of their own homes, are becoming increasingly common, particularly for patients with chronic conditions. These systems enable healthcare providers to track patient progress in real-time, adjust treatment plans as needed, and intervene before complications arise [2, 31]. As these technologies continue to evolve, the future of biomedical signal acquisition holds great promise for improving healthcare delivery, enhancing diagnostic accuracy, and empowering patients to take control of their health.

1.3 Introduction to Signal Processing in Biomedical Engineering

In the last few decades, biomedical engineering has emerged as a transformative field, revolutionizing the way healthcare professionals diagnose, monitor, and treat patients. Central to this transformation is the application of signal processing techniques, which enable the extraction of valuable insights from raw physiological data. Every organ system in the human body—whether the brain, heart, muscles, or lungs—produces electrical, mechanical, or biochemical signals that contain crucial information about a person's health. However, these signals are rarely clean or straightforward. They are often noisy, distorted, and influenced by numerous internal and external factors, making accurate interpretation a significant challenge [44].

Signal processing serves as the backbone of modern medical diagnostics and therapy, enabling raw signals to be filtered, analyzed, and converted into clinically meaningful parameters. The impact of signal processing can be seen across a wide range of applications: from ECG used to diagnose heart disease, to EEG that monitor brain activity in epilepsy and sleep disorders [45], to medical imaging modalities such as MRI and CT scans that provide detailed anatomical and functional insights [46]. As computational power increases and AI continues to integrate further into medical decision-making, signal processing is becoming increasingly sophisticated, enabling real-time, automated analysis and precision medicine tailored to individual patients.

The field of biomedical signal processing has undergone significant evolution over the past century. Initially, medical signal analysis relied on visual inspection by trained specialists, with physicians manually interpreting ECG paper printouts or listening to heart sounds through a stethoscope. However, with advancements in computational technology and digital signal processing (DSP), automated systems began to take over many aspects of signal analysis, improving both accuracy and efficiency [47].

- Early 20th Century: ECG was first recorded by Willem Einthoven, using a string galvanometer, which was large and impractical for clinical use.
- Mid-20th Century: The advent of digital computers allowed for more precise signal filtering and analysis, laying the groundwork for automated medical diagnostics.
- Late 20th Century: Signal processing techniques such as Fourier transforms, wavelet analysis, and principal component analysis (PCA) were introduced, significantly enhancing the ability to extract information from biomedical signals.
- 21st Century & AI Integration: The rise of machine learning, deep learning, and real-time signal analytics has transformed biomedical signal processing, enabling AI-driven diagnostics, predictive analytics, and personalized medicine (Fig. 1.8).

As the field continues to progress, several exciting trends are shaping the future of biomedical signal processing:

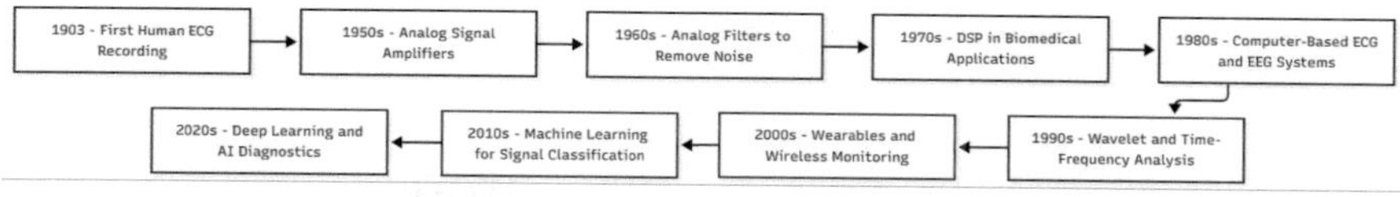

Fig. 1.8 The evolution of biomedical signal processing

- Wearable and Remote Healthcare Technologies: The development of smartwatches, portable ECG monitors, and remote patient monitoring systems is driving demand for real-time, cloud-based signal processing [48].
- AI-Enhanced Signal Analysis: Deep learning models are becoming increasingly sophisticated, enabling highly accurate ECG classification, brain signal decoding, and automated disease detection [49].
- Multimodal Signal Fusion: Combining multiple signals—such as ECG + PPG or EEG + fMRI—provides a more comprehensive view of physiological processes, improving diagnostic precision [50].
- Personalized and Adaptive Signal Processing: Future healthcare solutions will be tailored to individual patients, utilizing adaptive filtering algorithms that adjust according to each patient's unique physiological patterns [51].

1.3.1 Importance of Signal Processing in Diagnostics and Therapy

The human body is a complex system of electrical, mechanical, and biochemical processes that generate signals reflecting health and disease. Every heartbeat, every neural impulse, and every fluctuation in blood pressure carries invaluable diagnostic information—if only we can extract it. Biomedical signal processing enables healthcare professionals to translate raw physiological data into meaningful clinical insights [52]. Without signal processing, the wealth of information embedded in medical signals would remain buried under noise, artifacts, and variability, rendering diagnosis and treatment unreliable.

From the early days of manually recorded electrocardiograms to today's real-time artificial intelligence-driven diagnostics, biomedical signal processing has evolved into an indispensable tool for modern medicine. It underpins cardiology, neurology, respiratory medicine, radiology, intensive care, and rehabilitation, among other fields. Advances in computational techniques have enabled not only the early detection of diseases but also the prediction of their progression, the personalization of treatments, and the automation of complex clinical workflows.

Signal processing has had a particularly profound impact on non-invasive diagnostics. Where traditional medicine often relied on subjective clinical observations, today's healthcare systems leverage machine learning-enhanced signal analysis to diagnose conditions with remarkable precision. A simple wearable device can continuously monitor heart rhythm abnormalities, sending real-time alerts for arrhythmias

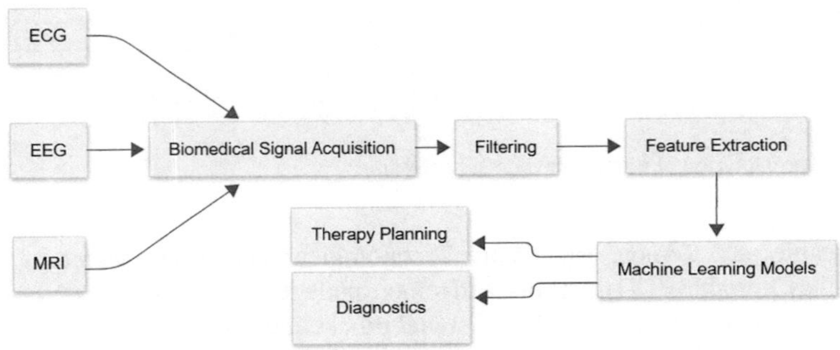

Fig. 1.9 Clinical decision-making

that might otherwise go unnoticed. Neurological disorders, which once required invasive procedures to diagnose, can now be detected through advanced EEG and MRI signal analysis techniques [2, 49]. Patients in intensive care units are continuously monitored using signal-processing-driven systems that alert physicians to subtle signs of deterioration before they become critical.

Beyond diagnostics, signal processing plays an equally crucial role in therapy. Implantable cardiac devices regulate the heart rhythm, neuroprosthetics restore lost function to individuals with paralysis, and real-time feedback systems enable brain-machine interfaces to communicate with assistive devices [53]. The convergence of signal processing, artificial intelligence, and real-time computing is ushering in a new era of medicine—one where data drives decisions, interventions are personalized, and treatment efficacy is continually optimized (Fig. 1.9).

Signal Processing in Disease Diagnosis

Few medical signals have had as profound an impact on diagnostics as the electrocardiogram (ECG). First recorded in the early 20th century using rudimentary string galvanometers, the ECG has since become a cornerstone of cardiovascular diagnostics. Each wave in an ECG tracing—the P wave, QRS complex, and T wave—represents a critical phase of the heart's electrical cycle. Deviations from standard wave patterns can indicate arrhythmias, myocardial infarction, ischemia, or conduction abnormalities [54]. However, raw ECG signals are rarely pristine. Motion artifacts, baseline drift, muscle noise, and powerline interference often contaminate them. Traditional visual interpretation of ECG printouts, while still practiced, is prone to human error and limited by inter-observer variability. Advanced digital signal processing techniques have significantly enhanced ECG interpretation by automatically filtering out noise, detecting abnormalities, and classifying arrhythmias with near-human accuracy. One of the most remarkable advancements in ECG analysis has been the application of machine learning and deep learning models. These algorithms, trained on massive datasets, can recognize complex arrhythmic patterns, predict the risk of sudden cardiac arrest, and even estimate cardiac output and myocardial strain—parameters previously requiring invasive testing [8, 55]. With

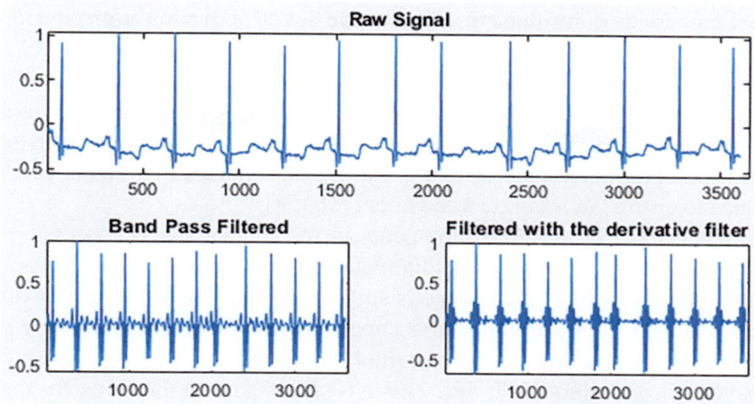

Fig. 1.10 A comparison of a raw ECG signal with artifacts and the signal after digital filtering

the advent of wearable ECG monitors, patients at risk of atrial fibrillation or ventricular tachycardia can now be continuously monitored, allowing for early intervention and reduced hospitalization rates (Fig. 1.10).

The human brain generates a symphony of electrical activity, captured through electroencephalography (EEG). This non-invasive technique has been instrumental in diagnosing epilepsy, sleep disorders, neurodegenerative diseases, and cognitive impairments. Unlike ECG, EEG signals are highly variable, reflecting the brain's constantly changing state of activity. The challenge of EEG interpretation lies in distinguishing meaningful neural patterns from background noise, muscle artifacts, and environmental interference [56]. Traditional EEG analysis involved visual inspection of waveforms by neurologists—a labor-intensive and error-prone process. However, modern signal processing approaches, including wavelet transforms, PCA, and independent component analysis (ICA), have revolutionized the field of EEG diagnostics. These methods enable the extraction of distinct frequency bands (delta, theta, alpha, beta, gamma) that correlate with different cognitive and pathological states. One of the most significant breakthroughs in EEG signal processing has been in the realm of seizure detection and prediction. Algorithms trained on EEG recordings can now identify seizure onset with remarkable accuracy, allowing patients to receive early warnings before an episode occurs [57]. Additionally, the use of deep neural networks (DNNs) in EEG processing is opening new frontiers in brain-computer interfaces (BCIs), enabling individuals with paralysis to control robotic limbs or communicate through thought alone.

Magnetic Resonance Imaging (MRI), another powerful tool in neurological diagnostics, benefits immensely from signal processing techniques. Raw MRI signals are reconstructed using Fourier Transform-based algorithms, and advanced image denoising, segmentation, and enhancement techniques enable radiologists to detect brain tumors, strokes, and multiple sclerosis lesions with unparalleled precision [33, 47]. Functional MRI (fMRI), which measures brain activity in real time, relies heavily on signal processing algorithms to separate true neural activation from physiological noise.

Respiratory diseases such as chronic obstructive pulmonary disease (COPD), asthma, and sleep apnea are commonly diagnosed using spirometry, pulse oximetry, and respiratory sound analysis. However, these signals are prone to motion artifacts and environmental noise, necessitating the use of sophisticated signal processing techniques to ensure an accurate assessment [50, 58].

One of the most exciting developments in respiratory signal processing is the application of machine learning for automated classification of lung sounds. Traditional auscultation with a stethoscope is subjective and depends on the physician's experience. AI-enhanced digital stethoscopes now record and analyze lung sounds in real time, distinguishing between normal breath sounds, wheezes, crackles, and other pathological patterns [42, 59]. This has significant implications for the early diagnosis of pneumonia, bronchitis, and pulmonary fibrosis.

Signal Processing in Therapy

Signal processing plays an indispensable role in intensive care units (ICUs), where real-time monitoring systems continuously analyze multiple physiological parameters. Patients admitted to an ICU often require constant surveillance of heart rate, blood pressure, oxygen saturation (SpO_2), respiratory rate, and brain activity to detect early signs of deterioration. In such settings, even a slight delay in recognizing a critical event can be life-threatening [52]. Continuous monitoring generates vast amounts of data, but without efficient signal processing techniques, extracting actionable insights from this deluge of information would be impossible. For instance, heart rate variability (HRV) analysis, derived from ECG signals, provides insight into the autonomic nervous system's function, enabling physicians to detect sepsis, myocardial dysfunction, or even impending cardiac arrest before overt clinical signs appear [51].

Respiratory monitoring is another area where advanced signal processing has transformed care. Capnography, which measures end-tidal CO_2 levels, is now an essential tool for detecting hypoventilation and apnea, particularly in sedated or ventilated patients. Algorithms filter out noise and ensure that alerts are triggered only when clinically relevant patterns emerge, reducing false alarms that could desensitize healthcare providers. Neonatal ICUs (NICUs) particularly benefit from signal processing advancements [33]. Premature infants are highly susceptible to apnea of prematurity, where breathing pauses for dangerously long periods. Machine learning models applied to chest impedance and nasal airflow signals can now predict apneic episodes before they occur, allowing for timely intervention.

Signal Processing in Medical Imaging and Therapy Planning

Medical imaging is perhaps the most transformative area where signal processing has reshaped diagnostics and therapy. Technologies such as MRI, CT, PET scans, and ultrasound rely heavily on image reconstruction algorithms, noise reduction techniques, and pattern recognition models. The goal is to enhance image clarity while preserving diagnostic accuracy. For example, in oncology, radiation therapy planning requires precise tumor delineation [46, 60]. Traditional methods of manual contouring by radiologists are time-consuming and susceptible to inter-observer variability. Advanced segmentation algorithms utilizing deep learning have significantly enhanced the accuracy of tumor identification in CT and MRI scans, facilitating personalized treatment planning.

Similarly, in cardiology, myocardial perfusion imaging (MPI) detects regions of ischemia and infarction. The interpretation of MPI scans is significantly improved by Fourier and wavelet-based image enhancement techniques, which eliminate background noise while preserving critical details.

Neuroprosthetics and brain-machine interfaces (BMIs) represent one of the most revolutionary applications of biomedical signal processing. These technologies bridge the gap between the nervous system and external assistive devices, restoring mobility, communication, and independence to individuals with paralysis or neuromuscular disorders. The core challenge in BMI development is accurately interpreting neural signals and converting them into actionable outputs. Electrocorticography (ECoG), EEG, and local field potentials (LFPs) serve as input signals, but these signals are inherently noisy, complex, and highly dynamic. Advanced machine learning models play a critical role in extracting meaningful patterns from high-dimensional neural recordings. These models enable real-time decoding of intended movements, allowing users to control robotic limbs, type on virtual keyboards, or even operate exoskeletons. For example, intracortical microelectrode arrays implanted in the motor cortex have allowed individuals with quadriplegia to regain control over robotic arms [61]. A combination of signal preprocessing, feature extraction, and deep learning classification refines neural activity into precise motor commands.

Similarly, cochlear implants utilize signal processing to convert sound waves into electrical stimuli, directly stimulating the auditory nerve in individuals with profound hearing loss. Speech enhancement algorithms improve sound clarity, making communication possible even in noisy environments [62].

Wearable medical devices, powered by advanced signal processing and AI, are reshaping preventive medicine and chronic disease management. Devices such as smartwatches, continuous glucose monitors (CGMs), and smart ECG patches allow real-time tracking of heart rate, glucose levels, oxygen saturation, and physical activity. One of the most impactful developments in wearable technology is automated atrial fibrillation (AFib) detection using PPG and ECG sensors. Deep learning classifiers, trained on massive datasets, can now distinguish between benign heart rate variability and clinically significant arrhythmias, enabling the early detection of stroke risk. In diabetes management, CGMs analyze interstitial glucose concentrations, and predictive modeling techniques forecast impending hypoglycemic episodes, alerting patients before dangerous lows occur (Figs. 1.11 and 1.12).

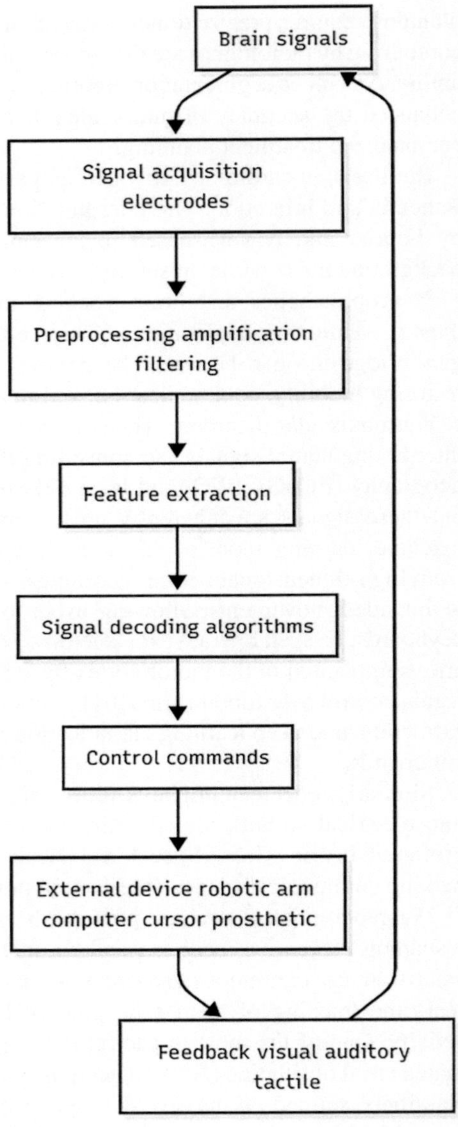

Fig. 1.11 A flowchart of a BMI system

1.3.2 Biomedical Signal Processing Stages

Biomedical signals are inherently complex, noisy, and multidimensional. The challenge lies in converting these raw signals into clinically relevant information while minimizing artifacts and maximizing accuracy. To achieve this, a structured signal

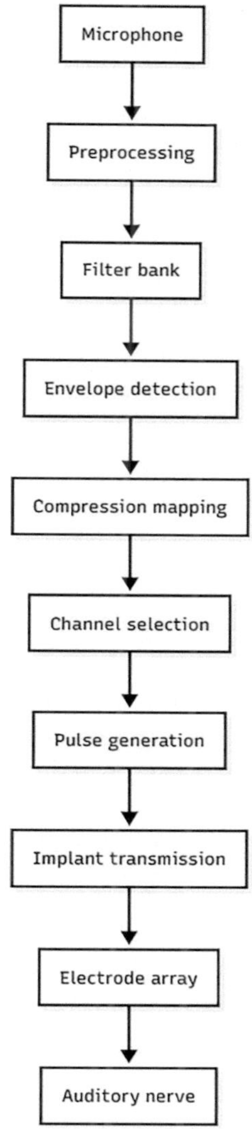

Fig. 1.12 A cochlear implant signal processing

processing pipeline is essential. Biomedical signal processing begins with acquisition and preprocessing, followed by feature extraction, transformation, filtering, and classification. These stages are the foundation of all modern medical diagnostic and therapeutic technologies [63] (Fig. 1.13).

The first step in biomedical signal processing is acquisition, where sensors and electrodes detect physiological activity and convert it into a measurable electrical signal. This process is far from straightforward due to the complex interplay of biological and environmental factors that influence the quality of recorded signals. For instance, an ECG sensor detects the electrical impulses generated by the heart. While

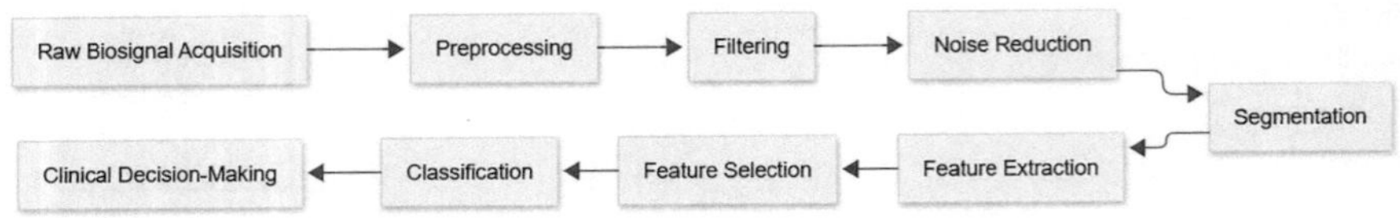

Fig. 1.13 A conceptual diagram of signal processing stages

the underlying physiology remains the same, the recorded signal varies significantly depending on:

- Electrode placement (chest vs. limb leads).
- Skin conductivity and impedance (affected by sweat, hair, or movement).
- External electromagnetic interference (from power lines, devices).

Consider the case of EEG, which records brain activity. Early EEG systems in the 1920s relied on large electrodes and produced low-resolution signals. Today's high-density EEG systems use hundreds of electrodes and sophisticated digital amplification to improve accuracy.

Common Modalities and Their Applications

Biomedical signals can be categorized based on the type of physiological process being recorded (Table 1.2).

One significant advancement in modern biomedical signal acquisition is the transition from wired systems to wireless and wearable devices. The rise of smartwatches, fitness trackers, and implantable biosensors has revolutionized the way data is collected, processed, and analyzed in real-time.

Equation 1.6 Generalized signal acquisition model

$$x(t) = s(t) + n(t) \tag{1.6}$$

where:

- x(t) is the recorded signal,

Table 1.2 Biomedical signal types and acquisition methods

Signal type	Measurement device	Medical application
ECG	Electrodes on skin	Diagnosing heart conditions
EEG	Scalp electrodes	Detecting epilepsy, brain activity
EMG	Surface or needle electrodes	Assessing muscle function
PPG	Optical sensor	Measuring heart rate, oxygen saturation
MRI	Radiofrequency coils	Soft tissue imaging
CT	X-ray detectors	3D imaging of bones and organs
Ultrasound	Piezoelectric transducer	Real-time imaging of organs and pregnancy

- s(t) is the true physiological signal,
- n(t) is the noise component.

Modern acquisition systems integrate AI-powered adaptive filtering, which can dynamically suppress noise without distorting the true signal, significantly improving accuracy.

Once acquired, biomedical signals undergo preprocessing to remove artifacts, noise, and baseline drift. Biomedical signals are highly susceptible to various types of noise, which must be filtered out before analysis. Some of the most common sources of noise include:

- Powerline Interference (50/60 Hz noise): Caused by nearby electrical devices.
- Motion Artifacts: Occurs when a patient moves, affecting signal stability.
- Muscle Noise (EMG contamination): Affects EEG and ECG recordings.

To remove these unwanted components, signal filtering techniques are employed. A Notch Filter is commonly used to eliminate powerline interference.

Equation 1.7 Notch filter transfer function.

$$H(f) = 1 - \frac{G}{1 + j\frac{f - f_0}{\Delta f}} \tag{1.7}$$

where:

- f_0 is the frequency to be eliminated
- Δf is the bandwidth.

Many biomedical signals, particularly ECG and EEG, are affected by baseline drift, where the signal's reference level shifts over time. This issue can obscure crucial information. A common technique used to correct this is wavelet-based detrending. Additionally, normalization ensures that signals from different patients can be compared on a standard scale. One of the most commonly used normalization methods is z-score normalization.

Equation 1.8 Z-score normalization

$$X' = \frac{X - \mu}{\sigma} \tag{1.8}$$

where:

- μ is the mean
- σ is the standard deviation of the signal.

After preprocessing, the biomedical signal remains a complex time-series waveform, necessitating further transformation to extract relevant features. This stage focuses on identifying and quantifying characteristics of the signal that hold diagnostic value. Feature extraction is critical in biomedical signal processing because raw signals often contain redundant, noisy, or uninformative data. Instead of feeding the entire waveform into a classifier or diagnostic algorithm, extracting meaningful features allows for more efficient and accurate analysis.

The most straightforward approach to feature extraction involves analyzing the signal directly in the time domain. Many physiological parameters are traditionally measured in this domain because they correspond closely to real-world clinical observations. For example, in ECG analysis, cardiologists examine:

- Peak-to-peak intervals (e.g., R-R intervals for heart rate variability).
- Amplitude variations (e.g., ST-segment elevation for myocardial infarction).
- Morphological changes (e.g., widened QRS complex in conduction disorders).

Equation 1.9 HRV from R-R intervals

$$HRV = \sqrt{\frac{1}{N-1}\sum_{i=1}^{N}(RR_i - \bar{RR})^2} \tag{1.9}$$

where:

- RR_i is the time interval between two successive R-peaks,
- RR^- is the mean R-R interval over the entire signal.

In EEG, time-domain features such as mean amplitude, root-mean-square (RMS), and zero-crossing rate help differentiate between normal and abnormal brain activity, particularly in epilepsy detection (Table 1.3).

Time-domain features, while intuitive, often fail to capture hidden periodic components in biomedical signals. Many physiological signals contain relevant information in specific frequency bands, making frequency-domain analysis a powerful tool (Figs. 1.14 and 1.15).

The Fourier Transform (FT) is the most widely used method to convert time-domain signals into their frequency components.

Equation 1.10 Discrete FT:

$$X(k) = \sum_{n=0}^{N-1} x(n)e^{-j2\pi kn/N} \tag{1.10}$$

Table 1.3 Time-domain features in common biomedical signals

Feature	Definition	Application
Mean value	Average amplitude over time	Baseline drift detection
Peak-to-peak amplitude	Difference between max and min values	ECG QRS detection
Zero-crossing rate	Number of times the signal crosses zero	EMG muscle activity analysis
RMS (Root mean square)	Measure of signal power	EEG signal analysis

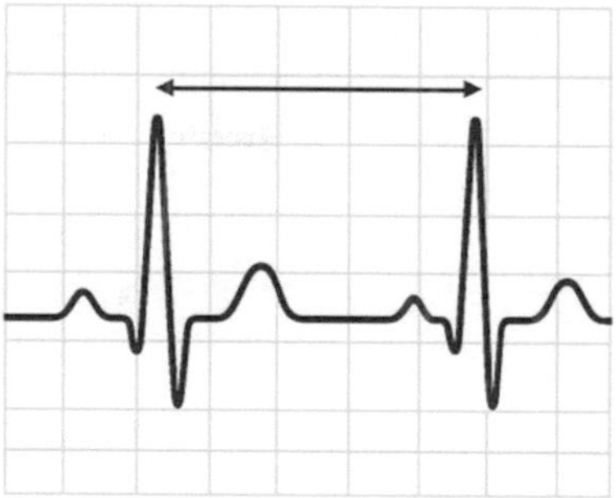

Fig. 1.14 Example of ideal ECG waveform

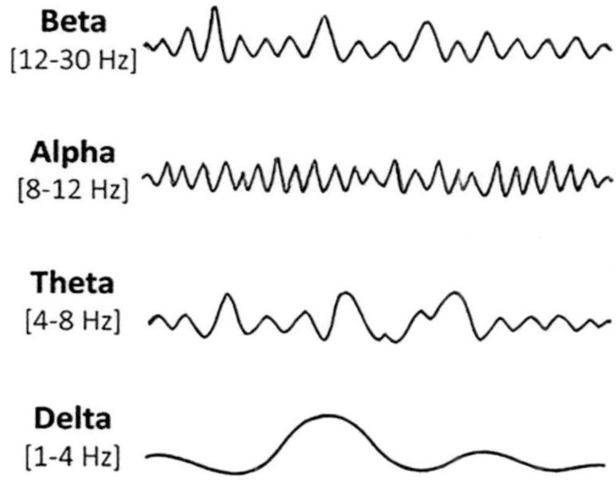

Fig. 1.15 EEG spectrum with different frequency bands

where:

- x(n) is the original signal,
- N is the number of signal samples,
- X(k) represents the frequency components.

For example, EEG signals contain specific frequency bands associated with different mental states:

- Delta (0.5–4 Hz): Deep sleep and brain injuries.
- Theta (4–8 Hz): Drowsiness and memory processing.
- Alpha (8–13 Hz): Relaxation and alertness.
- Beta (13–30 Hz): Active thinking and focus.
- Gamma (30–100 Hz): High-level cognitive function.

In heart rate analysis, frequency-domain features help assess autonomic nervous system activity:

- Low-frequency (LF, 0.04–0.15 Hz): Sympathetic and parasympathetic activity.
- High-frequency (HF, 0.15–0.40 Hz): Parasympathetic (vagal) activity.
- LF/HF Ratio: Indicator of stress levels and cardiovascular health.

Many biomedical signals exhibit non-stationary behavior, meaning their frequency content changes over time. Traditional Fourier Transform techniques assume stationarity and fail to capture these variations.

To address this limitation, time–frequency techniques such as the Short-Time Fourier Transform (STFT) and Wavelet Transform (WT) are used (Table 1.4).

Equation 1.11 Short-Time Fourier Transform (STFT)

$$STFT\{x(t)\} = \int_{-\infty}^{\infty} x(t)\mathrm{w}(\mathrm{t} - \tau)\mathrm{e}^{-j\omega t}\mathrm{dt} \tag{1.11}$$

where:

- w(t) is a sliding window function.

The Wavelet Transform (WT) offers a more flexible approach by analyzing the signal at multiple scales.

Equation 1.12 Continuous Wavelet Transform (CWT)

$$CWT(s, \tau) = \frac{1}{\sqrt{s}} \int_{-\infty}^{\infty} x(t)\psi^{*}\left(\frac{\mathrm{t} - \tau}{\mathrm{s}}\right)\mathrm{dt} \tag{1.12}$$

where:

- s represents scale
- τ represents translation.

Wavelet Transform has been particularly useful in:

- Seizure detection from EEG signals.
- ECG arrhythmia classification.
- Respiratory sound analysis for asthma diagnosis.

After extracting key features, the next step is classification, where machine learning (ML) or deep learning models assign meaning to the extracted data (Table 1.5).

Traditional classification approaches involve [55]:

Table 1.4 Frequency-domain features and their clinical applications

Feature	Frequency range	Medical application
Delta band (EEG)	0.5–4 Hz	Sleep studies, coma assessment
Alpha band (EEG)	8–13 Hz	Cognitive function analysis
LF power (HRV)	0.04–0.15 Hz	Sympathetic nervous system activity
HF power (HRV)	0.15–0.40 Hz	Parasympathetic nervous system activity

Table 1.5 Machine learning vs. deep learning in biomedical signal processing

Approach	Advantages	Limitations
Traditional ML	Explainable, less data-intensive	Requires manual feature extraction
Deep learning	High accuracy, automated feature extraction	Requires large datasets, computationally expensive

- Threshold-based methods (e.g., detecting bradycardia if HR < 60 bpm).
- Statistical classifiers (e.g., Linear Discriminant Analysis).
- Machine learning models (e.g., Support Vector Machines, Random Forests).
- Deep learning approaches (e.g., CNNs, LSTMs for ECG/EEG classification).

For instance, in ECG-based heart disease detection, ML models can classify:

1. Normal sinus rhythm
2. Atrial fibrillation
3. Ventricular tachycardia
4. Bundle branch blocks.

Deep learning techniques, particularly Convolutional Neural Networks (CNNs), have dramatically improved the accuracy of classification in:

- Diabetic retinopathy detection from fundus images.
- Brain tumor classification from MRI scans.
- Sleep stage classification from EEG recordings.

Once a biomedical signal has undergone preprocessing, feature extraction, and classification, the final and most crucial step is clinical decision-making and implementation. This stage involves interpreting the classified results, integrating them into clinical workflows, and ensuring that the processed signals directly contribute to patient care (Fig. 1.16).

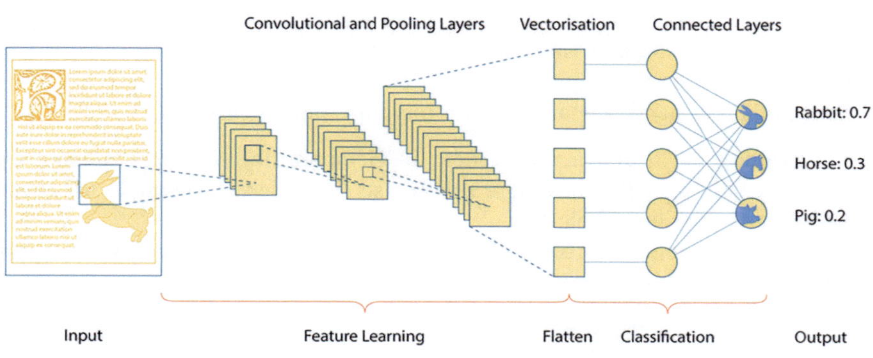

Fig. 1.16 Example of CNN architecture (https://commons.wikimedia.org/wiki/File:Convolutional_Neural_Network.png)

A Clinical Decision Support System (CDSS) is an intelligent software platform that uses processed biomedical signals to assist healthcare professionals in making accurate, timely, and data-driven diagnoses [16, 44]. These systems rely on:

1. Machine learning-based diagnostic models (e.g., ECG-based atrial fibrillation detection).
2. Rule-based expert systems (e.g., algorithms that flag abnormal EEG patterns).
3. Real-time monitoring and alert systems (e.g., automatic detection of apnea events in sleep studies).

For instance, in ICUs, CDSS continuously monitors EEG, ECG, and oxygen saturation signals, alerting clinicians when patterns indicate potential complications, such as sepsis, arrhythmias, or neurological deterioration.

A well-designed CDSS can reduce diagnostic errors, improve early disease detection, and optimize treatment planning.

Equation 1.13 CDSS Confidence Score Calculation

$$S_{\text{conf}} = \frac{\sum_{i=1}^{N} w_i P_i}{\sum_{i=1}^{N} w_i} \tag{1.13}$$

where:

- P_i is the probability of each diagnostic category,
- W_i is the weight assigned to each probability based on clinical significance.

For example, a CDSS analyzing ECG signals might assign:

- 70% probability of atrial fibrillation,
- 20% to ventricular tachycardia,
- 10% to normal rhythm, resulting in an automated confidence score for physician verification.

To enhance clinical usability, signal-processed data is increasingly being integrated into Electronic Health Records (EHRs). This allows:

- Automatic population of vital signs and trends.
- Comparison of current signals with past recordings.
- Sharing of data across departments for multidisciplinary analysis.

For example, in telemedicine, remote ECG monitoring devices continuously update a patient's EHR, allowing cardiologists in different hospitals to assess real-time heart function (Table 1.6).

Despite its benefits, implementing signal-processing-based decision support systems in real-world hospitals presents challenges:

1. Data Standardization Issues

    - Different medical devices produce signals in proprietary formats (e.g., Philips vs. GE ECG machines).

Table 1.6 Benefits of EHR integration with signal processing

Feature	Impact on clinical workflow
Real-time data updates	Faster intervention in emergencies
Historical signal comparison	Identifies disease progression
Automated reporting	Reduces physician workload
Remote accessibility	Enables telemedicine applications

- Lack of interoperability between hospital systems hinders seamless integration.

2. Accuracy and Trust in AI-Driven Diagnoses
   - Physicians may be hesitant to trust machine learning models without transparent explanations of how decisions are made.
   - False positives in automatic detection systems can lead to unnecessary interventions, while false negatives could result in missed diagnoses.
3. Computational Constraints in Real-Time Processing
   - Processing biomedical signals in real time requires high computational power, especially for deep learning models.
   - Hospitals must balance cost vs. benefit when implementing AI-driven hardware.
4. Regulatory and Ethical Considerations
   - Medical AI models must comply with regulations like the FDA (USA), CE Mark (Europe), or MDR (Medical Device Regulation).
   - Ethical concerns arise around patient data privacy, bias in AI algorithms, and the potential for misuse of automated decision-making.

The integration of advanced signal processing with AI and big data is paving the way for personalized medicine, where:

- AI algorithms tailor treatment plans based on real-time physiological signals.
- Wearable biosensors provide continuous health monitoring outside of hospital settings.
- Predictive modeling anticipates disease onset before symptoms appear.

For example, emerging "digital twin" models simulate an individual's cardiovascular system using their ECG, blood pressure, and oxygen saturation data, allowing doctors to predict how a patient will respond to specific treatments [27, 48].

Equation 1.14 Predictive Model for Personalized Medicine

$$Y = f(X) + \varepsilon \tag{1.14}$$

where:

- X represents a patient's real-time biomedical signals,
- f(X) is a trained AI model mapping signals to disease risk,
- ε represents unexplained variability in patient response.

1.4 Ethical and Legal Considerations

Biomedical signal acquisition and processing are fundamental to modern diagnostic medicine. Signals such as ECG, EEG, EMG, and blood oxygenation levels provide critical real-time data about a patient's health, allowing clinicians to diagnose, monitor, and treat various conditions. However, the process of acquiring and processing these signals is accompanied by a range of ethical and legal considerations that must be carefully navigated to protect patient rights, ensure accuracy, and comply with regulations [64].

Signal processing, including filtering, noise reduction, amplification, and feature extraction, plays a crucial role in making raw biomedical data usable for clinical purposes. These techniques can enhance the quality of the signals, reduce artifacts, and improve the precision of diagnostic outcomes. However, these methods must be applied in a way that respects patient privacy, ensures data integrity, and minimizes risks to patient safety [65]. This chapter examines the ethical and legal implications of biomedical signal acquisition and processing, with a focus on key issues including patient consent, data privacy, risk of harm, and regulatory compliance (Fig. 1.17).

1.4.1 Ethical Considerations in Biomedical Signal Acquisition and Processing

Ethical issues related to the acquisition and processing of biomedical signals are complex and often intersect with broader principles in healthcare, including autonomy, beneficence, non-maleficence, and justice. The following sections examine these principles and their relevance to biomedical signal processing.

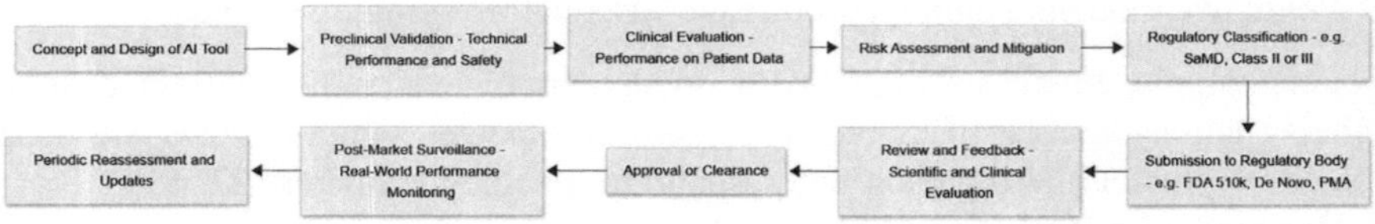

Fig. 1.17 CNN-based classification of ECG signals (https://commons.wikimedia.org/wiki/File:Convolutional_Neural_Network.png)

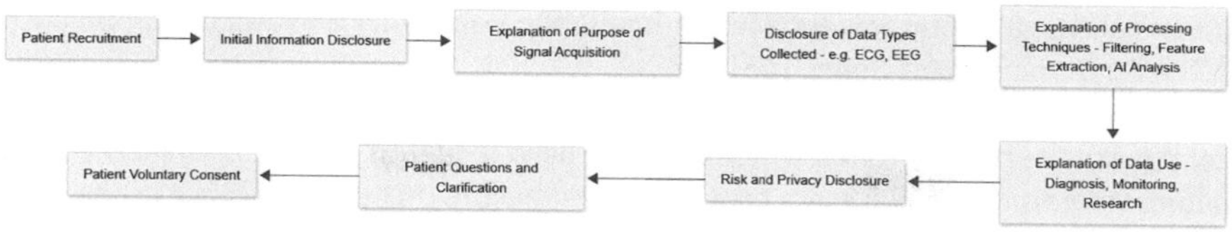

Fig. 1.18 The informed consent process for biomedical signal acquisition

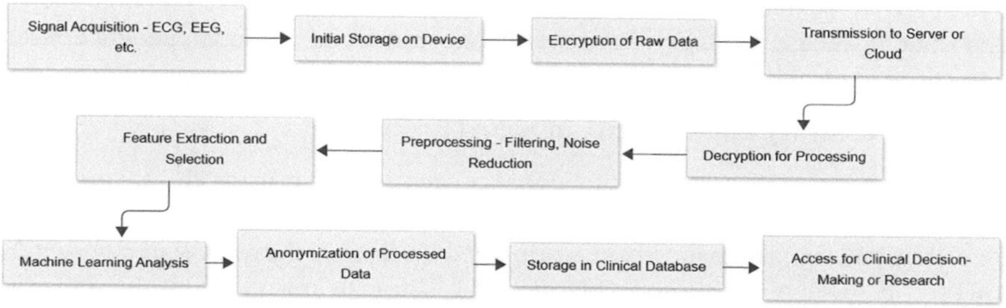

Fig. 1.19 The data flow in biomedical signal acquisition and processing

Informed Consent and Autonomy in Signal Acquisition

In healthcare, informed consent is a fundamental ethical requirement. Before acquiring biomedical signals, patients must be informed about the nature of the procedure, its purpose, and how the data will be used [66]. This includes explaining any processing techniques applied to the signals, such as filtering, noise reduction, or amplification. For instance, when a patient undergoes an ECG to monitor heart function, they must understand the procedure involved in acquiring the signal and any subsequent processing. While these procedures are generally non-invasive, the patient's autonomy must be respected by ensuring they voluntarily agree to the procedure (Figs. 1.18 and 1.19).

In the context of signal processing, patients should be informed about the types of processing that will be performed. For example, when an EEG signal is acquired, the raw data may contain noise from muscle activity or electrical interference. The signal is then processed to remove this noise and enhance the relevant neurological features. However, patients should be informed about this process and its potential impact on the interpretation of their data [16, 67].

Privacy, Confidentiality, and Data Protection

Biomedical signals are often considered highly sensitive data, as they provide detailed insights into a patient's health status. Consequently, data privacy and confidentiality are key ethical considerations in the acquisition and processing of biomedical signals. The data must be protected from unauthorized access, use, or disclosure, in accordance with privacy laws such as the General Data Protection Regulation (GDPR) in the European Union and the Health Insurance Portability and Accountability Act (HIPAA) in the United States [68].

For example, an ECG or EEG contains unique patient information, such as heart rate patterns or brain activity, which could be used to identify a patient or infer certain medical conditions. Healthcare providers must ensure that proper security measures, such as encryption and secure data storage, are in place to protect the data during the acquisition and processing stages. Additionally, the processed data must be handled with care. In some cases, data used for research purposes is anonymized to prevent patient identification, but patients must still be informed about the potential uses of their data. If the data is shared for research or commercial purposes, patient consent is required to ensure transparency and safeguard privacy [69].

Ensuring Accuracy and Integrity in Signal Processing

One of the core ethical concerns in signal processing is ensuring the accuracy and integrity of the data. Biomedical signal processing techniques, such as filtering, denoising, and signal enhancement, are applied to improve the quality of raw signals; however, if misapplied, these techniques can distort or misrepresent the data. For example, ECG signal filtering removes noise caused by muscle activity or power line interference. However, overly aggressive filtering could inadvertently remove important cardiac features or introduce errors into the data. Similarly, in EEG processing, improper artifact removal could lead to the loss of critical neurological signals, affecting diagnosis or treatment decisions.

The principle of non-maleficence (do no harm) is directly relevant here. The primary goal of signal processing is to enhance the quality of the data, making it more useful for diagnostic or therapeutic purposes. However, clinicians and researchers must exercise caution to ensure that the processing methods do not obscure critical information or introduce errors that could lead to inaccurate diagnoses or inappropriate treatment plans.

Risk of Harm and Patient Safety

While biomedical signal processing techniques are generally low-risk, there is still the potential for harm if these techniques are not applied correctly. For example, over-filtering or over-smoothing a signal might result in the removal of vital information that could affect the diagnosis or treatment [70]. This could lead to misdiagnosis or delayed treatment, ultimately harming the patient. Another example is the amplification of weak signals in some types of monitoring. If the signal is amplified excessively, it may cause artifacts or exaggerate patterns that do not exist in the raw data, resulting in misleading results. This could be particularly problematic in the context of telemedicine or remote monitoring, where the accuracy of processed signals is crucial for ensuring correct treatment. Patient safety must be a priority during the entire process of signal acquisition and processing. Healthcare providers and researchers should ensure that all signal processing techniques are thoroughly validated and that the risks of incorrect processing are minimized (Figs. 1.20 and 1.21).

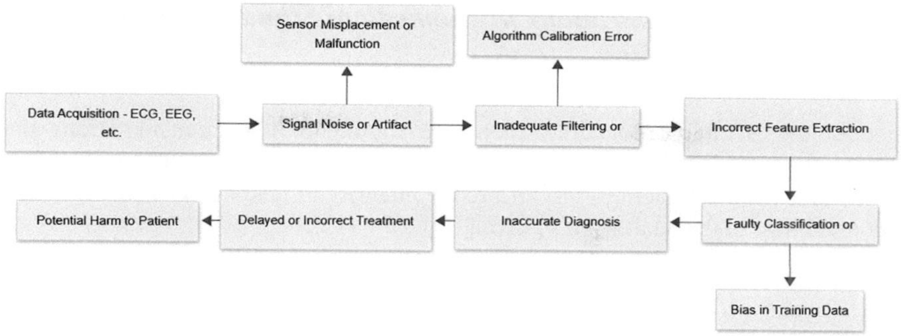

Fig. 1.20 A flowchart showing the potential risks of inaccurate signal processing and their impact on patient safety

Fig. 1.21 A diagram comparing the regulatory frameworks for biomedical signal processing in the EU and the US

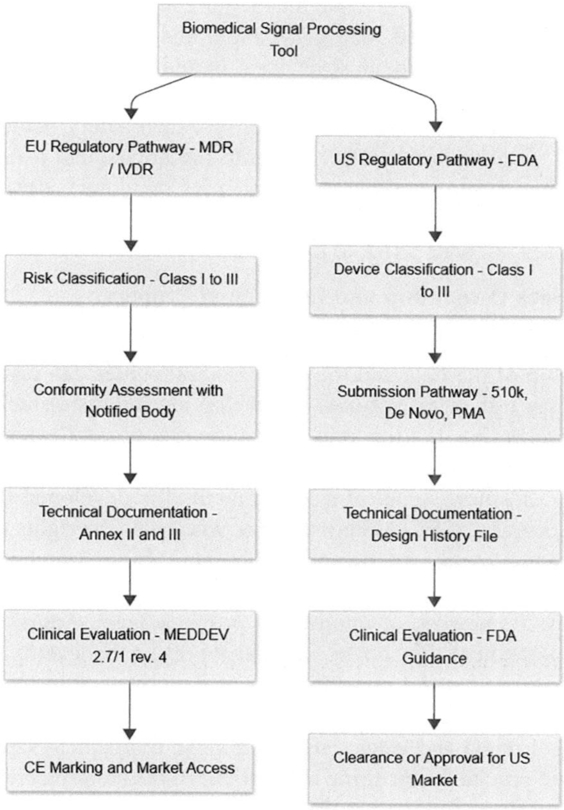

1.4.2 Legal Considerations in Biomedical Signal Acquisition and Processing

In addition to ethical concerns, there are significant legal considerations that govern the acquisition and processing of biomedical signals. These laws are designed to protect patient rights, ensure the integrity of medical data, and maintain accountability in the healthcare system [16].

Regulatory Compliance and Standards

Biomedical signal processing is subject to regulatory oversight by various governing bodies to ensure that medical devices and signal processing systems meet specific standards for safety and effectiveness. In the European Union, Medical Device Regulation (MDR) and In Vitro Diagnostic Regulation (IVDR) govern the use of medical devices, including those that acquire and process biomedical signals.

In the United States, the Food and Drug Administration (FDA) is responsible for regulating medical devices, including those used for acquiring and processing biomedical signals. Devices that are used to acquire and process data for clinical use must meet FDA standards to ensure their safety, accuracy, and reliability [15, 71].

For example, an ECG monitoring device that uses signal processing techniques to analyze heart rhythms must meet regulatory standards before it can be used in clinical practice. This ensures that the device provides accurate and reliable data and does not pose a risk to patient safety.

Data Ownership and Intellectual Property

As biomedical signal processing is often part of research and development, the ownership of the data and the processed outcomes can raise legal concerns. Researchers may utilize patient data to develop algorithms or technologies for enhanced signal processing, but this raises questions about intellectual property (IP) rights and data ownership [72, 73].

If a new signal processing method is developed that significantly improves the accuracy of ECG interpretation, who owns the rights to the algorithm? Is the healthcare provider the owner of the data? The researcher who developed the method? Or the company that funds the research? These questions must be addressed within the framework of intellectual property laws. Additionally, any commercial use of biomedical signal data, whether for research, development, or commercial purposes, requires careful consideration of data ownership, patient consent, and contractual agreements.

Ethical and legal considerations in biomedical signal acquisition and processing are crucial for ensuring that patient rights are protected, data integrity is maintained, and the safety and well-being of patients are prioritized. The process of acquiring and processing biomedical signals must be conducted with complete transparency, accuracy, and care, while adhering to strict legal frameworks that govern data privacy, patient consent, and regulatory compliance [16]. As technology continues to evolve and the use of biomedical signal processing becomes more integrated into clinical

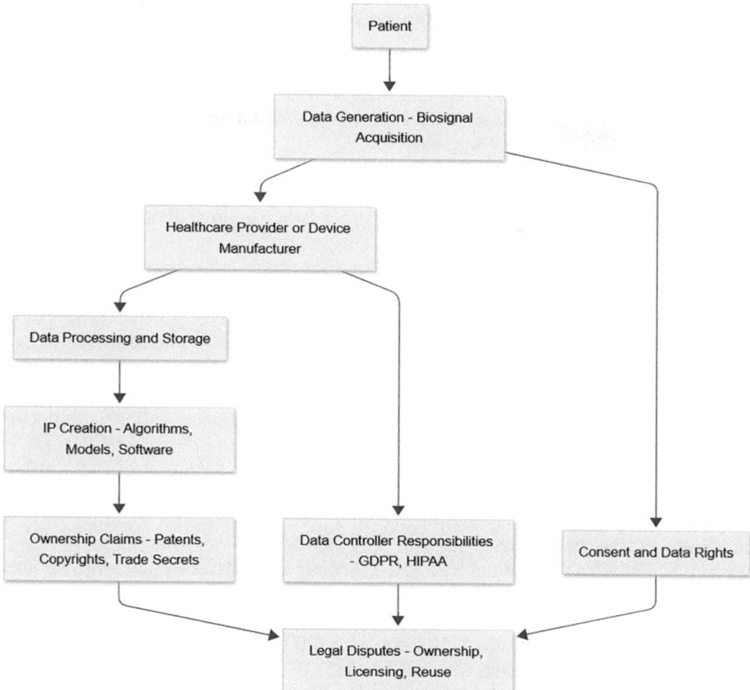

Fig. 1.22 A conceptual diagram showing the relationship between data ownership, intellectual property, and legal considerations in biomedical signal processing

practice and research, ongoing attention to ethical and legal concerns will be critical. Ensuring that these concerns are addressed will not only protect patients but also foster trust in medical technologies and support the continued development and advancement of healthcare (Fig. 1.22).

References

1. Jasim AD (2020) Review of remote ECG signal monitoring, preprocessing and arrhythmia detection. J Educ Pure Sci 10(2):167–178
2. Badnjevic A, Gurbeta Pokvić L (2021) Internet of Things (IoT) based smart healthcare system. In: 2016 international conference on signal processing, communication, power and embedded system (SCOPES). IEEE
3. Çelebi M (2020) Portable ECG monitoring device design based on Arduino. In: 2020 medical technologies congress (TIPTEKNO)
4. Zareba W (2020) Open access for the annals of noninvasive electrocardiology. Ann Noninvasive Electrocardiol 25(1):e12737
5. Singh DP, Pasupulla AP, Shanko A, Nithya P, Singh RM (2024) Performance analysis of ECG disease classification using SVM classifier. In: 2024 IEEE international conference on

interdisciplinary approaches in technology and management for social innovation (IATMSI), vol 2, pp 1–4
6. Mašetić Z, Gurbeta Pokvić L, Badnjević A (2021) Machine learning for prediction of electrocardiogram signal parameters. Acta Inform Med 29(1):21–26
7. Lin J (2024) Wearable emotion recognition technology based on heart rate variability and EEG rhythm. Trans Comput Sci Intell Syst Res
8. Sarić R, Jokić D, Beganović N, Pokvić LG, Badnjević A (2020) FPGA-based real-time epileptic seizure classification using artificial neural network. Biomed Signal Process Control 62:102106
9. Spahić L, Pokvić LG, Badnjević A (2021) Medical device conformity assessment: regulations and practice. Acta Inform Med 29(2):94–99
10. Subasi A (2019) Biomedical signal analysis and its usage in healthcare. Biomed Eng Appl Healthc 423–452
11. Khanna A, Selvaraj P, Gupta D, Sheikh TH, Pareek PK, Shankar V (2023) Internet of things and deep learning enabled healthcare disease diagnosis using biomedical electrocardiogram signals. Expert Syst 40(4):e12864
12. Badnjević A, Pokvić LG, Hasičić M, Bandić L, Mašetić Z, Kovačević Ž (2019) Evidence-based clinical engineering: machine learning algorithms for prediction of defibrillator performance. Biomed Signal Process Control 54:101629
13. Merletti R (2010) The electrode–skin interface and optimal detection of bioelectric signals. Physiol Meas 31(10):E01
14. Thomas P, Moni R (2016) Methods for improving the classification accuracy of biomedical signals based on spectral features. Technol 7(1):105–116
15. Badnjević A, Pokvić LG, Džemić Z, Bečić F (2020) Risks of emergency use authorizations for medical products during outbreak situations: a COVID-19 case study. Biomed Eng Online 19:1–14
16. Badnjević A, Pokvić LG, Deumić A, Bećirović LS (2022) Post-market surveillance of medical devices: a review. Technol Health Care 30(6):1315–1329
17. Olaniyan OM, Oyedeji AI, Ifeka OI (2021) Development of a microcontroller EEG-based system for diagnosis of autism spectrum disorder in developing countries. J Phys Conf Ser 1734(1):012033
18. Zarei R (2017) Developing enhanced classification methods for ECG and EEG signals. Doctoral dissertation, Victoria University
19. Kovačević Ž, Gurbeta Pokvić L, Spahić L, Badnjević A (2020) Prediction of medical device performance using machine learning techniques: infant incubator case study. Health Technol 10(1):151–155
20. Rabuske T, Silva M, Brito D, Silva J, Busse HI, Silva AV, Cardoso S (2021) A biomedical imaging system based on an integrated array of magnetoresistive sensors. In: 2021 19th IEEE international new circuits and systems conference (NEWCAS), pp 1–4
21. Zhou Y, Song S, Wang S, Wan Y, Yang T, Yu X, Zhao M (2023) A review on direct digital conversion techniques for biomedical signal acquisition. Electronics 12(12):2676
22. Choudhary SK, Bera TK (2024) Design and analysis of low-noise instrumentation amplifier for biomedical signal acquisition systems. In: 2024 4th international conference on computer, communication, control & information technology (C3IT), pp 1–6
23. Gresham B, Tang W (2023) Cross-correlation for dynamic predictive sampling signals in biomedical signal acquisition. In: Cross-correlation for dynamic predictive sampling signals in biomedical signal acquisition, pp 1–5
24. Chowdhury A, Das D, Cheung RC, Chowdhury MH (2023) Hardware/software co-design of an ECG-PPG preprocessor: a qualitative & quantitative analysis. In: 2023 international conference on electrical, computer and communication engineering (ECCE), pp 1–6
25. Coccia A, Amitrano F, Donisi L, Cesarelli G, Pagano G, Cesarelli M, D'Addio G (2021) Design and validation of an e-textile-based wearable system for remote health monitoring. Acta Imeko 10(2):220–229
26. Ng CL, Reaz MBI, Crespo ML, Cicuttin A, Shapiai MI, Ali SHBM, Chowdhury MEH (2024) A versatile and wireless multichannel capacitive EMG measurement system for digital healthcare. IEEE Internet Things J

27. Hrvat F, Spahić L, Pokvić LG, Badnjević A (2020) Artificial neural networks for prediction of medical device performance based on conformity assessment data: Infusion and perfusor pumps case study. In: 2020 9th Mediterranean conference on embedded computing (MECO), pp 1–4
28. Vityazeva T, Mikheev A (2020) Accuracy loss in multi-rate processing of biomedical signals. In: 2020 9th Mediterranean conference on embedded computing (MECO), pp 1–4
29. Wan Y, Song S, Zhao M, Li M, Wang S, Tan Z (2022) A review on direct digital conversion for biomedical signal acquisition ICs. In: 2022 IEEE 16th international conference on solid-state & integrated circuit technology (ICSICT), pp 1–3
30. Li Y (2024) Advancements in wearable smart devices for biomedical signal processing: technologies and applications. Highlights Sci Eng Technol 112:351–358
31. Chmielewski M, Pieczonka P, Kukiełka M, Gutowski T (2020) Software architecture optimization of mobile biomedical sensor-based tools providing analytical services for disease diagnostics and assistance. Procedia Manuf 44:575–582
32. Gu J (2024) AI-empowered neural processing for intelligent human-machine interface and biomedical devices. Open Access Gov
33. Liu J, Fan J, Zhong Z, Qiu H, Xiao J, Zhou Y, Zhou J (2023) An ultra-low power reconfigurable biomedical AI processor with adaptive learning for versatile wearable intelligent health monitoring. IEEE Trans Biomed Circuits Syst 17(5):952–967
34. Masihi S, Panahi M, Maddipatla D, Hanson AJ, Fenech S, Bonek L, Atashbar MZ (2021) Development of a flexible wireless ECG monitoring device with dry fabric electrodes for wearable applications. IEEE Sens J 22(12):11223–11232
35. Kumar S (2021) A review of transimpedance amplifiers used in biomedical applications. In: 2021 5th international conference on computing methodologies and communication (ICCMC), pp 1314–1321
36. Pullano SA, Kota VD, Kakaraparty K, Fiorillo AS, Mahbub I (2022) Optically unobtrusive zeolite-based dry electrodes for wearable ECG monitoring. IEEE Sens J 22(11):10630–10639
37. Song J, Zhou T, Liang Z, Liu R, Guo J, Yu X, Li J (2021) Electrochemical characteristics based on skin-electrode contact pressure for dry biomedical electrodes and application to wearable ECG signal acquisition. J Sens 2021(1):7741881
38. Resta FSA, Setiawan R, Hikmah NF, Amalia R (2024) Noncontact electrocardiogram system for mattresses using silver conductive fabric. In: 2024 11th international conference on electrical engineering, computer science and informatics (EECSI), pp 33–38
39. Tasneem N, Mahbub I (2020) Dry electrode based low-power ECG acquisition system with adaptive motion artifacts cancellation. In: 2020 IEEE 63rd international Midwest symposium on circuits and systems (MWSCAS), pp 913–916
40. Cai Y, Chen X, Jin Y (2022) Electrodes, front-end and downstream applications for the mobile ECG. In: 2022 international conference on electronics and devices, computational science (ICEDCS), pp 113–116
41. Shah SAA, Ali M, Altaf MAB (2021) A low-power bio-potential acquisition for the wearable shockable electrocardiogram system. In: 2021 1st international conference on microwave, antennas & circuits (ICMAC), pp 1–4
42. Gurbeta L, Badnjevic A, Maksimovic M, Omanovic-Miklicanin E, Sejdic E (2018) A telehealth system for automated diagnosis of asthma and COPD. J Am Med Inform Assoc 25(9):1213–1217
43. Qaisar SM, Fesquet L (2020) An effective QRS selection based on the level-crossing sampling and activity selection. In: 2020 6th international conference on event-based control, communication, and signal processing (EBCCSP), pp 1–5
44. AlHinai N (2020) Introduction to biomedical signal processing and artificial intelligence. Biomedical signal processing and artificial intelligence in healthcare. Academic Press, pp 1–28
45. Turovskii YA, Borzunov SV, Vakhtin AA (2021) Estimation method of medical and biological signal features in “human-computer” interface systems. In: 2021 14th international conference management of large-scale system development (MLSD), pp 1–4

46. Wei Y, Zhou J, Wang Y, Liu Y, Liu Q, Luo J, Huang L (2020) A review of algorithm and hardware design for AI-based biomedical applications. IEEE Trans Biomed Circuits Syst 14(2):145–163
47. Zhu Y, Zhang H, Liu B, Dou J. Exploring cutting-edge developments in deep learning for biomedical signal processing. Int J Adv Comput Sci Appl. 2024;15(5).
48. Srinivasagan R (2025) Advanced low-power filter architecture for biomedical signals with adaptive tuning. PLoS ONE 20(1):e0311768
49. Patel V, Shah AK (2021) Machine learning for biomedical signal processing. Elsevier E-Books, pp 47–66
50. Solis R, Jin Y, Zhang J (2023) Evolutionary deep learning using hybrid EEG-fNIRS-ECG signals for workload classification. Neurocomputing 540:125–135
51. Birhanu H, Kassaw A (2021) Cardiovascular signal processing: state of the art and algorithms. In: Future of information and communication conference. Cham: Springer International Publisher, pp 113–127
52. Jalagam MK, Mittal VK (2021) Recent studies on applications using biomedical signal processing: a review. In: 2021 2nd global conference for advancement in technology (GCAT), pp 1–11
53. Xu L, Xia Y (2023) EEG signal classification and feature extraction methods using deep learning: a review. Front Neurosci 17:1120456
54. Karthik S, Santhosh M, Kavitha MS, Paul AC (2022) Automated deep learning-based cardiovascular disease diagnosis using ECG signals. Comput Syst Sci Eng 42(1):183–199
55. Marwa SR, Shakir MA (2024) Improving ECG signals classification by using deep learning techniques: a review. ITM Web Conf 64:01023
56. Khalid M, Pluempitiwiriyawej C, Wangsiripitak S, Murtaza G, Abdulkadhem AA (2024) Applications of deep learning in ECG classification for disease diagnosis: a systematic review and meta-data analysis. Eng J
57. Khatar Z, Bentaleb D, Drissi S (2024) Bayesian methods for multi-objective optimization of hybrid numerical filters in ECG signal processing for accurate arrhythmia classification. Stat Optim Inf Comput 13(3):993–1012
58. Xue S, Geng X, Li D (2022) Feature extraction and classification of EEG signal based on deep learning
59. Merdjanovska E, Rashkovska A (2020) Respiration extraction from single-channel ECG using signal-processing methods and deep learning. In: 2020 43rd international convention on information, communication and electronic technology (MIPRO), pp 321–326
60. Hafizović L, Čaušević A, Deumić A, Bećirović LS, Pokvić LG, Badnjević A (2021) The use of artificial intelligence in diagnostic medical imaging: systematic literature review. In: 2021 IEEE 21st international conference on bioinformatics and bioengineering (BIBE), pp 1–6
61. Abhishek M, Rohit S (2023) AI for the detection of neurological conditions using EEG data. Int J Healthc Technol
62. Nalmpantis A, Passalis N, Tefas A (2023) End-to-end trainable Gaussian filtering for electrocardiogram signal classification using deep learning. In: 2023 31st European signal processing conference (EUSIPCO), pp 1060–1064
63. Catic A, Gurbeta L, Kurtovic-Kozaric A, Mehmedbasic S, Badnjevic A (2018) Application of neural networks for classification of Patau, Edwards, Down, Turner and Klinefelter syndrome based on first trimester maternal serum screening data, ultrasonographic findings and patient demographics. BMC Med Genomics 11:1–12
64. Kalina P (2020) Ethical and legal manifestations of informed consent. Technium Soc Sci J 8:753
65. Adeniyi AO, Arowoogun JO, Okolo CA, Chidi R, Babawarun O (2024) Ethical considerations in healthcare IT: a review of data privacy and patient consent issues. World J Adv Res Rev 21(2):1660–1668
66. Gera A, Sood J (2023) Informed consent—ethical doctrine and a legal mandate. Curr Med Res Pract 13(6):286–291
67. Chougule U, Sharma K, Kolekar S, Nerlekar HV, Gupta S (2025) Informed consent in surgery: legal and ethical considerations. J Neonatal Surg 14(2S):21–30

68. Williamson SM, Prybutok V (2024) Balancing privacy and progress: a review of privacy challenges, systemic oversight, and patient perceptions in AI-driven healthcare. Appl Sci 14(2):675
69. Risdawati I (2024) Legal aspects in implementing an informed consent system in patient health practices. Int J Soc Law 2(1):115–126
70. Moeini S, Shahriari M, Shamali M (2020) Ethical challenges of obtaining informed consent from surgical patients. Nurs Ethics 27(2):527–536
71. Albalawi AF, Yassen MH, Almuraydhi KM, Althobaiti AD, Alzahrani HH, Alqahtani KM (2024) Ethical obligations and patient consent in the integration of artificial intelligence in clinical decision-making. J Healthc Sci 4(12):957–963
72. Alhasan TK (2025) Managing legal risks in health information exchanges: a comprehensive approach to privacy, consent, and liability. J Healthc Risk Manag 44(4):12–24
73. Badnjevic A, Deumic A, Imamovic E, Pokvic LG (2023) A novel method for conformity assessment testing of defibrillators for post-market surveillance purposes. Technol Health Care 31(1):317–325

Chapter 2
Biomedical Signal Acquisition and Processing

This section explores the physiological origins, acquisition methods, and preprocessing requirements of biomedical signals. It begins with the biological basis of signal generation (Sect. 2.1), emphasizing electrodes, recording techniques, and the inherent limitations of measurement systems. Signal conditioning and amplification (Sect. 2.2) are then discussed, including filtering, band-limiting, impedance matching, and isolation to ensure signal fidelity. The principles of analog-to-digital conversion (Sect. 2.3) are introduced, covering architectures, performance metrics, and emerging AI-enabled smart ADC systems. Noise and artifact removal (Sect. 2.4) is analyzed, with methods for minimizing interference across different modalities. Core analytical methods, time-domain (Sect. 2.5), frequency-domain (Sect. 2.6), and time–frequency techniques such as wavelets (Sect. 2.7)—are reviewed, demonstrating their role in feature extraction and diagnostic interpretation. Finally, digital filtering methods (Sect. 2.8) are outlined, with low-pass, high-pass, band-pass, and notch filter designs tailored to biomedical applications. Collectively, this section provides a technical foundation for the accurate acquisition and processing of biomedical signals.

2.1 Physiological Origins of Biomedical Signals

The human body is a dynamic and intricate system, constantly generating physiological signals that provide insight into its inner workings. From the electrical impulses governing heart rhythms to the subtle fluctuations in blood volume detectable through optical techniques, these signals form the foundation of biomedical diagnostics and monitoring. Understanding the physiological origins of biomedical signals is crucial for developing technologies that accurately capture, process, and interpret these

© The Author(s), under exclusive license to Springer Nature Switzerland AG 2026
A. Badnjević and L. Spahić, *Biomedical Signals and Systems*, Series in BioEngineering,
https://doi.org/10.1007/978-3-032-10982-8_2

biological activities. The complexity of human physiology necessitates a broad spectrum of biomedical sensing techniques, each designed to detect and analyze specific types of signals.

Biomedical signals arise from various physiological activities, including electrical, mechanical, acoustic, and optical processes. Electrical signals, such as those recorded by electrocardiography (ECG) and electroencephalography (EEG), originate from the ionic movements within excitable tissues like the heart and brain. Mechanical signals, such as those captured by ultrasound transducers, stem from the movement of organs, tissues, and fluids within the body. Optical signals, exemplified by photoplethysmography (PPG), rely on the interaction of light with biological tissues to provide information about blood circulation and oxygenation [1]. Each of these modalities offers unique advantages and challenges, which dictate their applications in both clinical and research settings.

The origins of biomedical signals lie in the fundamental physiological processes that sustain life. The heart, as an electrically active organ, generates impulses that travel through specialized conduction pathways, resulting in characteristic waveforms observable in ECG recordings. Similarly, the brain's neural activity produces electrical fluctuations that can be detected via EEG. These signals not only reflect normal physiological function but also serve as indicators of pathological conditions [2]. For instance, abnormal cardiac rhythms in an ECG may suggest arrhythmias, while irregular EEG patterns can indicate neurological disorders such as epilepsy. Beyond electrical activity, mechanical forces within the body generate vital biomedical signals. The oscillations of arterial walls due to pulsatile blood flow can be measured through pulse wave analysis, providing critical information about vascular health. Respiratory movements create pressure changes in the lungs that can be captured by spirometry, aiding in the assessment of pulmonary function. Muscle contractions produce myoelectric signals, which are recorded by electromyography (EMG) to evaluate neuromuscular disorders and rehabilitation progress. The mechanical properties of tissues and organs can also be examined using ultrasound imaging, where high-frequency sound waves penetrate the body and reflect off internal structures to create detailed anatomical and functional maps [3].

Optical biomedical signals offer another dimension of physiological insight. The principles of light absorption, reflection, and scattering underlie technologies like PPG and optical coherence tomography (OCT). In PPG, light-emitting diodes and photodetectors measure variations in blood volume within microvascular tissues, enabling the estimation of heart rate and blood oxygen levels [4]. OCT utilizes low-coherence interferometry to generate high-resolution images of biological tissues, making it invaluable in ophthalmology and dermatology. The advantages of optical methods lie in their non-invasiveness and real-time capabilities, but their effectiveness can be influenced by factors such as tissue composition, motion artifacts, and ambient light interference [5, 6]. Despite the technological advancements in biomedical signal acquisition, each modality has its inherent limitations. Electrical recordings, though widely used, are susceptible to noise from external sources and require proper electrode placement to ensure accurate readings. Mechanical measurements often necessitate direct contact with the body, which can be uncomfortable for patients and introduce motion artifacts. Optical techniques, while providing high spatial resolution, may struggle with penetration depth and variability in tissue optical properties [7]. Understanding these limitations is essential for optimizing signal acquisition strategies and improving diagnostic reliability.

The physiological origins of biomedical signals are deeply rooted in the intricate interplay of biological systems. By studying these signals, researchers and clinicians gain access to a wealth of information that aids in disease detection, treatment planning, and patient monitoring. As technology continues to evolve, the ability to harness these signals with greater precision and efficiency will further advance the field of biomedical engineering, ultimately improving healthcare outcomes and patient well-being.

2.1.1 Electrodes

Electrodes are essential components in biomedical signal acquisition, as they serve as the interface between biological tissues and electronic measurement systems. They enable the detection of bioelectrical signals such as ECG, EEG, and EMG. The performance of an electrode is influenced by its material composition, design, and the specific biological tissue with which it interacts [8].

Principles of Electrode Function

Electrodes function by converting ionic currents in the body into measurable and analyzable electrical currents. When an electrode is placed on the skin or directly on tissues, it establishes an electrochemical interface where charge transfer occurs. The interface is governed by the half-cell potential, which is the voltage difference between the electrode and the surrounding electrolyte solution. The fundamental equation governing the voltage at this interface is the Nernst equation (Eq. 2.1):

$$E = E^0 + \frac{RT}{zF} \ln\left(\frac{[\mathrm{Ox}]}{[\mathrm{Red}]}\right) \tag{2.1}$$

where:

- E is the electrode potential,
- E^0 is the standard electrode potential,
- R is the universal gas constant,
- T is the temperature in Kelvin,
- z is the number of electrons transferred,
- F is the Faraday constant,

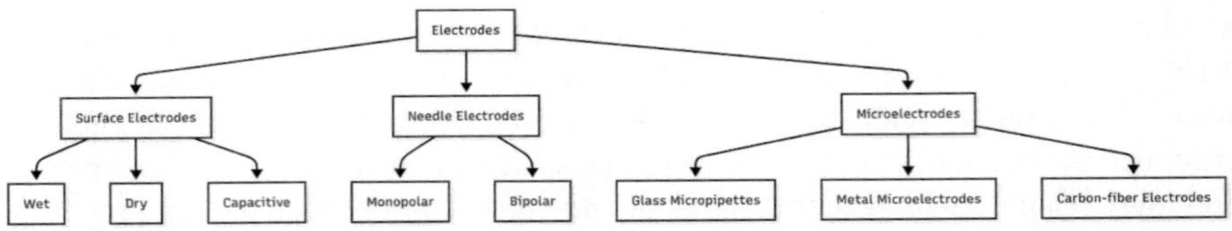

Fig. 2.1 The three main types of electrodes

- [Ox] and [Red] represent the concentrations of oxidized and reduced species, respectively.

In biomedical applications, electrodes should minimize polarization effects, which can distort the signal and introduce noise. Electrode impedance is a crucial factor in determining signal quality, particularly in high-frequency applications such as EEG. The skin–electrode contact, the electrode material, and the properties of the electrolyte influence the impedance. The impedance Z of an electrode can be modeled as Eq. 2.2:

$$Z = R_s + \frac{1}{j\omega C_d} \tag{2.2}$$

where:

- R_s is the series resistance of the electrode–skin interface,
- C_d is the double-layer capacitance,
- ω is the angular frequency of the signal,
- j is the imaginary unit.

Lower impedance is desirable to reduce signal loss and improve fidelity (Fig. 2.1).

Surface Electrodes: Non-Invasive Monitoring of Bioelectrical Signals

For many diagnostic and monitoring a pplications, a non-invasive approach is preferred, which is where surface electrodes play a crucial role. These electrodes are applied directly to the skin to measure signals from the heart (ECG), brain (EEG), and muscles (EMG). Because the skin acts as a natural barrier to electrical signals, a conductive interface is necessary to improve signal transfer, which is often achieved using electrolyte gels or hydrogels.

One of the most commonly used surface electrodes is the metal-plate electrode, typically made from silver or silver chloride (Ag/AgCl). The choice of Ag/AgCl is not arbitrary—it provides a stable half-cell potential, meaning that it minimizes unwanted electrical noise that could distort the signal. This stability makes it particularly useful for ECG and EEG applications, where even minor fluctuations can affect diagnostic accuracy. The primary advantage of Ag/AgCl electrodes lies in their low impedance and excellent signal quality, which is essential for long-term monitoring. Since they are easy to apply and relatively inexpensive, they remain a staple in clinical settings. However, they do have limitations: they are prone to motion artifacts, meaning that sudden movements or improper attachment can introduce noise [9]. Additionally, for optimal performance, the skin must be adequately prepared, which might not always be convenient in emergency settings.

Adhesive Gel Electrodes: Enhanced Contact for Reliable Readings

To overcome some of the challenges of metal-plate electrodes, adhesive gel electrodes were developed. These electrodes incorporate a layer of conductive gel or hydrogel that enhances contact with the skin, reducing impedance and improving signal consistency. This makes them a preferred choice for EEG and ECG recordings, particularly for long-term monitoring where stability is critical. Since the gel acts as a bridge between the skin and the electrode, it reduces the resistance at the electrode–skin interface, allowing for higher-quality signal acquisition. This reduces the likelihood of signal degradation over time [10]. However, there is a trade-off: the gel can dry out, diminishing its effectiveness, and some individuals may experience skin irritation with prolonged use.

Dry Electrodes: The Next Step in Wearable Technology

The reliance on gels in traditional electrodes presents challenges, especially for wearable and mobile health applications. Enter dry electrodes, which eliminate the need for any conductive gel by using advanced materials such as conductive polymers, carbon-based materials, or metallic coatings to establish electrical contact [11]. Their most significant advantage is convenience, since they do not require skin preparation or gel application, they are ideal for continuous monitoring in real-world settings, such as in smart textiles and wearable health devices. However, the downside is that they typically have higher impedance compared to gel-based electrodes, which can affect signal quality [12]. Additionally, they tend to be more susceptible to motion artifacts, requiring sophisticated signal processing techniques to clean the data [13].

Needle Electrodes: Penetrating Deeper for Precision

While surface electrodes are effective for capturing electrical activity at the skin level, they are not always suitable for measuring deeper bioelectrical signals. This is where needle electrodes come into play. These electrodes are specifically designed for insertion into tissues, making them ideal for applications such as intramuscular EMG and intraoperative neurophysiological monitoring. In its simplest form, a monopolar needle electrode consists of a thin metal needle that is inserted directly into the muscle or nerve. Since a monopolar electrode captures electrical activity from a relatively large area, a reference electrode is required on the skin to differentiate the signal from background noise. They offer high spatial resolution, making them ideal for pinpointing neuromuscular activity in specific regions. However, because they capture a broad electrical field, their readings can be affected by signals from neighboring tissues. Moreover, their invasive nature means that they can be uncomfortable or even painful for patients, and precise positioning is crucial for obtaining accurate results (Figs. 2.2 and 2.3).

To improve signal clarity and reduce external interference, bipolar needle electrodes were developed. Instead of a single recording surface, these electrodes feature

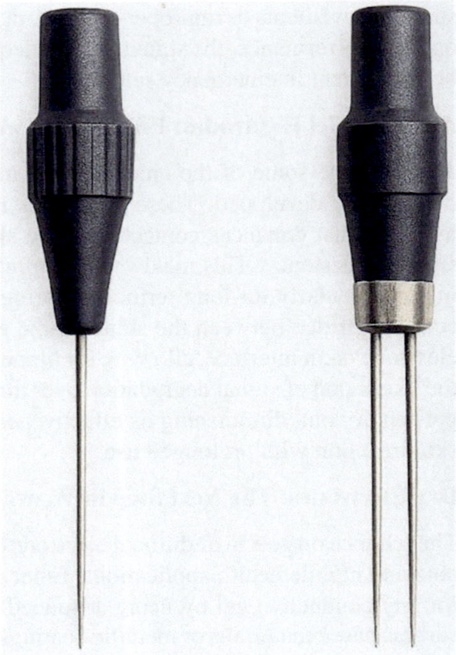

Fig. 2.2 Monopolar versus bipolar needle electrodes

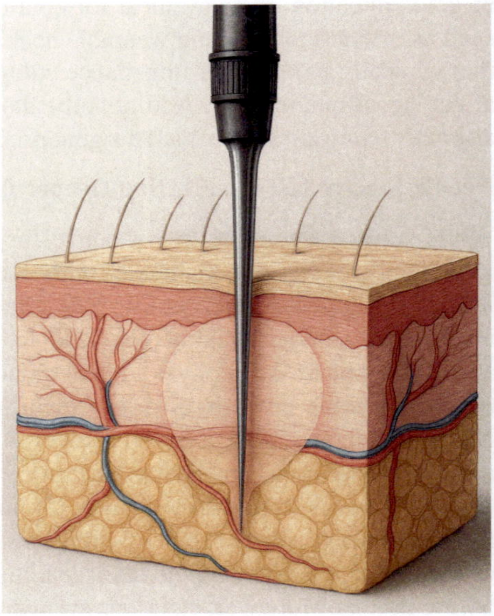

Fig. 2.3 Depth of signal capture in needle electrodes

two closely spaced recording points within the same needle, allowing them to detect local voltage differences while canceling out background noise. Because they are less susceptible to external interference, they provide cleaner, more localized signals, making them ideal for detailed nerve conduction studies and intraoperative monitoring. The trade-off is that their design is more complex and, consequently, they are more expensive than monopolar electrodes.

Advanced Microelectrodes

For applications requiring extreme precision, such as neuroscience research and brain-computer interfaces (BCIs), standard needle electrodes are not sufficient. Instead, microelectrodes—electrodes with tip diameters in the micrometer range—are used. These techniques enable researchers and clinicians to record activity from individual neurons or very small groups of cells, providing an unprecedented level of detail [6].

Among the earliest and most precise microelectrodes are glass micropipette electrodes. These are ultra-fine glass tubes filled with an electrolyte solution that can penetrate a single cell to measure intracellular activity. Their exceptionally high spatial resolution allows for direct measurement of intracellular potentials, making them indispensable in cellular electrophysiology. However, they are also fragile and difficult to manipulate, requiring highly specialized equipment and expertise.

A more modern alternative is silicon microelectrodes, which are fabricated using microelectromechanical systems (MEMS) technology. These electrodes can be precisely engineered with multiple recording sites, enabling simultaneous data collection from multiple neurons. They offer high durability and stability, making them ideal for long-term implantation in research and clinical applications. They are also compatible with advanced signal processing techniques, allowing researchers to analyze complex neural networks in unprecedented detail. However, their fabrication is complex and costly, making them less accessible than traditional electrode technologies.

2.1.2 Applications and Limitations

ECG records the heart's electrical activity through surface electrodes, typically using a 12-lead configuration. Recent advancements include dry electrodes and wireless wearable devices, which enable continuous monitoring. ECG is widely used in diagnosing arrhythmias, ischemia, myocardial infarction, and heart failure, as it can detect abnormal heart rhythms, identify changes associated with heart attacks, and monitor conduction abnormalities. However, ECG is prone to motion artifacts caused by physical movement, and the accuracy can be compromised by improper electrode placement and skin resistance, which introduces signal distortion [14].

EEG records brain electrical activity through electrodes placed on the scalp, typically following the 10–20 system of electrode placement. With the advent of wireless and high-density arrays, EEG now allows for more detailed brain mapping.

EEG plays a crucial role in detecting seizures, analyzing sleep stages and disorders, and facilitating brain-computer interfaces for individuals with disabilities [15]. However, EEG suffers from limited spatial resolution compared to imaging techniques like MRI, and the data can be affected by noise and artifacts from eye and muscle movements. Additionally, variations in scalp impedance can impact signal quality (Figs. 2.4, 2.5 and 2.6).

EMG records the electrical activity produced by skeletal muscles during contraction, typically using surface electrodes placed on the skin or intramuscular needle electrodes for deeper muscle assessment. Wearable EMG sensors now enable real-time monitoring of muscle activity, which is valuable in diagnosing neuromuscular disorders, such as muscular dystrophy, peripheral neuropathies, and conditions like carpal tunnel syndrome. Additionally, EMG plays a key role in rehabilitation and

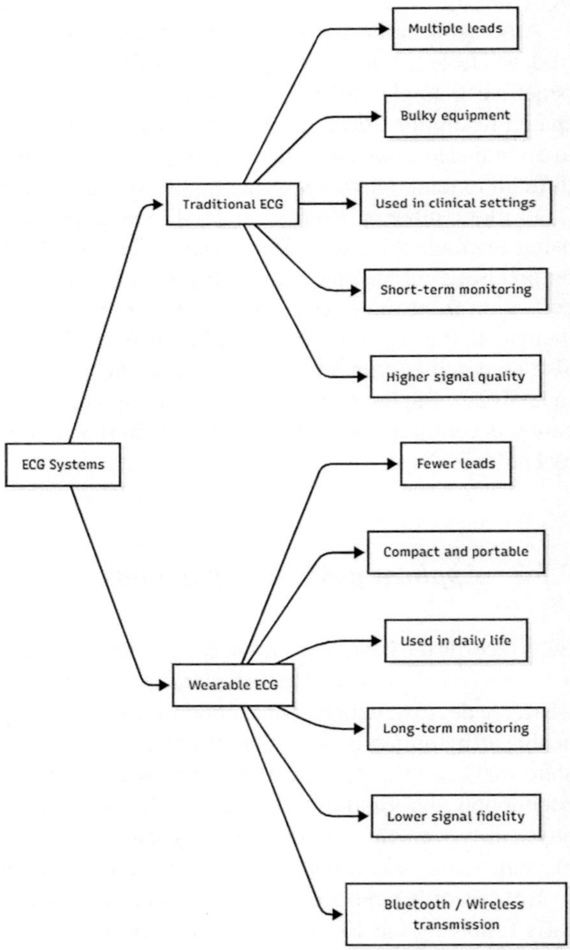

Fig. 2.4 Comparison of traditional versus wearable ECG systems

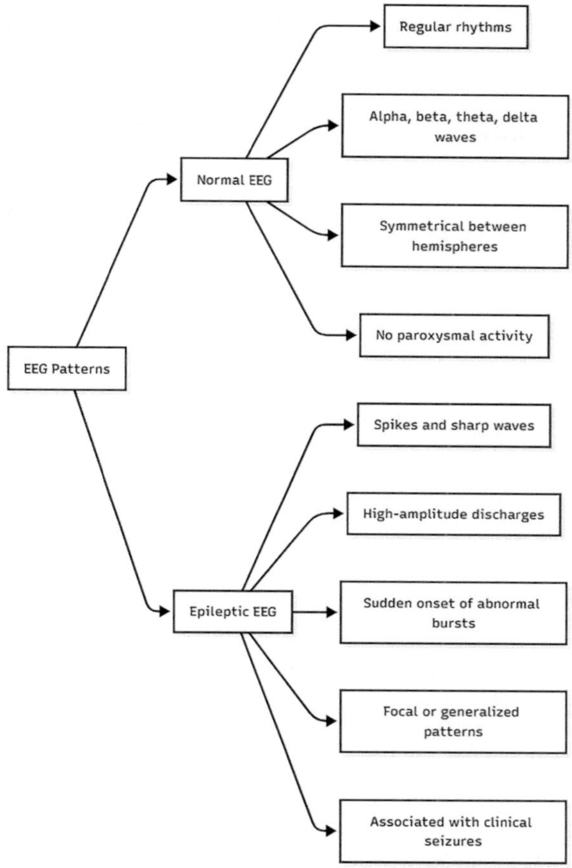

Fig. 2.5 Comparison of normal versus epileptic EEG patterns blood pressure is commonly measured using cuff-based systems (sphygmomanometers) or invasive methods like arterial catheters. Wearable cuffless devices, utilizing techniques such as PPG or ultrasound, are also being developed for continuous monitoring. BP measurement is essential in managing hypertension, assessing cardiovascular health preoperatively, and conducting ambulatory monitoring to track fluctuations over time. While cuff-based methods can be uncomfortable and may lack accuracy if not calibrated, cuffless devices may perform inconsistently under certain conditions, limiting their reliability [16, 17]

prosthetic control. However, EMG can be susceptible to interference from external noise, such as electrical equipment, and the placement of electrodes on the skin may not always provide accurate readings, particularly in patients with obesity or excessive body hair [18].

Respiratory signals are typically monitored using devices that measure airflow, respiratory effort, or oxygen saturation. Techniques such as spirometry and pulse oximetry are the most common methods for assessing lung function and detecting

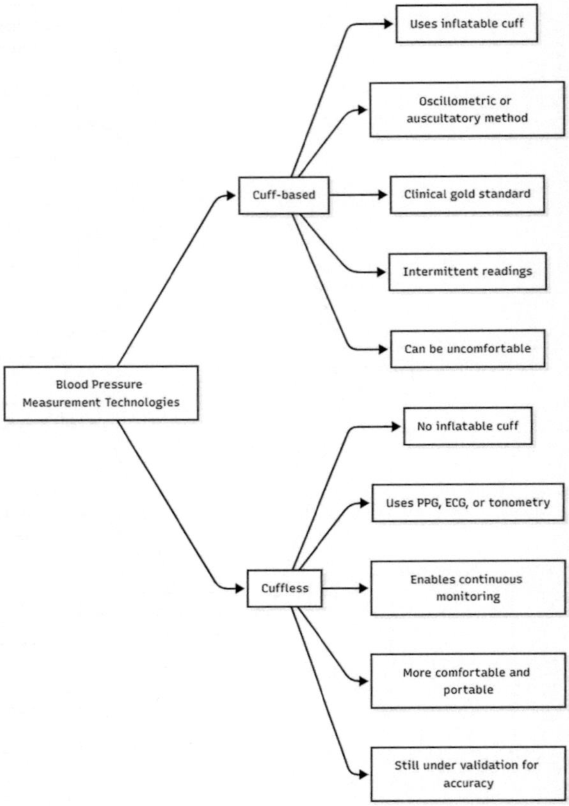

Fig. 2.6 Comparison of cuff-based versus cuffless BP technologies

conditions like asthma, chronic obstructive pulmonary disease (COPD), and sleep apnea. Advances in wearable sensors, utilizing technologies such as PPG, impedance pneumography, and capnography, enable continuous, non-invasive monitoring of respiratory health [19]. Despite their advantages, respiratory measurements can be influenced by external factors such as body position, movement, and environmental conditions. Additionally, the accuracy of non-invasive methods can be affected by factors such as skin tone, sensor calibration, and ambient light conditions.

2.2 Signal Conditioning and Amplification

In biomedical signal acquisition systems, the quality and reliability of the signal are paramount for correct analysis and interpretation. Biomedical signals, such as ECGs, EEGs, EMGs, and others, are inherently weak and often contaminated with various types of noise. These signals require careful preprocessing before they can be digitized and analyzed. Signal conditioning encompasses a range of operations performed on a signal to enhance its quality and suitability for further processing. Among these operations, amplification plays a central role. This chapter offers a comprehensive and in-depth examination of signal conditioning and amplification in biomedical systems, covering theoretical foundations, practical implementations, design challenges, and specific applications.

Biomedical signals represent physiological phenomena. These signals can be electrical, mechanical, optical, chemical, or thermal, and each type comes with its own set of challenges in terms of acquisition and processing. Electrical signals, such as ECG or EEG, are of particular interest because of their broad diagnostic relevance and the relative ease with which they can be acquired non-invasively. These signals often fall within specific frequency bands and amplitude ranges. For example, an EEG signal typically has an amplitude in the microvolt range and frequency content up to about 50 Hz, whereas EMG signals have higher frequencies and larger amplitudes. However, they are also susceptible to various types of noise, such as power line interference, motion artifacts, baseline drift, and other physiological or environmental sources.

Before these signals can be digitized by an analog-to-digital converter (ADC), they must undergo signal conditioning to ensure they are within the appropriate amplitude and frequency range and to remove unwanted noise components. Signal conditioning encompasses a range of processes, including amplification, filtering, impedance matching, isolation, and protection.

2.2.1 *Signal Conditioning*

Signal conditioning is a fundamental process in biomedical instrumentation, encompassing the set of operations applied to a biosignal to render it suitable for analysis, visualization, storage, or further computational processing. Biomedical signals, such as ECG, EEG, EMG, and others, are inherently low in amplitude, often contaminated with various forms of noise and interference, and can exhibit significant variability in terms of frequency content, baseline drift, and dynamic range. The analog electrical output from a biosensor or transducer typically reflects not only the physiological phenomenon of interest but also artifacts stemming from environmental factors, biological variability, and limitations of the acquisition hardware. Without appropriate signal conditioning, the raw signal is generally unsuitable for digitization and analysis, particularly in the context of diagnostic systems, real-time monitoring, or therapeutic control loops [20].

Signal conditioning bridges this gap by transforming the raw, analog biosignal into a form that is compatible with the subsequent stages of the signal processing chain—typically digitization by an ADC and processing by a microcontroller, microprocessor, or digital signal processor (DSP) (Fig. 2.7).

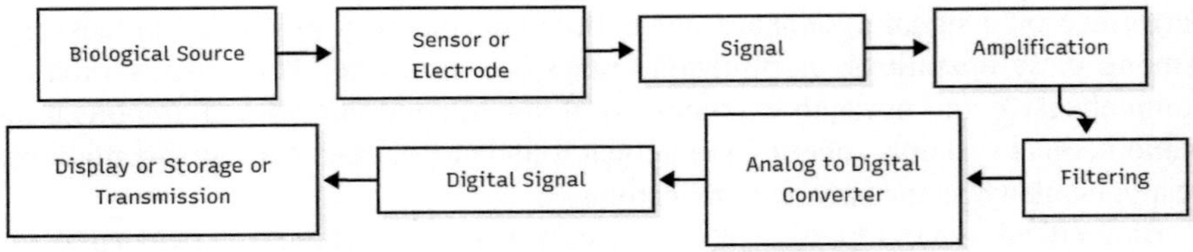

Fig. 2.7 Block diagram of a biomedical signal acquisition

Unlike industrial signals or controlled laboratory signals, biomedical signals are biologically variable, nonstationary, and embedded in a high-noise environment [21]. They present several challenges that must be addressed in the conditioning stage:

- Low Amplitude: Many bioelectrical signals range from microvolts to millivolts. For example, EEG signals typically range from 10 to 100 μV, while ECG signals may vary between 0.5 and 5 mV.
- Noise Sensitivity: Biosignals are susceptible to noise from multiple sources, including:

– Electromagnetic interference (EMI), especially from 50/60 Hz mains power
– Motion artifacts (electrode movement)
– Baseline drift due to changes in skin impedance or patient movement
– Crosstalk from other physiological signals

- Large Dynamic Range: Signals such as EMG can have significant variability, ranging from a few microvolts at rest to several millivolts during strong contraction.
- Safety Considerations: Direct electrical contact with the body mandates rigorous safety protocols to prevent current leakage or faults.

Signal conditioning addresses these challenges by employing a suite of hardware and sometimes hybrid analog–digital techniques tailored to the specific characteristics of each signal [22]. The signal conditioning stage typically involves the following operations, applied either sequentially or in integrated architectures.

As most biosignals are well below the input range of standard ADCs (which often require a 0–3.3 V or 0–5 V range), amplification is essential. The objective is to increase signal amplitude without introducing distortion, and while preserving the frequency and phase characteristics of the input. Amplification may also include differential input stages to improve common-mode noise rejection [23].

2.2.2 Fundamental Concepts of Amplification

Amplification is a foundational process in biomedical signal conditioning, responsible for scaling weak physiological signals to voltage levels suitable for further processing, including analog filtering, digitization, and computational analysis. The essence of amplification lies in increasing the amplitude of an electrical signal while preserving its temporal and spectral integrity [24]. This means that the waveform shape, frequency components, and timing information must remain undistorted during the amplification process. In practical terms, this requires a high degree of linearity and low noise operation across the relevant frequency band.

Many biopotential signals, such as ECG, EEG, EMG, and intracranial neural recordings, have amplitudes in the range of microvolts to low millivolts, which are often several orders of magnitude below the minimum input requirements of ADCs. Without appropriate amplification, such signals would be obscured by quantization noise, system noise, or even remain undetectable. Therefore, amplifiers form the front line of any biomedical acquisition chain [25]. The goals of signal conditioning are to:

- Improve the signal-to-noise ratio (SNR)
- Prevent aliasing and distortion
- Adapt signal levels for digitization
- Ensure patient safety through electrical isolation.

Amplifiers are characterized by a set of electrical properties that determine their suitability for biomedical applications: gain, input and output impedance, bandwidth, common-mode rejection, noise performance, and linearity. Each of these parameters must be carefully tuned to the characteristics of the physiological signal and the constraints of the measurement environment [26, 27]. Gain refers to the factor by which the amplitude of a signal is increased. It is a dimensionless quantity when expressed as a linear ratio, and it may also be expressed in logarithmic form as decibels (dB), which is often more convenient for describing signal levels across wide dynamic ranges.

Definitions: Let V_{in} be the input voltage and V_{out} the output voltage (Eq. 2.3):

$$\text{Gain (linear)} = \frac{V_{\text{out}}}{V_{\text{in}}}$$

$$\text{Gain (dB)} = 20 \cdot \log_{10}\left(\frac{V_{\text{out}}}{V_{\text{in}}}\right) \tag{2.3}$$

Example: If an EEG signal has an amplitude of 20 μV and needs to be brought up to 2 V for digitization, the required gain is:

$$\frac{2\text{V}}{20\mu\text{V}} = 100{,}000 \Rightarrow \text{Gain} = 100\text{dB}$$

This is a typical requirement in high-fidelity EEG or cortical recording systems. However, such high gain must be achieved across multiple amplification stages to preserve stability and minimize distortion. Gain can be:

- Fixed (determined by resistor values in the circuit)
- Programmable (digitally adjustable via external control)
- Or variable (using analog methods such as variable gain amplifiers, VGAs).

Overamplification can lead to clipping, where the output exceeds the amplifier's voltage rails, whereas underamplification results in poor resolution and increased susceptibility to noise. Therefore, gain must be designed in conjunction with the expected signal amplitude and the ADC range.

Impedance represents the opposition to current flow in AC circuits and is defined as (Eq. 2.4):

$$Z = R + j\omega L - \frac{j}{\omega C} \tag{2.4}$$

In biomedical instrumentation, the input impedance of the amplifier must be significantly higher than the impedance of the source (e.g., skin–electrode interface), often by a factor of 100 or more. This ensures that the amplifier does not load the source, which would cause signal attenuation and potential waveform distortion due to the voltage divider effect. Typical electrode–skin interface impedance values range from:

- 10 kΩ–100 kΩ for ECG wet electrodes
- Up to 500 kΩ or more for dry or capacitive electrodes.

Hence, input impedance of the amplifier is typically designed to be in the range of 10 MΩ–1 GΩ, depending on the application. Output impedance should be as low as possible to drive subsequent stages (e.g., filters, ADC inputs) without signal loss. An ideal voltage amplifier has infinite input impedance and zero output impedance.

Impedance Matching

In contrast to RF or power applications, where impedance matching maximizes power transfer, biomedical systems focus on voltage transfer and signal fidelity, favoring high-impedance inputs and low-impedance outputs. In biomedical recordings, differential amplifiers are often used to acquire signals from pairs of electrodes [28]. While the biopotential signal is expected to appear as a differential mode voltage, many unwanted interferences—such as electromagnetic fields from power lines—couple equally to both electrodes and appear as common-mode signals. The Common-Mode Rejection Ratio (CMRR) quantifies an amplifier's ability to suppress these common-mode components (Eq. 2.5):

$$\text{CMRR} = \frac{A_{\text{differential}}}{A_{\text{common}}};\ \text{CMRR}_{(\text{dB})} = 20 \cdot \log_{10}\left(\frac{A_{\text{differential}}}{A_{\text{common}}}\right) \tag{2.5}$$

where:

- $A_{differential}$ is the gain for differential signals.
- A_{common} is the gain for common-mode signals (ideally zero).

A CMRR of at least 80–100 dB is desired in EEG and ECG front-ends, particularly in urban environments with high levels of 50/60 Hz interference.

Impedance Matching and Isolation

Impedance matching refers to the careful design consideration that the input impedance of the signal conditioning circuit is suitably matched to the output impedance of the biosensor or the physiological signal source, and that the output impedance of the conditioning stage is compatible with the input impedance of subsequent processing blocks. The electrical impedance of a biosignal source, such as an electrode-tissue interface, can vary widely but is often very high (in the order of kilo-ohms to mega-ohms), especially in non-invasive surface electrodes [35]. If the amplifier or conditioning stage connected to the source has low input impedance, a loading effect occurs, leading to:

- Signal amplitude attenuation: The voltage measured across the amplifier input drops due to voltage division between the source impedance and the amplifier input impedance.
- Distortion of frequency content: The source impedance combined with input capacitance and parasitic elements may act as a filter, altering the spectral properties of the original signal.
- Reduced SNR: Because the effective signal voltage decreases, noise contributions from later stages become more significant.

Hence, high input impedance is essential to ensure minimal current draw from the source and to preserve signal fidelity.

Typical Input and Output Impedance Values

- Input impedance of biomedical amplifiers: Typically ranges from 1 MΩ to several GΩ. For example, EEG and ECG amplifiers often require input impedance above 10 MΩ to prevent attenuation of low-amplitude signals.
- Output impedance of amplifiers: Should be low, generally less than 100 Ω, to drive subsequent stages or long cables without signal loss or distortion.

The electrode–electrolyte-skin interface forms a complex impedance that is frequency-dependent and nonlinear. This impedance can be modeled as a combination of resistive and capacitive components, often approximated by the Randles circuit:

- Rs: Series resistance from skin and electrode leads.
- Cdl: Double-layer capacitance at the electrode–electrolyte interface.
- Rct: Charge transfer resistance.

The frequency-dependent impedance, Z(f), influences both signal amplitude and phase. High-input-impedance amplifiers reduce the loading effect on this complex impedance, thereby preserving the physiological signal's temporal and spectral characteristics [36].

Techniques for Achieving High Input Impedance

- FET input stages: Field-effect transistors (FETs) offer very high input impedance due to their insulated-gate structure.
- Instrumentation amplifiers: Integrated instrumentation amplifiers designed for biomedical applications combine high input impedance with excellent common-mode rejection.
- Buffer amplifiers: Voltage follower configurations using op-amps can present extremely high input impedance to the source while providing low output impedance.

Low output impedance is important for the following reasons:

- Driving long cables or loads: To maintain signal amplitude and reduce noise pickup.
- Impedance bridging: Ensures that the downstream ADC or filter stage receives the full voltage without losses.

Output impedance matching prevents signal reflections and power losses, which is especially critical in high-frequency or digital biomedical systems.

Electrical isolation in biomedical instrumentation separates the patient from the system's power and electronics, preventing electrical shock hazards and eliminating ground loops that introduce interference. Patients are inherently vulnerable to electrical currents due to compromised skin integrity or the presence of implanted devices. Unintended current flow from system faults, mains interference, or differential grounding can cause severe harm or even be fatal. Biomedical standards such as IEC 60,601-1 prescribe strict requirements for patient isolation in medical devices. Isolation amplifiers maintain signal integrity while providing galvanic separation between input and output. They transfer the signal across an isolation barrier via methods such as:

- Capacitive coupling: Signal modulates a high-frequency carrier transmitted across a capacitor barrier.
- Magnetic coupling: Transformer-based isolation.
- Optical coupling: Using LED-photodiode pairs within the amplifier package.

Isolation amplifiers are designed to have low noise, high linearity, and wide bandwidth while meeting isolation voltage ratings typically from 1.5 kV to 5 kV [37]. Optoisolators utilize an LED and a phototransistor or photodiode pair to transmit signals through light. They provide electrical isolation up to several kilovolts and are commonly used for digital signal isolation in telemetry or control circuits. While effective for digital signals, the bandwidth and linearity limitations of these devices

Table 2.1 Summary and best practices in impedance and isolation

Aspect	Design goal	Typical values	Notes
Input impedance	High to prevent source loading	>10 MΩ for ECG/EEG	FET input stages, buffers preferred
Output impedance	Low for efficient signal transfer	<100 Ω	Prevents signal loss and reflections
Isolation voltage	Highly focused on protecting patient safety	≥1.5 kV to 5 kV	Must comply with IEC standards
Patient leakage current	Minimal to prevent shock risk	<10 μA	Regular safety testing is required

typically preclude direct analog signal isolation without additional circuitry. Transformer coupling is used to isolate AC signals and power supplies. In some biomedical devices, small isolation transformers are integrated for power line decoupling or RF isolation (Table 2.1).

Isolation Requirements in Biomedical Systems

- Patient leakage current: Must be minimized to safe levels (microamps) under normal and fault conditions.
- Isolation voltage rating: Defines the maximum allowable voltage differential across the isolation barrier.
- Isolation resistance: High resistance (>100 MΩ) across the barrier to prevent leakage.
- Transient immunity: Ability to withstand electrical surges and spikes without breakdown.

Practical Implementation in Hospitals and Wearable Devices

- Hospital-grade ECG and EEG machines use isolation amplifiers integrated into the input stage to ensure patient safety during extended monitoring.
- Portable and wearable monitors incorporate galvanic isolation combined with battery power to minimize risks.
- Signal ground referencing: The use of driven right leg circuits and isolated grounds helps reduce common-mode voltages while maintaining isolation.

2.2.3 *Aims of Signal Conditioning*

The ultimate aim of signal conditioning is to prepare the analog biosignal for reliable and accurate digital processing, diagnosis, or visualization. The key goals are:

Improve Signal-to-Noise Ratio (SNR)

By amplifying the signal and suppressing noise sources, conditioning circuits enhance the SNR, which is critical for both qualitative (e.g., visual inspection of ECG waveforms) and quantitative analysis (e.g., automated arrhythmia detection). This is particularly important for weak signals like EEG or cortical recordings, where SNR may be < 0 dB without conditioning [38, 39].

Prevent Aliasing and Distortion

Aliasing occurs when higher-frequency components fold into the lower-frequency band during digitization due to insufficient sampling rates. To avoid this, low-pass anti-aliasing filters are implemented as part of the conditioning stage. Nonlinearities in amplification and saturation must also be avoided to preserve waveform shape [40].

Adapt Signal Levels for Digitization

As ADCs have fixed input ranges (e.g., 0–3.3 V for low-power microcontrollers), the signal must be scaled and possibly offset to utilize the full dynamic range of the converter. This enhances quantization resolution and minimizes digitization error [41].

Ensure Patient Safety Through Electrical Isolation

Medical devices must comply with standards such as IEC 60,601, which prescribe strict limits on patient leakage currents. Isolation barriers in the conditioning path are critical to ensuring that, even in the event of a fault, the patient remains electrically safe.

Recent advances in integrated circuit design have enabled the consolidation of signal conditioning functions into Application-Specific Integrated Circuits (ASICs) or System-on-Chip (SoC) platforms, particularly for wearable or implantable biomedical devices [6, 13]. These platforms may incorporate:

- Low-noise amplifiers with programmable gain
- Tunable analog filters
- 24-bit high-resolution ADCs
- Digital filtering engines
- Isolation interfaces
- Power management.

An example is the ADS1298 from Texas Instruments, which offers an 8-channel, 24-bit ADC with integrated instrumentation amplifiers, right-leg drive, and lead-off detection, optimized for ECG and EEG systems.

Real-World Effects

Even minor mismatches in electrode impedance or cable asymmetry can significantly degrade CMRR in practice. Therefore, circuit board layout, shielding, and electrode preparation are just as important as intrinsic amplifier CMRR. The bandwidth of an amplifier refers to the range of frequencies over which the amplifier maintains a consistent gain (within -3 dB of the nominal gain). Biomedical signals are inherently band-limited, but their precise frequency content varies depending on the signal type [29]:

- ECG: 0.05–100 Hz
- EEG: 0.1–50 Hz
- EMG: 20–500 Hz (up to 2 kHz in high-fidelity systems)
- EOG: 0.1–10 Hz
- Neural spikes: 300–10 kHz.

The amplifier's bandwidth must encompass the entire frequency content of the physiological signal of interest. Filtering should not occur in the amplifier itself unless specifically designed as part of an analog filter stage.

Gain-Bandwidth Tradeoff

Most amplifiers have a gain-bandwidth product (GBP) that is constant (Eq. 2.6):

$$\text{GBP} = \text{Gain} \cdot \text{Bandwidth} \tag{2.6}$$

Thus, higher gain reduces bandwidth unless compensated by multi-stage designs or wideband amplifier architectures.

Example: An op-amp with a GBP of 1 MHz configured for a gain of 100 will have a bandwidth of:

$$\frac{1\,\text{MHz}}{100} = 10\,\text{kHz}$$

This is sufficient for EMG, but not for neural spike recordings, which may require wideband, low-noise amplifiers with a bandwidth greater than 100 MHz (Table 2.2).

The instrumentation amplifier is the workhorse of biomedical signal amplification. It offers high input impedance, excellent CMRR, and adjustable gain. Unlike simple differential amplifiers, instrumentation amplifiers use three operational amplifiers (op-amps) to achieve precise gain control and improved performance.

The gain of a typical 3-op-amp configuration is (Eq. 2.7):

$$A = 1 + \frac{2R_1}{R_G} \tag{2.7}$$

Table 2.2 Typical amplitude, frequency range, and required gain for standard biomedical signals

Signal	Amplitude range	Frequency range	Required gain
ECG	0.5–5 mV	0.05–100 Hz	100–1000 ×
EEG	10–100 μV	0.5–50 Hz	10,000 ×
EMG	50 μV–5 mV	10–5000 Hz	100–1000 ×

Amplifiers introduce thermal, flicker, and bias current noise. The noise performance of an amplifier is often characterized by its input-referred noise density, given in $nV/\sqrt{Hz}$. Minimizing noise involves:

- Shielded cable
- Twisted pair wiring
- Ground loop elimination
- Low-noise amplifier selection.

2.2.4 Filtering and Band Limiting

Filtering is a foundational signal conditioning technique used to remove or attenuate undesired frequency components from biomedical signals while preserving the integrity of the physiological information of interest. The need for filtering arises because biosignals are often embedded in noise introduced by both internal (physiological) and external (environmental) sources. Effective filtering enables improved SNR, prevents aliasing during digitization, and enhances the interpretability and diagnostic value of signals [29].

Biomedical applications impose unique constraints on filter design: filters must maintain waveform morphology (i.e., phase linearity in some applications), minimize signal distortion, and conform to stringent safety and power consumption standards, especially in wearable or implantable devices [30]. Before discussing filter types, it is essential to recognize the typical sources of interference encountered in biomedical signal acquisition:

- Powerline interference: 50 Hz or 60 Hz sinusoidal noise induced via capacitive or inductive coupling from electrical equipment or building infrastructure.
- Baseline drift: Low-frequency fluctuations (<0.5 Hz), often due to respiration, electrode–skin impedance variation, or patient movement.
- EMG noise: High-frequency muscular activity superimposed on signals such as ECG and EEG, particularly problematic during ambulatory monitoring.
- Motion artifacts: Abrupt, nonstationary shifts in the baseline caused by electrode movement or mechanical vibration.
- Electrode polarization noise: Particularly significant in high-impedance interfaces, leading to low-frequency drift.

These noise sources overlap with parts of the frequency spectrum occupied by the actual signal. Hence, filtering must be carefully designed to preserve diagnostically relevant information while rejecting interference [28]. The frequency generally categorizes the filters that range they attenuating to passing. In biomedical signal processing, the following types of filters are most commonly employed:

High-pass filters (HPFs) attenuate frequencies below a specified cutoff frequency fc. They are primarily used to remove baseline drift, electrode DC offsets, and very low-frequency components that are not of interest [31]. For example, in ECG acquisition, a high-pass filter with a cutoff frequency ($fc\approx$) of approximately 0.05–0.5

Hz is commonly used. Mathematically, an analog first-order high-pass filter has a transfer function (Eq. 2.8):

$$H(s) = \frac{sRC}{1 + sRC} \tag{2.8}$$

where R and C define the cutoff frequency (Eq. 2.9):

$$f_c = \frac{1}{2\pi RC} \tag{2.9}$$

Low-pass filters (LPFs) attenuate frequency components above a specified cutoff frequency. They are used to suppress high-frequency noise, including EMG interference and radio-frequency (RF) signals. In ECG systems, LPFs with cutoff frequencies between 100 and 150 Hz are common.

Band-pass filters (BPFs) combine HPF and LPF stages to pass only a specific range of frequencies. For instance, EEG signals typically range from 0.5 to 70 Hz, and an EEG amplifier may implement a band-pass filter with these cutoff frequencies [1].

Notch filters (band-stop filters) attenuate a narrow range of frequencies and are used extensively in biomedical systems to eliminate powerline interference (typically centered at 50 Hz or 60 Hz, depending on the region). These filters are designed with a very narrow bandwidth to avoid suppressing adjacent physiological frequency content [32].

A typical notch filter transfer function can be represented as (Eq. 2.10):

$$H(s) = \frac{s^2 + \omega_0^2}{s^2 + \frac{\omega_0}{Q}s + \omega_0^2} \tag{2.10}$$

where $\omega_0 = 2\pi f_0$ is the quality factor controlling the bandwidth.

Filters can be realized in either analog circuitry (pre-digitization) or through digital signal processing (post-digitization). The selection depends on the system's architecture, timing constraints, power budget, and intended clinical application. Analog filters are typically implemented using either passive components (such as resistors, capacitors, and inductors) or active components (such as operational amplifiers). They are crucial in front-end conditioning to suppress noise and prevent aliasing before ADC sampling [33].

- Passive Filters: Constructed using RC (resistor–capacitor) or RLC (resistor-inductor-capacitor) networks. These are simple, low-cost, and require no power, but can introduce loading effects if not properly buffered.
- Active Filters: Utilize op-amps to achieve gain and buffering. They offer precise control over filter characteristics, thereby avoiding impedance mismatch issues. Common topologies include Sallen–Key and multiple-feedback configurations.

Table 2.3 Summary of regulatory parameters for amplifier systems in medical devices

Parameter	Limit (IEC 60,601-1)
Leakage current	<10 μA
Isolation voltage	>1.5 kV RMS
CMRR	>100 dB
Input impedance	>10 MΩ

Advantages of analog filters:

- Immediate signal modification (no delay)
- Low power consumption in discrete circuits
- Crucial for pre-ADC anti-aliasing.

Limitations:

- Limited tunability
- Susceptibility to component tolerances and drift
- Less effective for adaptive or nonstationary filtering.

Digital filters are applied after the ADC and provide exceptional flexibility, stability, and accuracy. They are particularly suitable for real-time or offline processing in embedded systems, DSPs, or software-defined biomedical devices (Table 2.3).

Two broad classes:

- Finite Impulse Response (FIR): Linear phase, consistently stable, but requires more computation.
- Infinite Impulse Response (IIR): More efficient (fewer coefficients), but may introduce phase distortion.

Digital filtering enables:

- Adaptive filtering (e.g., removing EMG noise from ECG in real time)
- Multiband decomposition (e.g., wavelet-based EEG analysis)
- Noise cancellation using reference signals (e.g., adaptive noise cancellation for ECG with accelerometer reference).

Several application-specific criteria must be considered when designing filters for biomedical use [32]:

- Phase Linearization: For applications such as ECG or PPG waveform analysis, preserving the time-domain shape of the signal is crucial. Linear-phase FIR filters are preferred in such cases.
- Delay Constraints: Real-time monitoring systems (e.g., defibrillators or anesthesia monitors) must use filters that do not introduce unacceptable delay.
- Tunable Cutoffs: Some wearable systems include tunable or adaptive filters to respond to changing noise environments or patient activity.
- Power Budget: In implantables or battery-powered wearables, filter implementation must consider power efficiency.
- Component Tolerance.

An ECG acquisition system typically incorporates the following filter stages:

- High-pass filter (0.05 Hz): Removes baseline wander due to respiration.
- Low-pass filter (100–150 Hz): Suppresses EMG and RF noise.
- Notch filter (50/60 Hz): Eliminates powerline interference.

These filters may be implemented as part of a compact analog front-end IC with programmable parameters. In digital post-processing, additional filters may be applied to smooth, enhance the QRS complex, or reduce noise using techniques such as wavelet denoising [34].

All biomedical signal conditioning systems must adhere to medical device regulations. Key standards include:

- IEC 60,601-1: General requirements for basic safety and essential performance.
- IEC 60,601-2-47: Particular requirements for ECG systems.
- ISO 14971: Risk management in medical devices.

Safety requirements:

- Leakage currents must be <10 μA for patient-applied parts.
- Isolation barriers must withstand >1.5 kV RMS.
- Systems must fail safely under fault conditions.

2.3 Analog-to-Digital Conversion

In the realm of biomedical engineering, ADC is not merely a technical mechanism—it is a foundational enabler of nearly every digital health technology we rely on today. From hospital bedside monitors to wearable heart rate sensors and remote diagnostic systems, ADC forms the critical link between the analog biological world and the digital devices that interpret, process, and act upon its signals. As healthcare systems worldwide increasingly pivot towards data-driven and connected solutions, understanding the principles, roles, and implications of ADC in biomedical systems is more critical than ever.

Physiological signals—such as the electrical activity of the heart (ECG), brain (EEG), muscles (EMG), and measurements like blood oxygen saturation (SpO2), blood pressure, and glucose levels—are analog by nature. They fluctuate continuously over time, reflecting the complex dynamics of human biology. These signals are typically low in amplitude, susceptible to noise, and highly variable in frequency, which introduces unique challenges in capturing them with fidelity. To perform meaningful digital processing, whether it is filtering noise, detecting abnormal patterns, or storing data for longitudinal analysis, these analog signals must first be converted into digital form. ADC is the process that performs this translation, capturing the signal at discrete time intervals (sampling) and converting its amplitude to a finite number of levels (quantization). This process, while conceptually straightforward, is critically sensitive in biomedical applications. Minor errors in ADC can lead to significant clinical consequences. A misread ECG wave, a false saturation alarm, or a flawed insulin pump reading can be the result of a poorly performing ADC stage.

In high-stakes environments such as intensive care units (ICUs) and surgical suites, the margin for error is minimal. Patient monitors rely on continuous, real-time digitization of biosignals to trigger alarms, guide interventions, and document patient status. In wearable health devices, which operate under strict power and size constraints, the ADC must be not only accurate but also highly efficient. These devices are often deployed in remote or ambulatory settings, where reliable signal acquisition and digitization are essential for telemedicine and home-based care [42, 43].

The Role of ADC in Medical Devices and Safety–Critical Systems

Medical devices span a broad spectrum, from portable diagnostic tools to complex life-support systems, and in all these, ADC plays a defining role. The digitized signal serves as the basis for automated analysis, control loops, user feedback, and long-term data archiving. For example, ECG machines must sample at sufficiently high rates (usually $\geq$500 Hz) and provide at least 12-bit resolution to accurately capture the QRS complex and subtle waveform features. In infusion and perfusion pumps, precision in ADC is vital for monitoring fluid delivery rates. A quantization error in pressure readings could lead to an overdose or underdose. In conformity assessment testing of infusion pumps, the resolution of the ADC plays a pivotal role in accurately detecting critical thresholds in flow rate and pressure measurements. Higher ADC resolution enhances the system's sensitivity, enabling more precise identification of subtle variations that may indicate performance deviations or safety concerns. Similarly, dialysis machines rely on accurate digitization of biochemical and pressure parameters to ensure effective blood purification.

Mechanical ventilators, which have become especially prominent during respiratory crises such as COVID-19, depend on ADC to translate sensor outputs related to airflow, pressure, and oxygen levels into actionable data. Without a reliable ADC, ventilators cannot maintain closed-loop control systems that respond in real time to a patient's needs [44].

From Discrete Circuits to Integrated Intelligence

Over the past few decades, the evolution of ADC technology has paralleled advances in microelectronics and signal processing. In earlier biomedical devices, analog circuits dominated. The digitization process was often external or peripheral, leading to bulky systems with limited signal processing capability. Today, ADCs are deeply embedded in mixed-signal microcontrollers and application-specific integrated circuits (ASICs) that form the brains of compact and wearable medical electronics [45, 46].

Modern ADC systems in biomedical devices are increasingly intelligent. They include adaptive sampling strategies that adjust resolution and speed based on signal conditions, low-power modes for wearable use, and integrated filtering to reject noise. Research has shown that integrating noise-aware sampling with delta-sigma architectures significantly enhances the accuracy of biosignal digitization while conserving power [47]. AI-enhanced biomedical systems take this further by combining ADC with real-time classification and predictive analytics. For instance, a wearable ECG monitor might not only digitize a heartbeat but also pre-process and analyze it locally to detect arrhythmias before transmitting data to the cloud [48].

Regulatory Requirements and Post-market Surveillance

Medical devices, especially those involving life-critical decisions, are subject to rigorous international regulatory frameworks. These include the IEC 60,601 series for electrical safety and performance, ISO 13485 for quality management, and FDA regulations for design and testing. ADC components, while often buried inside larger systems, are central to compliance with these standards.

Post-market surveillance (PMS) has gained increasing attention, especially as devices become more complex and integrated. The findings reveal that many performance issues in the field, previously attributed to mechanical faults, were rooted in ADC drift, resolution degradation, or signal timing errors [49, 50]. To support these surveillance efforts, machine learning is being applied to long-term signal recordings to detect anomalies that suggest ADC degradation.

As digital healthcare systems become more distributed, autonomous, and intelligent, the demands on ADCs will only increase. Future ADC systems will need to adapt in real time, adjust sampling based on context, detect noise or artifact patterns, and even self-calibrate using feedback from AI systems. Concepts such as compressed sensing and adaptive quantization are already being explored to optimize data throughput without sacrificing fidelity [51]. Next-generation biomedical devices will integrate ADC as part of a dynamic ecosystem, where sensing, digitization, analysis, and decision-making form a seamless loop. This vision aligns with efforts to embed intelligence not just in processing but in the very act of digitization itself. ADC will no longer be a passive step in the signal chain, it will be an active participant in intelligent sensing, forming the bedrock of a predictive, proactive digital healthcare landscape.

2.3.1 Fundamental Concepts of Analog-to-Digital Conversion

To understand the behavior and performance of ADC in biomedical systems, it is crucial to unpack the foundational principles that underpin this technology. This includes the core processes of sampling, quantization, and encoding, as well as the critical concepts of resolution, noise, signal fidelity, and system error. In biomedical engineering, these elements are not only academic, they translate directly into the efficacy and safety of diagnostic and therapeutic tools.

Sampling: The First Step in Digitization

Sampling is the process of measuring the value of an analog signal at discrete time intervals. This forms the temporal skeleton of any digitized signal and determines the granularity with which time-varying biomedical phenomena are captured. According to the Nyquist-Shannon sampling theorem, to reconstruct a continuous signal accurately, the sampling frequency must be at least twice the highest frequency component present in the signal (Nyquist rate).

For example, in electrocardiography, where significant waveform features can have frequency components up to 150 Hz, a sampling rate of at least 300 Hz is required. However, modern systems often employ sampling rates between 500 and 1000 Hz to ensure robustness against signal aliasing and to capture transient events such as arrhythmias or pacemaker spikes [52]. Aliasing occurs when the sampling frequency is insufficient, causing high-frequency signal components to masquerade as lower-frequency artifacts in the digitized output. In clinical settings, aliasing can lead to misdiagnosis, for instance, mistaking muscular artifacts for cardiac anomalies. To mitigate this, analog low-pass filters are applied before the sampling stage to eliminate frequencies above the Nyquist limit.

Quantization: Mapping Analog Amplitudes to Discrete Values

Once sampled, the analog signal's amplitude must be quantized—mapped to a set of finite numerical values. The number of quantization levels is determined by the resolution of the ADC, usually expressed in bits. An n-bit ADC can represent $2^{\wedge n}$ distinct levels. Thus, a 12-bit ADC provides 4096 levels, allowing finer distinctions between signal amplitudes than an 8-bit ADC, which provides only 256 levels.

Quantization introduces an error known as quantization noise—the difference between the actual analog input and its digital approximation. This error is bounded within ± 0.5 least significant bits (LSBs) and becomes a significant source of noise, particularly in low-amplitude signals standard in biosensing [53]. Higher resolution ADCs reduce quantization noise and improve signal-to-noise ratio (SNR). For example, in EEG systems where microvolt-level resolution is required, 16-bit or even 24-bit ADCs are often necessary. However, increased resolution also demands more power, processing speed, and data storage, which must be balanced in portable or battery-powered devices [52, 54].

Encoding: Preparing Digital Data for Processing

Encoding refers to the method by which the quantized signal is formatted into binary or digital code for processing. This can include binary encoding, two's complement representation for bipolar signals, and digital filtering schemes. Some ADC architectures also include serial or parallel data outputs, affecting the design of the downstream processing pipeline. In high-integrity medical systems, error-detecting and error-correcting codes may be embedded within the ADC output stream to enhance reliability, particularly in noisy environments such as MRI suites or surgical theaters. This ensures that even if digitized data is corrupted during transmission, errors can be detected and corrected without compromising patient safety.

Signal Fidelity and Noise Considerations

Signal fidelity—the degree to which the digitized signal matches the original analog input—is influenced by sampling accuracy, quantization resolution, and system noise. Biomedical signals are especially susceptible to distortion from ambient electromagnetic interference, electrode artifacts, and motion-induced noise.

Various techniques are used to enhance signal fidelity:

- Shielding and grounding in hardware design
- Anti-aliasing and digital low-pass filters
- Oversampling combined with decimation to improve effective resolution
- Differential signal acquisition to cancel common-mode noise.

Digital imaging systems were shown to benefit significantly from ADC systems with real-time noise analysis and compensation, demonstrating how precision in digitization impacts diagnostic accuracy in modalities such as CT and MRI [15].

Performance Metrics in ADC Evaluation

Several performance metrics define ADC suitability for biomedical applications:

- SNR: Indicates the ratio of signal power to noise power. Higher SNR is essential for clear biosignal interpretation.
- Effective Number of Bits (ENOB): Reflects real-world resolution accounting for noise and distortion. A 12-bit ADC may exhibit an ENOB of only 10 bits due to imperfections.
- Integral Non-Linearity (INL) and Differential Non-Linearity (DNL): Measure how actual conversion deviates from ideal behavior.
- Total Harmonic Distortion (THD): Particularly relevant in frequency-domain biomedical analysis, such as auditory evoked potentials.

These parameters must be carefully matched to the requirements of specific biomedical devices. For example, ECG machines demand high ENOB and low DNL, while pulse oximeters may prioritize SNR and low power consumption [49].

2.3.2 ADC Architectures and Their Role in Biomedical Devices

As biomedical systems continue to evolve and diversify, the architectural design of analog-to-digital converters becomes increasingly important. The choice of ADC architecture affects not only the accuracy and resolution of signal digitization but also influences power consumption, speed, noise immunity, and integration complexity. In this section, we explore the most common ADC architectures used in biomedical devices, analyze their advantages and limitations, and discuss how they align with specific medical applications. ADC architectures vary in how they perform sampling, quantization, and encoding. The most commonly used ADC types in biomedical systems include:

- Successive Approximation Register (SAR) ADCs
- Delta-Sigma (ΔΣ) ADCs
- Flash ADCs
- Pipeline ADCs
- Integrating ADCs (dual-slope or multi-slope).

Each of these has distinct performance trade-offs that make them suitable for specific signal characteristics and device constraints.

Successive Approximation Register (SAR) ADCs

SAR ADCs are widely used in biomedical devices due to their moderate speed, low power consumption, and relatively high resolution. They operate by performing a binary search through all possible quantization levels to converge on a digital output. This makes them efficient for applications requiring up to 16-bit resolution at moderate sampling rates (up to a few MHz). In wearable biosensors, such as portable ECG monitors, PPG-based heart rate sensors, and glucose monitors, SAR ADCs are often preferred due to their compact size and energy efficiency [55]. These devices benefit from the SAR ADC's ability to balance performance with battery longevity, which is crucial in ambulatory settings.

Recent innovations in SAR design, such as capacitor-splitting and asynchronous conversion techniques, have further reduced energy per conversion while maintaining accuracy. Studies have shown that SAR ADCs can operate effectively at sub-microwatt power levels, enabling continuous monitoring in devices like fitness trackers and smartwatches [47, 51].

Delta-Sigma (ΔΣ) ADCs

Delta-Sigma ADCs are dominant in applications where high resolution and low noise are essential, such as EEG, EMG, and other biopotential measurements. These ADCs use oversampling and noise shaping to push quantization noise out of the frequency band of interest, followed by decimation filtering to achieve high effective resolution. In EEG systems, where signal amplitudes are often in the microvolt range and frequencies rarely exceed 100 Hz, ΔΣ ADCs with resolutions of 24 bits are ideal. They enable precise capture of subtle neural events and allow for artifact discrimination through post-processing.

Flash ADCs

Flash ADCs provide extremely high-speed conversion by using a parallel bank of comparators, one for each quantization level. While this allows for nanosecond-scale conversions, it also results in high power consumption and large silicon area requirements. In biomedical imaging applications such as CT and MRI, where fast sampling is essential to prevent motion artifacts and synchronize high-speed detectors, flash

ADCs are sometimes used [56]. However, their high power usage and limited resolution (typically up to 8 bits) restrict their utility in portable or low-power biomedical systems.

Pipeline ADCs

Pipeline ADCs offer a compromise between speed and resolution, typically achieving 10 to 14-bit resolution at high sampling rates. They are structured as a cascade of low-resolution stages, each processing part of the input signal and passing residuals down the pipeline. These ADCs are often used in digital radiography and ultrasound systems, where real-time imaging demands both fast conversion and moderate-to-high resolution. The latency introduced by pipelining is acceptable in these contexts, and advanced calibration techniques have reduced errors associated with stage mismatches [41]. In therapeutic ultrasound systems, pipeline ADCs were identified as integral components in achieving real-time echo signal digitization, enabling accurate depth and tissue differentiation [49].

Integrating ADCs

Integrating ADCs, such as dual-slope or multi-slope architectures, excel in applications where accuracy is more critical than speed. They work by integrating the input signal over a fixed time and comparing the result to a reference. This inherently averages out noise and is resistant to fluctuations in supply voltage. Blood analyzers, dialysis machines, and point-of-care testing devices frequently employ integrating ADCs due to their high immunity to interference and ability to provide precise measurements over longer time scales.

Selecting the appropriate ADC architecture requires a careful balance of several factors:

- Signal bandwidth: Low-bandwidth biosignals (EEG, SpO_2) may favor $\Delta\Sigma$ ADCs; high-speed applications (imaging) may require Flash or Pipeline ADCs.
- Resolution needs: Diagnostic tools need higher resolution; control systems can tolerate less.
- Power constraints: Wearables demand ultra-low-power SAR or $\Delta\Sigma$ ADCs.
- System size and integration: Compact designs benefit from architectures that minimize external components.

It is also essential to consider the downstream digital processing pipeline. For example, SAR and $\Delta\Sigma$ ADCs produce serial outputs that are easy to interface with microcontrollers, while Flash ADCs may require additional data handling circuitry. In the context of regulatory and performance testing, the choice of ADC architecture must be validated through conformity assessments and stress testing under clinical conditions [49].

2.3.3 ADC Performance Parameters in Medical Devices

Once a specific ADC architecture is selected for a biomedical application, the next critical step is to evaluate and ensure its performance under operational and clinical conditions. The quality of an ADC cannot be judged by architecture alone; instead, a set of performance parameters must be quantified to determine its suitability for safety–critical biomedical systems. These parameters collectively define how accurately and reliably the ADC represents the original analog signal and whether it can meet regulatory and clinical standards.

Signal-to-Noise Ratio (SNR)

Signal-to-noise ratio is one of the most essential indicators of ADC performance. It measures the ratio between the power of the actual signal and the power of the background noise introduced during the digitization process. A high SNR is critical in biomedical applications where signals are often weak, such as in EEG or fetal ECG monitoring. An SNR of at least 60–70 dB is typically required for reliable signal interpretation in these contexts [57]. Modern ΔΣ ADCs used in EEG systems have demonstrated SNR values exceeding 100 dB, supporting precise signal reconstruction. Devices with low SNR may produce distorted waveforms, potentially leading to misdiagnosis or missed clinical events. In clinical trials, SNR has been directly linked to the diagnostic quality of ECG signals in portable monitoring systems [58].

Effective Number of Bits (ENOB)

The Effective Number of Bits quantifies how many bits of an ADC's resolution are truly usable in the presence of noise and non-linearities. While a 12-bit ADC theoretically provides 4096 levels of resolution, its ENOB may be lower due to noise, clock jitter, or analog imperfections. ENOB is critical in medical systems where precise measurements are essential, such as in infusion pumps and dialysis machines [49]. An ENOB of at least 10.5 is generally acceptable for ECG applications, while EEG and EMG systems often demand ENOB values closer to 13–15 to achieve diagnostic-grade resolution [59].

Total Harmonic Distortion (THD)

THD measures the harmonic content introduced during the ADC process, expressed as a percentage or decibels. High THD can distort the spectral profile of the original signal, making frequency-domain analyses less reliable. In auditory evoked potential monitoring and electromyography, THD must be minimized to preserve clinical accuracy. ADCs with THD below -90 dB are typically considered acceptable for high-resolution audio and biosignal acquisition. THD also plays a role in functional brain monitoring using steady-state visual evoked potentials (SSVEP), where harmonic interference can corrupt analysis. In such systems, low-distortion ΔΣ ADCs are often employed.

Integral and Differential Non-Linearity (INL/DNL)

Integral Non-Linearity and Differential Non-Linearity describe how the actual ADC conversion deviates from ideal linear behavior. INL represents the cumulative deviation across the input range, while DNL measures the deviation between adjacent code transitions. Both affect the monotonicity and linear accuracy of the conversion. For medical devices like infusion pumps and ventilators, non-linear digitization can lead to unstable control feedback loops. High DNL contributed to calibration drift in post-market evaluations of mechanical ventilators, emphasizing the need for strict DNL limits in device specifications. Acceptable values for INL are typically below ± 0.5 LSB, and for DNL below ± 1 LSB, especially in closed-loop systems where precision feedback is essential [60].

Sampling Rate and Bandwidth

Sampling rate and bandwidth define the temporal resolution of the ADC and its ability to track fast signal changes. High sampling rates are necessary for rapidly changing physiological events, such as defibrillation shock capture, high-frequency EMG bursts, or ultrasound signal acquisition. In contrast, slowly varying signals like temperature or glucose concentration may be adequately digitized at lower rates. Ultrasound systems and electrocardiographs (ECGs) typically require sampling rates of 1–5 MHz and 500–1000 Hz, respectively. In contrast, wearable temperature monitors may operate with sub-Hz sampling frequencies to conserve power. Inadequate sampling rates in defibrillators led to under-detection of ventricular fibrillation patterns during PMS testing [49].

Jitter and Timing Accuracy

Jitter refers to the variation in the sampling time intervals, often caused by clock instability. In time-sensitive biomedical applications such as pulse transit time measurement or neurostimulation timing, jitter can corrupt temporal accuracy, leading to misinterpretation or therapeutic error. Minimizing jitter requires robust clock design and shielding from electromagnetic interference. ADCs integrated with temperature-compensated crystal oscillators (TCXOs) and phase-locked loops (PLLs) offer improved timing precision. In wearable ECG and pulse oximetry devices, clock jitter below 50 ps RMS is typically targeted [41, 61].

Power Consumption and Efficiency

Power consumption is a critical parameter in wearable and implantable devices. ADC power draw influences battery life, thermal behavior, and overall system operation. Designers often face trade-offs between resolution, sampling rate, and power. Ultra-low-power ADCs have been developed using SAR and $\Delta\Sigma$ architectures, consuming as little as 1–10 μW while maintaining acceptable SNR and ENOB. These advances support multi-day or even multi-week continuous monitoring in wearable systems [55]. In clinical settings, reducing power also minimizes heating effects in devices that contact patients for prolonged durations. Innovative neonatal incubators and continuous glucose monitors are among devices that benefit from low-power ADC implementations [50].

2.3.4 Advanced Trends—Smart ADC Systems and AI Integration

The rapid evolution of healthcare technologies, especially in the domains of wearable health, remote diagnostics, and AI-driven decision support, has brought ADC design to the forefront of medical innovation. Traditional ADCs, while effective for general signal conversion, are now being outpaced by the growing demand for smarter, more adaptive, and context-aware digital systems. Section 2.5 explores the cutting-edge developments in smart ADC systems and their integration with artificial intelligence (AI), machine learning (ML), and advanced signal analytics in biomedical engineering.

Smart ADCs: Concept and Capabilities

A "smart ADC" refers to an ADC system that includes built-in intelligence for context adaptation, self-calibration, power management, and sometimes edge-level signal interpretation. Unlike conventional ADCs, which perform fixed-rate sampling and standard quantization, smart ADCs can dynamically adjust parameters such as sampling frequency, resolution, and filtering based on signal characteristics and application requirements. For example, in a wearable heart rate monitor, a smart ADC can lower its sampling rate during resting periods and increase it during physical activity or abnormal rhythm detection. This adaptivity not only conserves power but also ensures high-fidelity signal acquisition when it is most needed [54].

Modern ADC systems are increasingly being designed to integrate with edge computing platforms. In such architectures, basic signal processing, such as baseline correction, noise filtering, and anomaly detection, is performed locally, often on the same microcontroller that houses the ADC. This reduces data transmission loads, enhances response time, and allows for real-time clinical decisions at the point of care. For instance, in smart insulin pumps, edge-level digitization and analysis can trigger immediate dosage adjustments in response to glycemic trends, without needing to communicate with cloud platforms. AI-enhanced edge ADC systems have been shown to lower false alarm rates in ICU monitors and to improve response times in neurostimulation feedback loops [62].

AI-Driven Signal Classification and Prediction

One of the most transformative applications of smart ADC systems is their role in feeding high-quality, pre-processed data into AI and ML pipelines. These systems are designed to ensure that only clinically relevant, artifact-free signals reach higher-level diagnostic engines. AI algorithms, in turn, can use this data for classification (e.g., arrhythmia detection), prediction (e.g., seizure likelihood), or recommendation (e.g., adjusting ventilator settings). Similar systems have been proposed for monitoring chronic conditions such as congestive heart failure or COPD, using smart ADC frameworks that integrate seamlessly with predictive models [63].

In long-term monitoring systems, calibration drift and environmental changes can degrade ADC performance over time. Smart ADCs equipped with self-calibrating algorithms can detect these drifts and apply digital compensation automatically. This is particularly important in devices used in home settings, where regular manual recalibration is impractical. Researchers have also introduced self-healing ADCs that use redundancy and machine learning to identify and isolate faulty components or channels. For example, if one analog path develops increased noise or bias, the system can flag the anomaly and reroute signal acquisition through alternative circuitry or apply corrective filters [64].

Artificial intelligence is also revolutionizing post-market surveillance (PMS) of ADC-equipped medical devices. By continuously analyzing digitized signals for anomalies, machine learning models can detect early signs of ADC malfunction, degradation, or incompatibility with evolving clinical conditions. A case study using AI for real-time monitoring of patient monitors, where deviations in signal digitization patterns triggered maintenance alerts. These systems are aligned with modern regulatory paradigms that emphasize continuous device validation and adaptive safety assurance, especially in high-risk medical technologies [50].

Emerging areas in smart ADC design include:

- Compressed sensing ADCs that reduce the number of samples required for accurate signal reconstruction.
- Neuromorphic ADCs inspired by biological systems, enabling event-based sampling.
- Quantum-enhanced ADCs leveraging qubit states for ultra-sensitive biosignal digitization.
- Biofeedback-linked ADCs that adapt in real time to physiological trends or therapeutic outcomes.

These technologies promise to push the boundaries of what is possible in medical signal acquisition, paving the way for ultra-personalized, autonomous care systems. As AI continues to mature, the synergy between adaptive ADC and intelligent analytics will form a foundational layer of next-generation healthcare.

2.4 Noise and Artifact Removal in Biomedical Signals

Biomedical signals are prone to various forms of noise, which can significantly hinder their diagnostic value and compromise patient safety. In this chapter, we focus on understanding these disturbances and equipping students with strategies to mitigate them.

In biomedical signal processing, noise refers to undesirable, often random signal components that obscure meaningful physiological data, typically arising from environmental or electronic origins [65]. Artifacts, on the other hand, may result from physiological sources such as muscle movement or eye blinking [49]. Although "artifact" is commonly used interchangeably with noise, it connotes signals that overlap the target frequency bands and can mimic actual physiological events, making them especially problematic in a clinical context.

Biomedical recordings work with low-amplitude signals—ECG typically spans just a few millivolts, while EEG often contains signals in microvolts. Even low-level noise can conceal critical waveform features. Faulty ECG readings due to motion artifacts can mislead heart rate and arrhythmia detection. In EEG, artifacts like eye blinks or muscle movements can dominate genuine brain activity, complicating neurological diagnostics [66, 67].

In classrooms, these sources are grouped as either environmental or physiological:

- Electrical/environmental noise: includes power-line interference (50/60 Hz), thermal noise, and electromagnetic pickup (Brownian, shot, pink noise)
- Physiological artifacts: from sources like muscle (EMG), cardiac (ECG contamination in EEG), respiration, and eye movements
- Motion artifacts occur frequently with wearable devices, as electrode displacement or body movement interferes with contact stability.

2.4.1 Common Types of Noise in Biomedical Signals

In biomedical engineering, understanding the various types of noise is essential before implementing noise removal methods. Each noise type presents unique challenges and affects specific physiological signals differently. Below, we explore the most prevalent categories, their causes, and their effects on signal integrity.

Power-line interference, also known as mains hum, originates from the electromagnetic fields generated by electrical grids and devices operating at 50 or 60 Hz. This noise typically manifests as a sinusoidal waveform superimposed onto the signal, making it challenging to detect slow-changing components in ECG or EEG signals. For instance, alpha wave detection in EEG, which occurs in the 8–13 Hz range, can be obscured by this interference. This form of noise can enter the acquisition system through capacitive coupling with nearby electrical devices, inadequate shielding, or shared grounding. It is particularly problematic in older hospital environments or poorly insulated mobile devices [68]. Engineers often combat power-line noise using narrowband notch filters, adaptive line enhancement algorithms, or digital band-stop filters that target 50/60 Hz and its harmonics without distorting adjacent physiological signals.

Baseline wander is a low-frequency shift in the signal's mean voltage level, often caused by patient movement, respiration, or electrode impedance fluctuations. It typically affects ECG recordings by distorting the isoelectric line, making it challenging to interpret ST segments and detect myocardial infarction. Respiratory cycles (~0.15–0.3 Hz) and movement artifacts contribute significantly to baseline shifts in wearable and ambulatory settings. Polynomial detrending, high-pass filtering (cut-off around 0.5 Hz), and cubic spline interpolation are among the most effective techniques to restore the accurate baseline without degrading the QRS complex.

Muscle-generated signals or EMG noise infiltrate recordings when voluntary or involuntary muscle contractions occur near the recording electrodes. These signals have frequency content overlapping with ECG (30–100 Hz) and EEG (20–300 Hz), which makes simple low-pass filtering insufficient. EMG contamination is particularly problematic in EEG, where cognitive activity needs to be distinguished from facial or jaw movements. In high-density EEG systems, independent component analysis (ICA) and supervised classifiers are employed to separate and remove EMG components while preserving brain signals [44]. Motion artifacts result from relative movement between the sensor and the skin. They occur during walking, speaking, or any postural change. These artifacts introduce transient distortions or abrupt shifts in the recorded signal and can completely obscure waveform features. In wearable PPG systems, motion artifact is the leading cause of erroneous heart rate detection. Techniques such as accelerometer-based noise referencing and adaptive filtering using real-time inertial data have shown significant success in wearable biosensor applications [69, 70].

This noise arises due to fluctuations in electrode–skin impedance, especially in dry or reusable electrodes. Such instability introduces erratic signal jumps or drifting baselines, particularly in ambulatory settings or long-duration monitoring scenarios. Proper electrode placement, skin abrasion, and use of conductive gels can reduce this interference. Novel solutions include dry electrodes with built-in impedance tracking and smart electrodes with automatic gain adjustment [32]. Systemic noise includes internal circuit noise such as thermal, shot, and flicker noise, as well as quantization noise from analog-to-digital conversion. These are mainly dependent on circuit design and the intrinsic properties of the electronics used. In low-power biosignal acquisition systems, engineers must trade off between sampling resolution, speed, and energy efficiency. Techniques such as oversampling, dithering, and analog shielding are employed to reduce the impact of these systemic sources [71].

2.4.2 Techniques for Noise and Artifact Removal

Effective removal of noise and artifacts in biomedical signals is essential for ensuring signal clarity, clinical reliability, and accurate diagnosis. This section introduces students to a wide range of traditional and modern techniques for noise mitigation, categorizing them by underlying methodology and application domain.

Before a signal is digitized, analog front-end (AFE) circuitry plays a critical role in noise suppression. Analog filters—typically low-pass, high-pass, and band-pass—are implemented to remove out-of-band noise components. For instance, anti-aliasing filters are crucial in ECG systems to avoid spectral folding during sampling.

Electromagnetic shielding and grounding practices are also important. Differential amplifiers in AFE circuits enhance common-mode rejection, thereby suppressing power-line noise and motion artifacts. These analog techniques form the first line of defense and are indispensable for high-fidelity signal capture [72].

Digital filters are widely used for real-time and offline noise removal. Finite Impulse Response (FIR) filters offer linear phase characteristics, while Infinite Impulse Response (IIR) filters are preferred for computational efficiency.

- Notch filters: Typically centered at 50 or 60 Hz to remove power-line interference.
- High-pass filters: Used to eliminate baseline wander (e.g., 0.5 Hz cut-off for ECG).
- Low-pass filters: Attenuate high-frequency EMG contamination in EEG or ECG.
- Adaptive filters: Adjust their parameters in real time using reference signals—such as accelerometer data—to cancel motion artifacts.

These filters are implemented in hardware DSPs, microcontrollers, or post-processing pipelines. Adaptive filter algorithms like LMS (Least Mean Squares) and RLS (Recursive Least Squares) are commonly used in ECG and PPG systems with motion sensor integration.

Wavelet transform (WT) techniques have gained popularity due to their ability to isolate nonstationary signal features at multiple scales. Discrete Wavelet Transform (DWT) decomposes a signal into coefficients that represent different frequency bands and time localizations. In noise removal, thresholding of wavelet coefficients can selectively suppress noise-dominated components. For ECG and EEG denoising, wavelets such as Daubechies, Symlets, and Coiflets are widely used. Multiresolution analysis also makes wavelet techniques well-suited for capturing motion artifacts or sudden signal changes, which are common in wearable applications. Hybrid methods combining WT with neural networks have shown superior performance in recent studies [15].

Blind Source Separation (BSS) methods aim to decompose multichannel signals into statistically independent or uncorrelated components. The two main techniques are:

- Independent Component Analysis: Assumes non-Gaussian, statistically independent sources. ICA is highly effective in EEG signal cleaning, where it can isolate eye blink artifacts, muscle noise, or ECG leakage.
- Principal Component Analysis (PCA): Focuses on maximizing variance and is used when source signals are Gaussian. Though less selective than ICA, PCA can be effective in EMG and PPG preprocessing for dimensionality reduction and trend separation.

Recent improvements include automated ICA with classification engines that detect artifact-related components [73]. These techniques are used primarily in event-related signals, such as evoked potentials in EEG or QRS detection in ECG.

- Ensemble Averaging: Enhances signal-to-noise ratio by averaging repeated measurements assumed to have consistent signal content but uncorrelated noise.

- Template Subtraction: Applies a known artifact model and subtracts it from the contaminated signal.

These methods work well in controlled environments but are limited in dynamic settings like wearables or mobile health applications [68]. Recent advances have enabled data-driven approaches to noise removal. These models learn features of noise and clean signals from labeled datasets and apply them in real-time or batch processing.

- Supervised Learning: Support Vector Machines (SVMs), Random Forests, and Convolutional Neural Networks (CNNs) trained on annotated datasets can classify and correct noisy segments.
- Unsupervised Learning: Clustering and autoencoders detect and reconstruct clean signals without prior labels.
- Hybrid Models: Combine traditional filtering (e.g., wavelets) with neural networks for adaptive denoising pipelines.

These techniques are especially impactful in wearable systems and ICU monitors, where noise characteristics change rapidly. For example, deep learning models have been trained to correct ECG during exercise or detect motion artifacts in ambulatory EEG [74].

2.4.3 Noise Reduction in Specific Modalities

1. Noise and artifact characteristics vary widely across different types of biomedical signals. As such, noise mitigation techniques must be adapted to the unique demands of each modality. This section explores practical applications of noise reduction strategies in four primary signal types: ECG, EEG, EMG, and PPG. Students will gain insight into both modality-specific challenges and how the techniques from Sect. 8.3 are applied in real-world clinical and wearable systems.

Electrocardiography (ECG)

ECG signals capture the electrical activity of the heart, with waveform components such as P-waves, QRS complexes, and T-waves providing crucial diagnostic information. ECG is susceptible to baseline wander, muscle noise, and motion artifacts. Because the signal operates primarily in the 0.05–100 Hz frequency range, precise filtering is essential.

- Baseline wander: Removed using high-pass filters (>0.5 Hz), spline interpolation, or adaptive filtering.
- Power-line interference: Suppressed using notch filters or adaptive line cancellation.
- Motion artifacts: Managed using accelerometer-assisted adaptive filtering, especially in wearable ECGs.

Wavelet-based denoising and template matching improve QRS detection accuracy, even under high noise conditions. AI models trained on large ECG datasets have also demonstrated excellent artifact classification and removal.

Electroencephalography (EEG)

EEG records electrical activity of the brain from the scalp. Frequencies range from delta (0.5–4 Hz) to gamma (30–100 Hz), with amplitudes in the microvolt range. Artifacts include eye blinks, muscle noise, and cardiac interference.

- Eye blink artifacts: Typically removed using ICA or template subtraction methods.
- Muscle artifacts (EMG): Suppressed with band-pass filters (1–40 Hz) or wavelet shrinkage.
- Power-line interference: Addressed with notch filters, but care is needed to avoid distorting EEG rhythms.

Advanced ICA variants like adaptive mixture ICA and automated labeling of artifact components are used in modern systems [50]. Real-time EEG processing for brain-computer interfaces (BCIs) also employs deep learning for noise-robust feature extraction [63].

Electromyography (EMG)

EMG signals reflect the electrical activity of muscles and are typically broadband (10–500 Hz). Noise can arise from crosstalk, electrode impedance variation, and power-line interference.

- Crosstalk: Reduced through optimal electrode placement and differential amplification.
- Baseline drift: Corrected using high-pass filters (~20 Hz cut-off).
- Signal enhancement: Achieved via wavelet analysis or PCA to isolate true muscle activation from noise.

Noise mitigation is particularly critical in prosthetic control and muscle fatigue assessment, where precision is key. Machine learning classifiers trained on noise-free EMG signals improve motion classification in robotic systems.

Photoplethysmography (PPG)

PPG signals are obtained optically and reflect changes in blood volume. They are commonly used in pulse oximetry and heart rate monitoring. The main challenges include motion artifacts and ambient light interference.

- Motion artifacts: Tackled using adaptive filters, sensor fusion with accelerometers, and wavelet decomposition.
- Ambient light interference: Managed with analog filtering and differential photodetector configurations.

Smartwatches and fitness trackers employ AI-based algorithms that correct motion-induced noise in real time using PPG and inertial sensor data fusion [75].

2.4.4 Future Trends in Artifact Removal

As biomedical signal acquisition becomes more mobile, intelligent, and embedded in everyday life, traditional noise mitigation approaches are being enhanced—and in some cases replaced—by adaptive, AI-driven, and neuromorphic systems. This section explores emerging trends that represent the next frontier of artifact removal in both clinical and wearable biomedical technologies.

AI-enhanced noise reduction involves dynamic models that adapt to context, individual physiology, and environmental changes. Unlike static filters, these systems can modify parameters like filter bandwidth, threshold levels, and window sizes in real time, based on signal quality indicators. Recent research integrates deep learning with traditional digital filtering pipelines. For instance, CNNs trained to detect and segment motion artifacts can trigger real-time signal correction. These hybrid models are now implemented in commercial smartwatches, wearable ECGs, and telemedicine platforms. With the advent of powerful edge computing platforms, machine learning models can now run locally on microcontrollers embedded in wearable health monitors. This enables real-time noise detection and correction without requiring cloud access, thus reducing latency and preserving privacy. Systems like neural signal processors (NSPs) and tinyML platforms support local AI inference for artifact classification and quality scoring. In ambulatory EEG and fetal monitoring, edge AI has been shown to improve signal clarity and reduce false alarms significantly [62].

Another major trend is multi-sensor fusion, where physiological signals (e.g., ECG, PPG) are interpreted in conjunction with contextual data from accelerometers, gyroscopes, and temperature sensors. By identifying specific motion signatures or postural states, filtering systems can better isolate fundamental signal components from movement-induced noise. This approach is beneficial in PPG-based heart rate monitoring during physical activity. Context-aware models have been demonstrated to achieve 20–30% better artifact rejection in activity-based datasets. Inspired by the structure of biological neurons, neuromorphic chips perform spike-based or event-driven computation with extremely low power consumption. These systems are ideal for wearable and implantable medical devices that require long battery life and continuous monitoring. Neuromorphic ADCs and event-based processors have been shown to perform real-time QRS detection and motion artifact suppression using less than 10 μW, making them promising tools for future biomedical applications [76].

Quantum signal processing is an emerging area that could revolutionize biosignal acquisition. By leveraging quantum bits (qubits), these systems promise ultra-sensitive detection with theoretically noise-free reconstruction. Although still in early research, quantum-enhanced filtering and quantum Fourier analysis are being explored for applications in magnetoencephalography (MEG) and high-resolution EEG [50]. Bio-inspired adaptive filtering—modeled on retinal adaptation and auditory gain control—is also being applied to dynamically adjust gain and filter coefficients based on ambient noise and signal features.

2.5 Time-Domain Analysis

Time-domain analysis is one of the most fundamental approaches in biomedical signal processing. Unlike frequency-domain or time–frequency techniques, time-domain methods focus on analyzing signal behavior as it evolves—without transforming it into another domain. For many physiological signals, such as ECG, EMG, and respiratory waveforms, time-domain features provide direct, interpretable insights into underlying physiological events.

In biomedical engineering, time-domain features are essential not only because they are computationally efficient but also because they capture clinically relevant trends with minimal preprocessing. For example, analyzing the average heart rate from an ECG, identifying muscle activation periods in EMG, or estimating breath cycles from respiratory traces—all rely on time-domain characteristics such as mean, variance, amplitude, or zero-crossings [48, 77]. While frequency analysis excels at identifying periodicity and spectral signatures, time-domain analysis is often more suitable for real-time monitoring, especially in wearable and edge-computing systems where resource constraints limit the feasibility of spectral decomposition [78].

Moreover, time-domain features serve as foundational building blocks for more complex machine learning and AI-based classification systems. Many models, especially those designed for diagnostic support, begin with raw time-domain statistics before moving into higher-level representations [79]. From a clinical perspective, changes in time-domain features can indicate pathological conditions or physiological events. In ECG, for instance, increased beat-to-beat interval variability may signal autonomic dysfunction. In EEG, sudden amplitude shifts in specific time windows can denote epileptic spikes or cognitive evoked potentials [48].

2.5.1 Key Time-Domain Features

Time-domain features form the backbone of many biomedical signal analysis systems. They provide a way to characterize the amplitude, shape, and variability of a signal over time, without requiring any transformation into the frequency domain. These features are especially useful in real-time monitoring environments and are favored in computationally constrained systems such as wearable devices. The most basic and widely used time-domain features include statistical descriptors like the mean, variance, and standard deviation, as well as dynamic characteristics like root mean square (RMS), zero-crossing rate, slope sign changes, and waveform length. More advanced measures such as signal energy, entropy, skewness, and kurtosis add richness to the feature set and help in distinguishing normal from pathological signal patterns.

The mean of a signal represents its average value and provides a general sense of the signal's baseline or offset. In practical terms, tracking the running mean of an ECG signal can help identify baseline drift caused by respiration or electrode movement. The median, a more robust metric, is less sensitive to outliers and is often used in EEG to assess baseline shifts across trials or epochs.

Variance and standard deviation measure how much a signal deviates from its mean. High variance in EMG signals is commonly associated with muscular fatigue or tremor, while low standard deviation in PPG signals may indicate stable cardiovascular function. These metrics are essential for quantifying variability, which is often a diagnostic indicator.

The root mean square value of a signal provides an estimate of its power. It is calculated by squaring each sample, averaging the result, and then taking the square root. In electromyography, RMS is one of the most reliable indicators of muscle contraction intensity. It is also used in analyzing neural signals, where the amplitude may correlate with activation levels.

The zero-crossing rate (ZCR) counts how frequently the signal changes sign from positive to negative or vice versa. In muscle signal analysis, a high ZCR can suggest spastic activity or tremor. In voice diagnostics or sleep staging, shifts in ZCR have been used to detect arousals or apnea events.

Slope sign changes and waveform length are features that capture the rate and extent of variation in signal slope. These features are handy in detecting abrupt transitions, such as epileptic spikes in EEG or arrhythmias in ECG. Waveform length also serves as a proxy for signal complexity, with higher values indicating more erratic or burst-like activity.

Energy and entropy are two features that delve into the magnitude and unpredictability of signals, respectively. Signal energy is the sum of squared amplitude values and reflects the overall strength of the signal. It is widely used in heart sound segmentation and gait analysis. Entropy, on the other hand, quantifies signal complexity or randomness. For example, a significant drop in EEG entropy may indicate seizure activity or anesthesia-induced unconsciousness.

Finally, skewness and kurtosis are higher-order statistical moments that describe the shape of the signal's amplitude distribution. Skewness assesses asymmetry, which can highlight abnormal waveform deflections such as those found in ischemic ECG. Kurtosis measures how peaked a distribution is, and high kurtosis values in EEG may signal epileptiform activity or sharp waveforms. Taken together, these time-domain features create a rich descriptive profile of a signal that can be used in both manual inspection and automated classification systems.

2.5.2 *Practical Applications in Biomedical Signals*

Time-domain analysis is extensively used in a variety of biomedical signal processing applications. By extracting meaningful features such as the mean, variance, RMS, and entropy from physiological signals, clinicians and researchers can monitor health, detect abnormalities, and support diagnosis in both hospital and wearable settings.

This section explores specific use cases across four key signal types: ECG, EMG, EEG, and respiratory waveforms.

Electrocardiography (ECG)

In ECG analysis, time-domain features help assess the temporal structure of heartbeats, particularly in heart rate variability (HRV) studies. HRV analysis relies on inter-beat intervals (RR intervals) extracted from ECG signals. Simple statistical features like mean RR interval, standard deviation of NN intervals (SDNN), and root mean square of successive differences (RMSSD) are standard indicators of autonomic nervous system function [69]. Time-domain analysis is also used to evaluate QRS complex morphology. Features such as RMS value, waveform length, and amplitude variation help in arrhythmia detection and cardiac stress assessment [80]. These metrics are employed in clinical ECG monitors and mobile heart rate devices.

Electromyography (EMG)

In EMG, time-domain features like RMS and variance are key indicators of muscle activity, fatigue, and neuromuscular health. During sustained contractions, a gradual decrease in RMS and an increase in signal variance indicate muscle fatigue, which is crucial for ergonomic assessments and rehabilitation monitoring [81]. Zero-crossing rate and slope sign changes are also employed to detect muscle onset timing and assess tremor severity in Parkinson's disease [67]. These features have been widely adopted in prosthetic control systems and biofeedback applications.

Electroencephalography (EEG)

EEG signals are often analyzed in the time domain for event-related potential (ERP) detection. Key features include amplitude, latency, and waveform energy within specific time windows following a stimulus. These parameters are used in cognitive neuroscience to assess attention, memory, and language processing [15, 62]. Entropy and higher-order statistics like kurtosis help detect epileptic seizures or abnormal neural activity. Automated seizure detection systems often rely on time-domain features extracted over short epochs, coupled with machine learning classifiers [82].

Respiratory and PPG Signals

Time-domain features are equally critical in analyzing respiratory waveforms and photoplethysmography (PPG). For respiration, metrics such as breath duration, signal amplitude, and variability help in diagnosing sleep apnea and monitoring mechanical ventilation [49]. In PPG, peak-to-peak intervals and pulse amplitude variability are extracted to monitor cardiovascular health and detect conditions like hypotension and hypovolemia [83]. Entropy measures and RMS are also used to quantify motion artifacts or assess signal quality in wearable PPG sensors. These applications demonstrate how time-domain features are embedded in both diagnostic algorithms and real-time monitoring systems across biomedical domains. In the next section, we will discuss how time-domain features are segmented, extracted, and organized in signal processing pipelines.

2.5.3 Signal Segmentation and Feature Extraction

In biomedical signal processing, raw signals must be segmented into meaningful intervals before features can be extracted. This step is critical because it aligns analysis with physiological events and ensures statistical reliability. Time-domain feature extraction typically depends on segmenting the signal into fixed-size or event-related windows.

One common approach is fixed-window segmentation, where signals are divided into equally sized time intervals (e.g., 1-s or 10-s windows). This method is suitable for long-term monitoring and facilitates consistent comparison across time. However, the choice of window size directly influences the quality of feature extraction. Smaller windows provide better temporal resolution but may be more sensitive to noise, whereas larger windows offer stability at the cost of responsiveness [50]. In contrast, event-based segmentation aligns signal windows with specific physiological events, such as R-peaks in ECG or stimulus onset in ERP studies. This approach ensures that extracted features are tied directly to clinically relevant phenomena. For example, HRV metrics are computed from RR intervals identified within the ECG, while ERP features are measured within stimulus-locked windows in EEG [84].

Another crucial decision is whether to use overlapping or non-overlapping windows. Overlapping windows increase the density of extracted features and can capture rapidly changing dynamics, which is especially important in EEG or EMG burst detection. However, they also increase computational load. Non-overlapping windows are more computationally efficient and are often used in low-power wearable devices [85]. After segmentation, the process of feature extraction involves computing descriptive statistics (e.g., mean, RMS, entropy) for each segment. In real-time systems, this step is optimized for speed and reliability. Many wearable biosignal systems now use embedded processors to perform on-device feature extraction, minimizing data transmission and improving battery life [86]. Segmentation and feature extraction pipelines also form the foundation for machine learning models used in classification and anomaly detection. Features derived from time-domain segments are used as inputs to classifiers like support vector machines (SVM), random forests, or deep neural networks to detect conditions such as sleep stages, stress, seizure onset, or arrhythmias [87]. Understanding how and when to segment a signal, and which features to extract, is fundamental to the design of robust biomedical analysis systems. In the next section, we will explore how these time-domain features are integrated with machine learning and artificial intelligence to build predictive and adaptive health monitoring systems.

2.5.4 Time-Domain Trends and AI Integration

In recent years, the role of time-domain features has expanded significantly with the advent of machine learning (ML) and artificial intelligence (AI) in biomedical signal analysis. These developments have made it possible to move beyond fundamental statistical analysis and toward predictive, adaptive systems that improve diagnostic accuracy, reduce clinician workload, and enable real-time health monitoring.

Time-domain features are often used as inputs to AI models due to their simplicity, interpretability, and compatibility with low-power processing systems. In wearable devices, for instance, features like RMS, mean, and signal energy are computed in real time and used to trigger alerts or adjust system behavior without relying on remote cloud processing [88]. Machine learning models, such as SVM, decision trees, k-nearest neighbors (k-NN), and random forests, frequently begin their analysis pipeline with time-domain feature vectors. These models have been employed to classify arrhythmias from ECG, detect motor intention from EMG, and differentiate sleep stages from EEG. Moreover, time-domain features are increasingly integrated with frequency- and time–frequency domain metrics to create hybrid models. This approach enhances classifier robustness and improves detection accuracy in noisy or dynamic environments. In seizure prediction, for example, combining EEG energy, entropy, and wavelet-based features with time-domain statistics improves performance compared to any single feature type [89]. Deep learning architectures such as convolutional neural networks (CNNs) and recurrent neural networks (RNNs) are also leveraging time-domain features, particularly for lightweight edge computing. Preprocessing pipelines often compute basic time-domain metrics that are then passed to neural networks for pattern recognition, reducing the burden of raw signal processing [90]. Another emerging trend involves explainable AI (XAI), where the interpretability of time-domain features supports transparency in clinical decision support tools. Features like heart rate variability, RMS amplitude, and signal variance have clear physiological correlates, making them attractive in regulatory and clinical settings where explainability is essential.

Finally, AI-enhanced systems are being developed for adaptive sensing, where the signal acquisition process adjusts dynamically based on the real-time evaluation of time-domain features. These self-optimizing systems reduce power consumption, minimize false alarms, and personalize monitoring for each patient [70].

2.6 Frequency Domain Analysis

In the study of various signals, particularly those related to biological systems, analyzing the frequency content is crucial for understanding the underlying processes that drive the signal's behavior. Whether investigating electrical impulses in the heart, brain, or muscles, the frequency characteristics of a signal often reveal far more information than time-domain analysis alone. Frequency-domain analysis provides a framework that allows researchers and clinicians to assess how the energy of a signal is distributed across different frequencies. This method of analysis plays an essential role in deciphering the complexities of physiological signals and provides a more comprehensive understanding of the phenomena being observed. While time-domain analysis, where signals are observed and interpreted as functions of time, offers valuable insights, it can often be limited in its ability to fully characterize specific systems. Many biological signals, such as those produced by the heart, brain, or muscles, consist of periodic or quasi-periodic components that are best understood through their frequency structure. Time-domain analysis, by its nature, can obscure the periodicity or frequency-specific components of a signal that may be essential for diagnosis or understanding a biological process. By shifting the analysis to the frequency domain, it becomes possible to isolate and examine these frequency components, which may not be apparent in a time-domain representation [20].

Frequency-domain analysis involves transforming a signal from the time domain into a domain where its frequency components represent it. This transformation enables the study of signals in a way that uncovers hidden periodic patterns, noise components, and underlying physical processes that might otherwise go unnoticed. It is beneficial in fields such as medical diagnostics, where signals like ECG, EMG, and EEG play an essential role in patient monitoring and treatment planning. These intrinsically time-based signals can often be better understood and interpreted when analyzed through their frequency characteristics, leading to more accurate diagnoses and deeper insights into the body's physiological state.

A fundamental concept in frequency-domain analysis is the concept of frequency components. A signal, whether from the heart, brain, or muscles, can typically be decomposed into a series of oscillations or periodic components. These components correspond to specific frequencies, with each frequency component representing a different aspect of the system's behavior. For example, the heart's rhythm is governed by low-frequency components. At the same time, muscle activity may exhibit a broader spectrum of frequencies that vary depending on the intensity and type of contraction. Respiratory signals exhibit frequency components related to the rate and rhythm of breathing. By analyzing the frequency content, medical professionals can gain deeper insights into a patient's physiological state, potentially identifying abnormalities such as arrhythmias, muscle disorders, or respiratory irregularities.

In many cases, frequency-domain analysis is used to isolate the frequency bands most relevant to a particular physiological process. For example, certain types of analysis may focus on identifying low-frequency components in heart signals to assess heart rate or detect arrhythmias. Similarly, muscle activity can be analyzed through its high-frequency components to study muscle fatigue or neuromuscular disorders. By focusing on specific frequency bands, researchers can narrow their analysis to only those aspects of the signal that are most informative for a particular medical condition. One of the key benefits of frequency-domain analysis is its ability to remove noise from the signal. Biological signals are often contaminated by external interference, such as power-line noise or equipment artifacts, which can obscure the accurate underlying signal. In the time domain, these artifacts can be difficult to distinguish from the real signal, making accurate analysis challenging. Frequency-domain analysis enables the identification of these extraneous frequency components, which can then be filtered out, leaving the cleaner, more relevant aspects of the signal intact. This process enhances the accuracy and reliability of the analysis, making it a powerful tool in medical diagnostics [40].

Moreover, frequency-domain analysis allows for the detection of hidden periodicity. Many biological signals are not perfectly periodic but exhibit periodic elements or quasi-periodic behavior that can be difficult to identify in the time domain. For instance, the electrical activity of the heart may exhibit irregularities that are not immediately apparent in the time domain; however, when the signal is viewed in the frequency domain, these irregularities may appear as peaks or anomalies at specific frequencies [31]. Similarly, EEG signals from the brain may reveal patterns that suggest the presence of neurological disorders such as epilepsy, even when those patterns are not evident in the time domain. Additionally, frequency-domain analysis is invaluable in analyzing complex signals that may contain overlapping frequency components. Signals from the body are often a mix of various frequencies, each corresponding to distinct processes or activities. In the time domain, these overlapping components can be challenging to separate and analyze. However, in the frequency domain, these components can be isolated, enabling a more detailed and precise understanding of the signal's composition [30]. This is particularly useful when trying to assess the contribution of various sources to a signal, such as identifying different muscle groups in an EMG signal or distinguishing between different types of brainwave activity in EEG data.

In medical applications, real-time monitoring is often a key consideration. Frequency-domain analysis can be instrumental in settings where continuous or near-continuous monitoring of biological signals is required. For example, real-time analysis of ECG signals can be performed to continuously monitor the heart's rhythm and detect any abnormalities as they occur. Similarly, real-time monitoring of respiratory signals allows for the detection of issues like apnea or abnormal breathing patterns, which can be vital for patients in critical care or sleep studies [91]. One of the key challenges in frequency-domain analysis, however, is dealing with the trade-off between time and frequency resolution. While the frequency domain offers advantages in identifying periodic components and filtering noise, it often has a limitation in pinpointing exact timing information about transient events within a signal. For instance, a sharp, transient change in the signal may be important, but frequency-domain analysis may not provide clear information about the exact timing of this event. This challenge is particularly relevant when dealing with biological signals that exhibit both periodic components and sudden, non-periodic events, such as heart attacks or epileptic seizures.

Despite these challenges, frequency-domain analysis remains an essential tool for researchers and clinicians working with biological signals. Its ability to reveal periodicities, uncover hidden frequency components, and filter out noise makes it invaluable in understanding the body's underlying physiology. Furthermore, its integration with other signal processing techniques, such as time-domain analysis or machine learning, has led to significant advances in the diagnostic and monitoring capabilities of medical systems.

2.6.1 Fourier Transform and Power Spectral Density

The Fourier Transform is a mathematical technique that decomposes a signal into its constituent sinusoids of varying frequencies, amplitudes, and phases. It is fundamental to signal processing and has wide applications in many fields, particularly in biomedical signal analysis. In this section, we will examine the significance of the Fourier Transform in the context of biological systems and its applications in biomedical engineering.

The Fourier Transform

F of a continuous signal (t) is given by the following Eq. 2.11:

$$X(f) = \int_{-\infty}^{\infty} x(t)\mathrm{e}^{-\mathrm{j}2\pi \mathrm{ft}}\mathrm{dt} \tag{2.11}$$

where:

- X(f) is the frequency-domain representation of x(t)x(t)x(t)
- f is the frequency variable
- t is time
- j is the imaginary unit
- $\mathrm{e}^{-\mathrm{j}2\pi\mathrm{ft}}$ is the complex sinusoid, which represents the basis function of the Fourier Transform.

This equation essentially decomposes the time-domain signal into an infinite sum of sinusoids of varying frequency. In biomedical signal analysis, the Fourier Transform is used to analyze the frequency content of biological signals, such as ECG, EEG, and EMG.

In practical applications, signals are often sampled at discrete intervals, necessitating the use of the Discrete Fourier Transform (DFT). The DFT is a sampled version of the continuous Fourier Transform and is used to analyze signals that have been discretized in both time and frequency [13].

The Discrete Fourier Transform (DFT) is given by Eq. 2.12:

$$X(k) = \sum_{n=0}^{N-1} x(n)e^{-j\frac{2\pi kn}{N}}, k = 0,1, \ldots, N - 1 \tag{2.12}$$

where:

- X(k) is the DFT of the discrete signal x(n)
- N is the total number of samples
- k is the index corresponding to the frequency component
- n is the index for time samples.

This discrete transformation enables the efficient computation of a signal's frequency components and serves as the foundation for algorithms such as the Fast Fourier Transform (FFT) (Fig. 2.8).

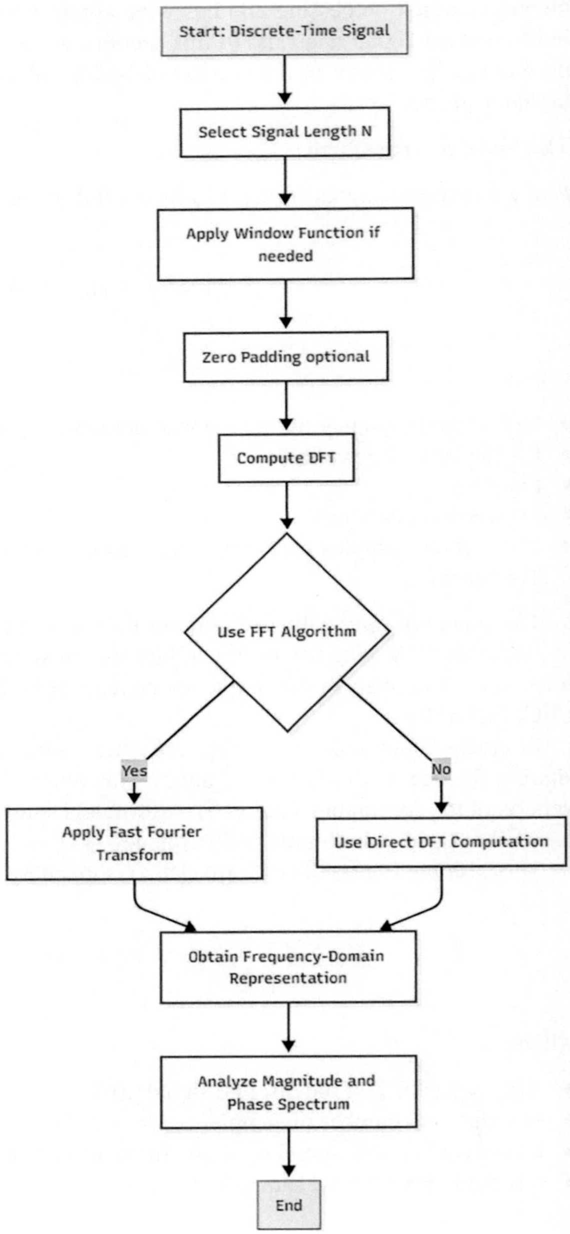

Fig. 2.8 A flow diagram showing the process of converting a discrete-time signal into its DFT representation, along with the FFT algorithm

The FFT is an efficient algorithm for calculating the DFT in a computationally efficient manner. The FFT drastically reduces the time complexity of DFT calculations, making it suitable for real-time biomedical signal analysis. The FFT algorithm works by recursively breaking down the DFT into smaller DFTs, reducing the time complexity from $O(N_2)$ to O(NlogN).

- Divide-and-conquer approach: The FFT splits the DFT computation into smaller subproblems, thus optimizing performance.
- Applications: The FFT is widely used in real-time signal analysis, such as monitoring heart rhythms or brain wave patterns.

While the Fourier Transform provides a powerful tool for analyzing signals, it has limitations when dealing with non-stationary signals, such as those commonly found in biomedical applications. Windowing is a technique used to overcome these limitations by segmenting the signal into smaller windows, each of which is analyzed independently.

The windowed Fourier Transform is given by Eq. 2.13:

$$X(f,t) = \int_{-\infty}^{\infty} x(\tau)\mathrm{w}(\mathrm{t}-\tau)\mathrm{e}^{-\mathrm{j}2\pi \mathrm{f}\tau}\mathrm{d}\tau \tag{2.13}$$

where:

- $w(t - \tau)$ is the window function
- $x(\tau)$is the signal.

Windowing enables time–frequency analysis, which is particularly useful for signals that change over time, such as EEG and EMG signals, which contain both high- and low-frequency components (Fig. 2.9).

The Power Spectral Density (PSD) is a critical concept in signal processing, as it describes how the power of a signal is distributed across different frequency bands. In biomedical signals, PSD analysis provides insights into the dominant frequencies and the relative power of each frequency component. The PSD S(f) of a signal x(t)is given by the square of the magnitude of the Fourier Transform (Eq. 2.14):

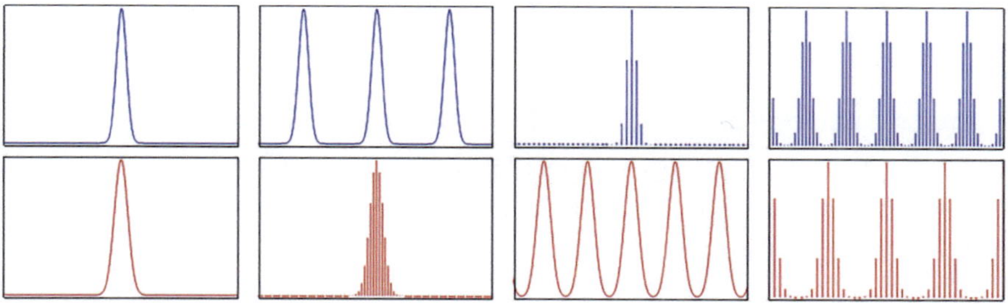

Fig. 2.9 Signal analysis using the Fourier Transform (https://commons.wikimedia.org/wiki/File:From_Continuous_To_Discrete_Fourier_Transform.gif)

$$S(f) = | X(f)|^2 \tag{2.14}$$

This equation represents the power of the signal as a function of frequency. The PSD is useful for identifying characteristic frequency components of signals, such as detecting arrhythmic events in ECG or identifying brainwave patterns in EEG.

For non-stationary signals, spectrogram analysis is a powerful technique. A spectrogram represents the signal in both the time and frequency domains, allowing for the observation of how the frequency content of a signal evolves. The spectrogram S(t,f) of a signal x(t) is computed using the STFT (Eq. 2.15):

$$S(t,f) = \left| \int_{-\infty}^{\infty} x(\tau)\mathrm{w}(\mathrm{t} - \tau)\mathrm{e}^{-\mathrm{j}2\pi\mathrm{f}\tau}\mathrm{d}\tau^2 \right|^2 \tag{2.15}$$

where:

- $\mathrm{w}(\mathrm{t} - \tau)$ is the window function.
- The magnitude squared provides the energy content at each time–frequency point.

Spectrograms are widely used in biomedical fields, particularly for monitoring EEG signals to detect seizures or identify sleep stages.

The Inverse Fourier transform enables the reconstruction of the original signal from its frequency-domain representation. This is essential when the signal is modified in the frequency domain (such as filtering out noise) and needs to be transformed back to the time domain. The inverse Fourier Transform F^{-1} (Eq. 2.16):

$$x(t) = \int_{-\infty}^{\infty} X(f)\mathrm{e}^{\mathrm{j}2\pi\mathrm{ft}}\mathrm{df} \tag{2.16}$$

By applying the inverse transform to the frequency-domain signal X(f), we can recover the original time-domain signal x(t).

2.6.2 Application in Identifying Signal Frequency Components

Example: Analyzing an ECG Signal Using the Fourier Transform

In this example, we will take an ECG signal, apply the DFT, and then analyze its frequency components. We will calculate the PSD and apply windowing for time–frequency analysis.

Step 1: Represent the ECG Signal.

We will simulate a simple ECG signal in time. This can be done by combining multiple sinusoidal waves that simulate the QRS complex, P-wave, and T-wave in an ECG. Assume the ECG signal can be represented as Eq. 2.17:

$$x(t) = A_1\sin(2\pi f_1 t + \phi_1) + A_2\sin(2\pi f_2 t + \phi_2) + A_3\sin(2\pi f_3 t + \phi_3) \tag{2.17}$$

where:

- A_1, A_2, and A_3 are the amplitudes for the QRS complex, P-wave, and T-wave.
- f_1, f_2, and f_3 are the frequencies for the respective components of the ECG.
- ϕ_1, ϕ_2, and ϕ_3 are phase shifts.

Let us simulate an ECG with the following parameters:

- $A_1 = 1$, $f_1 = 1\text{Hz}$ (QRS complex)
- $A_2 = 0.3$, $f_2 = 0.2\text{Hz}$ (P-wave)
- $A_3 = 0.2$, $f_3 = 0.1\text{Hz}$ (T-wave)

The ECG signal appears as follows in time (you would plot this in a real implementation). The frequencies correspond to the heartbeat frequency, as well as the frequencies of the P-wave and T-wave.

Step 2: Apply the Discrete Fourier Transform (DFT)

Next, we calculate the DFT to analyze the frequency components of the signal. The DFT equation is (Eq. 2.18):

$$X(k) = \sum_{n=0}^{N-1} x(n)e^{-j\frac{2\pi kn}{N}}, k = 0,1, \ldots, N - 1 \tag{2.18}$$

In this case:

- x(n)x is the ECG signal sampled at discrete intervals (say, at a sampling frequency of 500 Hz).
- N is the number of samples, for example, N = 1000 samples over 2 s.
- k corresponds to the index of the frequency bin.

Calculation: Suppose we sample the signal at 500 Hz for 2 s, resulting in a total of 1000 samples. Applying the DFT to this signal yields the frequency-domain representation.

Step 3: Plot the Frequency Spectrum.

After applying the DFT, we obtain the frequency-domain representation of the ECG signal, X(f). This will show us the different frequencies present in the signal, corresponding to the QRS complex, P wave, and T wave.

Step 4: Power Spectral Density (PSD).

Now, let us calculate the PSD, which represents how the power of the signal is distributed across different frequencies. The equation for PSD is (Eq. 2.19):

$$S(f) = |X(f)|^2 \tag{2.19}$$

Calculation:

- Calculate the magnitude of the DFT at each frequency, f.
- Square the magnitude to obtain the power at each frequency.

For instance, for $f_1 = 1$ Hz, you would calculate $|X(1)|^2$ and similarly for other frequencies.

Step 5: Windowing and Time–Frequency Analysis.

To analyze the signal in both time and frequency, we can apply windowing. For non-stationary signals like ECG, windowing helps break the signal into smaller segments (or windows), which can then be analyzed using the Fourier Transform. Let us use a Hamming window, which is a commonly used window function in signal processing. The window function w(t) is (Eq. 2.20):

$$w(t) = 0.54 - 0.46\cos\left(\frac{2\pi t}{N-1}\right), t = 0,1,\ldots,N-1 \tag{2.20}$$

This window is multiplied by the signal, and the Short-Time Fourier Transform (STFT) is calculated for each window. The equation for the STFT is (Eq. 2.21):

$$S(t,f) = \left|\int_{-\infty}^{\infty} x(\tau)\mathrm{w}(\mathrm{t} - \tau)\mathrm{e}^{-\mathrm{j}2\pi \mathrm{f}\tau}\mathrm{d}\tau\right|^2 \tag{2.21}$$

Calculation:

- Divide the ECG signal into smaller segments (e.g., 50 ms windows).
- Apply the Fourier Transform to each window.
- Plot the spectrogram, which shows how the frequency components evolve.

Example: Application of FT in Medical Imaging

The Fourier Transform plays a crucial role in medical imaging by enabling the conversion between spatial and frequency domains. It is an indispensable tool in various imaging modalities, particularly in Magnetic Resonance Imaging (MRI) and Computed Tomography (CT). The ability to move between these domains enables the reconstruction of high-quality medical images from raw data captured during the imaging process. In medical imaging, the raw data collected from patients, such as signals received by MRI or CT scanners, is often in the frequency domain. To visualize the data and obtain actionable diagnostic images, we need to convert this frequency-domain information back into the spatial domain. This process involves the use of the Fourier Transform and its inverse.

Magnetic Resonance Imaging is one of the most prominent imaging techniques that relies heavily on the Fourier Transform for reconstructing images. In MRI, the signals received from the body are in the frequency domain due to the nature of the data acquisition process. The image reconstruction process involves converting this frequency-domain data back into a spatial image that healthcare professionals can analyze.

1. Raw Data Acquisition: MRI scanners acquire raw data in the k-space, which is essentially the frequency domain. The k-space represents different frequency and spatial components of the object being scanned, with each point containing information about a particular spatial frequency.
2. Fourier Transform for Image Reconstruction: Once the raw k-space data is obtained, the Inverse Fourier Transform (IFT) is used to convert it back into the spatial domain, resulting in a two-dimensional or three-dimensional image.

This conversion is crucial because the MRI scanner initially acquires data in the frequency domain, as it is easier to capture oscillations and variations in signal frequency than direct spatial information. The Fourier Transform enables the extraction of this frequency-domain data and maps it onto a spatial representation of the body. To elaborate on the process further, the MRI raw data (k-space data) is often represented by a two-dimensional array of complex numbers. Each point in this array corresponds to a different frequency component. The Fourier Transform process, when applied to the k-space data, converts it into an image that can be interpreted in terms of anatomical features. In practice, the MRI reconstruction process involves applying the FFT, an efficient algorithm for computing the Fourier Transform. This method significantly accelerates the reconstruction, particularly for high-resolution scans, and is crucial in clinical settings where time is of the essence.

The Fourier Transform is also used in CT scans, albeit in a slightly different manner. In CT imaging, the raw data is acquired as a series of X-ray projections from multiple angles around the patient. These projections are initially collected in the frequency domain. The data is then processed using the Radon Transform, which is related to the Fourier Transform, to create cross-sectional images. The Inverse Radon Transform is applied, similar to the inverse Fourier Transform, to reconstruct the image in the spatial domain. The use of Fourier techniques in CT enables the extraction of detailed internal body structures, which is beneficial for detecting tumors, fractures, and internal bleeding (Fig. 2.10).

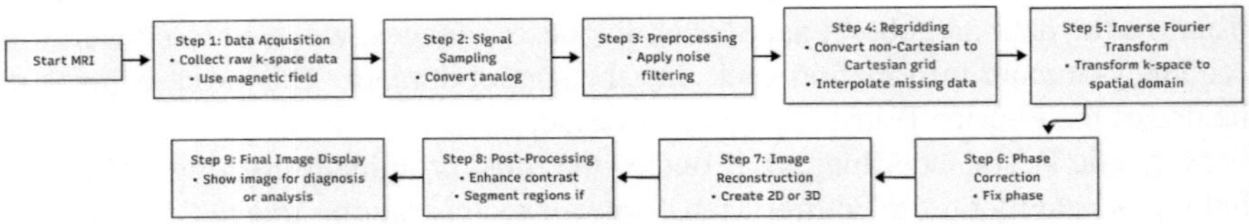

Fig. 2.10 A step-by-step process of MRI image reconstruction using the Fourier Transform

2.7 Time-Frequency Analysis

In the analysis of signals, particularly biological signals, time–frequency analysis offers a robust framework for understanding the dynamic behavior of signals that evolve both in time and frequency. Unlike traditional time-domain analysis, which focuses solely on how a signal changes over time, time–frequency analysis provides a means to examine the frequency content of a signal at various points in time. This approach becomes particularly relevant for complex biological systems, where signals may undergo rapid changes or contain both transient and oscillatory components that are critical for understanding underlying physiological processes.

In biological systems, such as the heart, brain, and muscles, signals often exhibit a combination of periodic and non-periodic elements. For example, the electrical activity in the heart (ECG), the brain (EEG), or the muscles (EMG) may involve sudden shifts or slow-wave patterns that cannot be fully captured using traditional time-domain analysis. Time–frequency analysis bridges this gap by providing a dual representation of signals, allowing for the simultaneous examination of both time and frequency characteristics [92, 93]. This method reveals insights into how the signal evolves across both dimensions, making it easier to identify transient events, periodic patterns, and frequency-specific behaviors that are essential for understanding the underlying biological phenomena. Time–frequency analysis is especially valuable in medical diagnostics, where it can be used to monitor complex signals in real-time, track changes over time, and detect subtle abnormalities that might otherwise go unnoticed. By analyzing signals in both time and frequency domains, researchers and clinicians can identify different types of pathologies, such as arrhythmias in ECG signals, abnormal brainwave activity in EEG signals, or muscle dysfunction in EMG data. This comprehensive approach enables a more accurate diagnosis and informed treatment planning, as time–frequency analysis can pinpoint specific frequency bands associated with particular physiological or pathological states [67, 94].

One key advantage of time–frequency analysis is its ability to offer both high time and frequency resolution, depending on the chosen method. Techniques such as the Short-Time Fourier Transform (STFT) and wavelet transform enable the decomposition of signals into localized frequency components, revealing both fast transients and periodic oscillations. This ability to zoom in on specific events in both time and frequency is essential for studying signals that are not strictly periodic but exhibit non-stationary behavior, like epileptic seizures or irregular heartbeats. Moreover, time–frequency analysis is instrumental in isolating specific components of mixed signals. Biological signals often consist of multiple overlapping frequencies that represent different underlying processes. For instance, in EEG signals, various brainwave patterns, such as alpha, beta, and delta rhythms, may coexist, each corresponding to different brain states or activities. Time–frequency analysis can separate these components, allowing researchers to study each one individually and gain a deeper understanding of the underlying processes at play.

In addition to its application in diagnostics, time–frequency analysis is instrumental in fields such as bioengineering and neuroscience, where complex signals from sensors and devices are analyzed to model biological systems. This can include everything from studying muscle fatigue through EMG signals to understanding brain function through EEG signals, where identifying specific frequency bands can reveal critical information about neural activity, cognitive states, and disease states.

Despite its powerful capabilities, time–frequency analysis is not without its challenges. One limitation lies in the trade-off between time and frequency resolution, often referred to as the uncertainty principle. While high-frequency resolution can provide more detailed information about frequency components, it can result in a loss of time resolution, making it more challenging to pinpoint exact moments of transient events. Conversely, improved time resolution may result in a loss of frequency information. This challenge can be mitigated through careful selection of analysis methods and optimization of parameters, but it remains an important consideration when applying time–frequency analysis to biological signals [95]. Nevertheless, time–frequency analysis is an indispensable tool in the study of biological signals, offering a unique and valuable perspective on the complex dynamics that drive physiological processes. Its ability to combine both time and frequency information enables the extraction of critical insights that would otherwise remain hidden in a purely time-domain or frequency-domain analysis. By facilitating the identification of abnormalities, uncovering hidden patterns, and enhancing real-time monitoring, time–frequency analysis has become a crucial technique in the fields of medical diagnostics, bioengineering, and neuroscience.

2.7.1 Short-Time Fourier Transform

In the realm of signal processing, both the Wavelet Transform (WT) and STFT are utilized to analyze non-stationary signals, particularly those that exhibit time-varying characteristics, such as biomedical signals. These methods are instrumental for investigating signals that cannot be adequately represented in either the time or frequency domain alone. Unlike traditional methods, such as the Fourier Transform, which decompose a signal into a set of frequency components for the entire duration, the STFT and WT enable both time and frequency analysis for signals whose frequency content changes over time [96, 97].

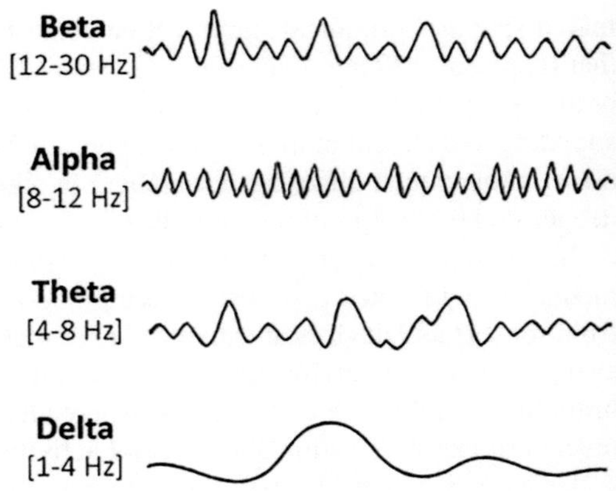

Fig. 2.11 Different EEG signals

STFT and Wavelet Transforms are two distinct approaches that achieve time–frequency localization. While both approaches provide insight into the time–frequency structure of a signal, they differ in their mathematical foundations, the way they localize both time and frequency, and their applications. These tools are crucial for analyzing physiological signals, such as ECG, EEG, EMG, and ECoG, providing valuable insights into heart rhythms, brain activity, and muscle contractions, among others [49]. The Short-Time Fourier Transform is an extension of the Fourier Transform, designed to analyze signals that change over time. The key idea behind the STFT is to divide the signal into small, overlapping segments (also called windows) and apply the Fourier Transform to each segment. This results in a two-dimensional representation of the signal in both time and frequency domains (Fig. 2.11).

Mathematically, the STFT of a signal x(t) is defined as Eq. 2.22:

$$STFT\{x(t)\} = X(t,f) = \int_{-\infty}^{\infty} x(\tau)\mathrm{w}(\tau - \mathrm{t})\mathrm{e}^{-\mathrm{j}2\pi \mathrm{f}\tau}\mathrm{d}\tau \qquad (2.22)$$

where:

- X(t,f) is the **STFT** of the signal x(t) representing the signal's frequency content at time t and frequency f
- w(τ − t) is a **window function** that selects a small portion of the signal around time t.
- $e^{-j2\pi f\tau}$ is the **Fourier kernel** that extracts frequency components from the signal.

The choice of window function w(t) is critical to the performance of the STFT. Standard window functions include the rectangular window, Hamming window, Hanning window, and Gaussian window, each having specific properties related to the trade-off between time resolution and frequency resolution [96].

The window function determines the trade-off between time and frequency localization: more expansive windows provide better frequency resolution but poorer time

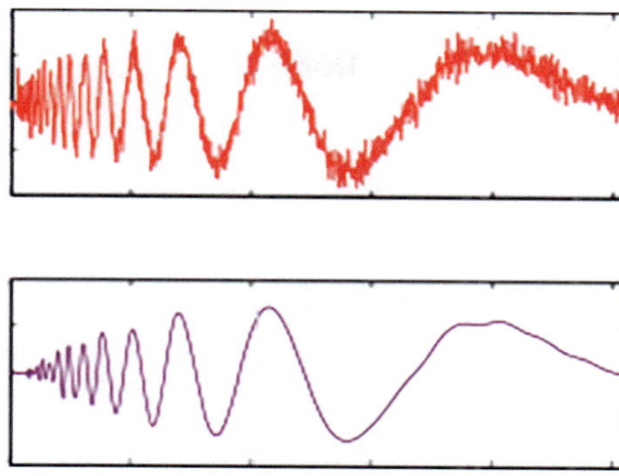

Fig. 2.12 Wavelet denoising (https://commons.wikimedia.org/wiki/File:Wavelet_denoising.svg)

resolution, and narrower windows provide better time resolution but poorer frequency resolution. The choice of window function depends on the specific characteristics of the signal being analyzed [93, 98].

The STFT is particularly useful in analyzing biomedical signals that are non-stationary or have time-varying frequency content. For example, EEG signals from the brain contain alpha, beta, delta, and theta waves, each of which corresponds to different brain states. The STFT can be used to identify these frequency components at different times, aiding in the diagnosis of conditions like epilepsy or sleep disorders. Similarly, in ECG signals, STFT can be used to track changes in heart rhythm and detect abnormalities, such as arrhythmias. The STFT, while powerful, suffers from a fixed time–frequency resolution. That is, the window size determines the time and frequency resolutions, and this can be limiting when dealing with signals that require varying time–frequency resolution. For instance, a signal that contains both low-frequency components (such as heart rate) and high-frequency components (like muscle contraction signals) may not be optimally captured using a fixed window size (Fig. 2.12).

2.7.2 Wavelet Analysis

The Wavelet Transform is another time–frequency analysis technique that overcomes some of the limitations of the STFT by allowing for a variable window size. In the WT, signals are analyzed using a family of wavelet functions that are dilated (scaled) or translated (shifted) to provide both time and frequency information.

Unlike STFT, where the window size is fixed, Wavelet Transform uses a scaled window (smaller at high frequencies and larger at low frequencies) to achieve a better time–frequency resolution across different frequency ranges.

The Continuous Wavelet Transform (CWT) of a signal x(t) is defined as Eq. 2.23:

$$CWT\{x(t)\} = W_x(a,b) = \int_{-\infty}^{\infty} x(t)\psi^*\left(\frac{t - b}{a}\right)dt \tag{2.23}$$

where:

- $W_x(a,b)$ is the **wavelet transform** of the signal x(t) at scale a and translation b.
- $\psi(t)$ is the **wavelet function**.
- a is the scale factor, which determines the frequency resolution.
- b is the translation factor, which determines the time resolution.

The mother wavelet $\psi(t)$ is the prototype of the wavelets that are used for decomposition. Common mother wavelets include:

- Morlet wavelet: Best suited for frequency analysis.
- Mexican Hat wavelet: Offers sound time localization.
- Haar wavelet: Discrete and used in cases where sharp transitions are important.

The choice of wavelet depends on the signal's characteristics and the desired resolution in time and frequency.

The Wavelet Transform is particularly well-suited for non-stationary signals, as it allows for the analysis of signals with local features at different scales. For example, EEG signals exhibit varying frequency components over time, such as spikes during seizures or bursts of activity during cognitive tasks. The WT can capture these non-stationary phenomena by using a localized frequency analysis with variable resolution. This capability is critical for detecting transient events, which are essential for monitoring conditions such as epilepsy, neural oscillations, or sleep stages. Another key biomedical application of the WT is in the analysis of ECG signals. The WT can detect rapid changes in the heart's electrical activity, such as arrhythmias, by breaking the signal down into components that capture both slow and fast changes in the waveform. The ability of the WT to localize transient events makes it a valuable tool for studying heart conditions, where sudden changes in the signal can be indicative of a pathological event [82, 99].

Advantages:

- Multiresolution analysis: The WT provides both time and frequency localization at different scales, making it particularly effective for analyzing non-stationary signals that contain both high-frequency transients and low-frequency oscillations.
- Adaptability: The Wavelet Transform is highly adaptable and can be customized to suit the signal's features by choosing an appropriate wavelet function.

Limitations:

- Computational complexity: The continuous wavelet transform (CWT) can be computationally intensive, particularly when working with large datasets or in real-time applications.
- Window selection: The choice of wavelet function and scale can significantly influence the results, requiring expertise and domain-specific knowledge.

Example: STFT and Wavelet Analysis of ECG Signal

Short-Time Fourier Transform on ECG Signal

Let us assume we have a sampled ECG signal, denoted as x(t), with a sampling frequency of Fs = 500 Hz. This means the signal is sampled at a rate of 500 samples per second. The signal duration is T = 10 s. We will compute the STFT of the ECG signal using a sliding window approach.

Step 1: Define Parameters.

- Window Function: We will use a Hamming window for the analysis.
- The window length is 200 samples (N = 200), meaning each window represents 200 samples of the ECG signal.
- The overlap between successive windows is 50% so that each window will overlap by 100 samples.

Step 2: Calculate STFT.

STFT is calculated using the Eq. 2.24:

$$X(t,f) = \sum_{n=-\infty}^{\infty} x(t+n)\, w(t-n)\, e^{-j2\pi fn} \tag{2.24}$$

where:

- x(t + n) is the ECG signal at time t + n
- w(t − n) is the window function.
- $e^{-j2\pi fn}$ is the **Fourier transform** of the windowed signal.

Step 3: Spectrogram.

- The STFT results in a time–frequency representation of the ECG signal, typically visualized as a spectrogram. This will have:
- Time on the x-axis.
- Frequency on the y-axis.
- Intensity (or magnitude) represented by the color scale.

Wavelet Analysis of ECG Signal

- The ECG signal is the same as before, sampled at Fs = 500 Hz and with a duration of T = 10 s.

Step 1: Choose a Wavelet Function.

We choose the Morlet wavelet as the mother wavelet for its good frequency localization. The Morlet wavelet is defined as Eq. 2.25:

$$\psi(t) = \frac{1}{\sqrt{\sigma}} \exp\left(-\frac{t^2}{2\sigma^2}\right) \exp(j2\pi f_0 t) \tag{2.25}$$

where σ controls the width of the wavelet, and f_0 is the center frequency.

Step 2: Apply Continuous Wavelet Transform (CWT).

The CWT of a signal x(t) with a wavelet $\psi(t)$ is given by Eq. 2.26:

$$W(t,f) = \int_{-\infty}^{\infty} x(t)\psi^{*}(\mathrm{t} - \tau)\mathrm{d}\tau \tag{2.26}$$

where $\psi * (t - \tau)$ is the **complex conjugate** of the wavelet function, and the integral computes how well the wavelet matches the signal at each scale (or frequency).

Step 3: Scale and Frequency Relationship.

The scales in CWT correspond to different frequencies. A low scale corresponds to high frequency, and a high scale corresponds to low frequency (Table 2.4).

Comparison of STFT and WT on ECG Signal

STFT is useful for stationary signals but faces challenges in non-stationary signal analysis, as it requires a fixed window size that compromises either time resolution or frequency resolution. The Wavelet Transform offers a flexible, adaptive approach, particularly well-suited for analyzing non-stationary signals, such as ECG, where both sharp transitions (high frequencies) and low-frequency components require different resolutions. Both techniques are valuable for time–frequency analysis, but

Table 2.4 Comparison of STFT and WT

Feature	Short-time Fourier transform	Wavelet transform
Time–frequency resolution	Fixed resolution determined by the window length	Adaptive resolution: high-frequency components are analyzed with better time resolution, and low-frequency components with better frequency resolution
Best for	Stationary signals. Struggles with non-stationary signals like ECG	Non-stationary signals, such as ECG, exhibit both fast transients and slower rhythms
Performance on sharp transitions (e.g., QRS complex)	Not ideal for analyzing sharp transitions due to fixed window size	Tracks rapid changes (e.g., QRS complexes) efficiently due to adaptive resolution
Handling of Frequency Content	Poor handling of changing frequency content over time in signals like ECG	Effectively captures changing frequency content in signals like ECG, with varying time–frequency resolution
Effectiveness for ECG	Struggles to capture fast transients and longer heart rhythms accurately	More effective for analyzing ECG due to its ability to capture both rapid changes and slow rhythms

the Wavelet Transform is generally preferred for analyzing complex biomedical signals, such as ECG, due to its multi-resolution capabilities [32, 100].

Example: Analyzing EEG Signal Using STFT and Wavelet Transform

EEG signals are frequently utilized to investigate brain activity and identify conditions such as seizures, sleep disorders, and various neurological disorders. These signals are inherently non-stationary, meaning their frequency components change over time [64, 87]. Thus, analyzing EEG signals requires methods that can capture both fast transients (such as spikes or seizures) and slower rhythms (such as alpha or delta waves).

Short-Time Fourier Transform (STFT) on EEG Signal

Let us assume we have a segment of an EEG signal sampled at 1000 Hz for 10 s:

- EEG Signal: A typical signal might contain components such as alpha waves (8–12 Hz), beta waves (13–30 Hz), and rapid spikes (above 30 Hz).
- Window length: 1 s.
- Overlapping windows: 50% overlap (0.5 s).

Step 1: Split the signal into overlapping windows.

- For a window length of 1 s, with 50% overlap, the signal is divided into two halves, creating windows every 0.5 s.

Step 2: Apply the Fourier Transform to each window, producing the frequency spectrum for that window.

- For each 1-s window, the Fourier Transform gives us the amplitude of the signal at various frequency bins (e.g., 0–5 Hz, 5–10 Hz, etc.).

The STFT provides a spectrogram of the signal, showing how the frequency content of the EEG changes over time. The result may show an alpha wave activity (8–12 Hz) for several seconds, followed by a sudden spike (30 + Hz) if the person experiences a brief seizure.

The limitations of STFT on EEG signals stem from the fact that the time–frequency resolution is fixed due to the choice of window length. For example, if we use a short window, we get better time resolution but poorer frequency resolution. A longer window improves frequency resolution but reduces the time resolution, which can make it harder to detect fast transitions in the EEG signal.

Wavelet Transform (WT) on EEG Signal

Using Wavelet Transform to analyze the same EEG signal:

- Wavelet Transform provides better time–frequency localization compared to STFT because of its adaptive resolution.
- High-frequency components (such as spikes or seizures) are analyzed with better time resolution, while low-frequency components (such as alpha or delta waves) are analyzed with better frequency resolution.

Step 1: Choose a mother wavelet:

- A common choice is the Morlet wavelet, which is well-suited for EEG analysis.

Step 2: Apply the CWT to the EEG signal.

- The CWT computes a set of coefficients for each time point in the signal, giving us the amplitude of the signal in various frequency bands (such as 0–5 Hz, 5–10 Hz, etc.), with better time resolution at higher frequencies and better frequency resolution at lower frequencies.

Step 3: Visualize the result as a time–frequency plot (similar to a spectrogram) but with adaptive resolution across the frequencies.

The Wavelet Transform produces a time–frequency plot where alpha waves (8–12 Hz) can be visible in the low-frequency bands, and spikes (above 30 Hz) can be precisely detected with high time resolution. This enables us to track both rapid events (such as seizures) and slower rhythms (like brain oscillations during sleep) (Table 2.5).

Comparison of STFT and WT on EEG Signal

The Wavelet Transform adapts to the signal's features, providing much better detection of fast transients (spikes) and slow rhythms in EEG signals compared to STFT. The Wavelet Transform is particularly effective in handling non-stationary signals, such as EEG, as it can simultaneously track both rapid changes and slow trends [62, 101].

Table 2.5 Comparison: STFT vs. wavelet transform for EEG analysis

Aspect	Short-time Fourier transform	Wavelet transform
Resolution	Fixed time–frequency resolution determined by window size	Adaptive resolution, high-time resolution for high frequencies, low-time resolution for low frequencies
Best for	Stationary signals with constant frequency content	Non-stationary signals with varying frequency content, like EEG
Sharp transitions (e.g., spikes)	Less effective at detecting sudden spikes or seizures	Detects sudden changes in the signal (spikes, seizures) with better time resolution
Low-frequency activity	Struggles to provide good frequency resolution at low frequencies	Provides better frequency resolution at low frequencies (alpha, delta waves)
Use case	Good for signals with relatively stable frequency content over time	Ideal for analyzing complex, non-stationary signals like EEG, especially for detecting seizures, sleep patterns, etc.

2.8 Digital Filtering Techniques

Signal processing is at the heart of modern technologies, offering profound impacts across diverse fields, from telecommunications to biomedical engineering. Within signal processing, digital filtering is one of the most vital techniques. Digital filters are employed to manipulate and process signals by removing unwanted noise, extracting useful features, or shaping the signal in a way that it becomes more informative for further analysis. Filters are indispensable when it comes to the processing of real-world signals, which are often corrupted by noise or distorted in various ways during measurement, transmission, or storage. At its core, digital filtering refers to the process of modifying or altering the characteristics of a digital signal in order to achieve a specific desired outcome, such as noise reduction or signal enhancement. These operations are performed through the use of mathematical algorithms that manipulate a signal's frequency components. The development of digital filters stems from the need to address real-world challenges faced when working with signals that are collected and processed in a digital form, such as audio signals, image data, or biomedical signals like ECG and EEG. The digital filter is based on the principles of both frequency-domain and time-domain signal processing. While analog filtering relies on continuous-time signals, digital filters are designed to operate on discrete-time signals, which are obtained by sampling continuous signals at specific intervals. The transition from analog to digital filtering is made possible through sampling and quantization, which are key concepts in DSP. Sampling is the process of converting a continuous-time signal into a discrete one by capturing its values at regular intervals, while quantization refers to the process of approximating those sampled values to a finite set of levels [64, 102].

At the heart of digital filtering are two fundamental types of filters: Finite Impulse Response (FIR) and Infinite Impulse Response (IIR) filters. Each of these filter types has distinct characteristics and is suitable for different applications. Understanding their differences is crucial for selecting the most suitable filter for a specific problem.

- FIR filters are characterized by a finite number of taps or coefficients, meaning the output of the filter depends only on a finite number of input samples. The key benefit of FIR filters is their inherent stability, as they lack feedback loops, unlike IIR filters. They are easy to design, and the lack of feedback ensures that they do not produce oscillations or instability. However, they often require a higher filter order (more coefficients) to achieve sharp frequency response characteristics, which results in increased computational costs.
- IIR filters, on the other hand, have an infinite number of taps, which means the output depends not only on the current and past input samples but also on past output samples. This recursive nature allows IIR filters to achieve the same performance as FIR filters with fewer coefficients, making them computationally more efficient. However, this efficiency comes at the cost of stability concerns. If not correctly designed, IIR filters may become unstable and lead to unwanted oscillations.

In both FIR and IIR filters, the primary goal is to manipulate the signal's frequency components. This is accomplished by setting specific frequencies to pass through the filter (passband) and attenuating others (stopband). This frequency manipulation can occur either in the time domain or the frequency domain. In the time domain, the filter operates by applying a series of coefficients to the input signal's samples, while in the frequency domain, the filter operates by modifying the frequency components of the signal directly [62, 103].

Digital filters are widely used across various fields. In biomedical signal processing, filters are often employed to remove noise from physiological signals such as ECG and EEG. For instance, in ECG analysis, filters help eliminate high-frequency noise and baseline wander, which can significantly distort the heart signal. In EEG signal analysis, digital filters remove artifacts such as eye movements or muscle contractions that can interfere with the analysis of brain activity. Beyond biomedical applications, digital filters are also widely used in fields such as audio signal processing, communications, and image processing. For example, in audio processing, filters are used for noise reduction, echo cancellation, and equalization. In image processing, filters play a crucial role in tasks such as edge detection, image sharpening, and blurring. Real-world signals, such as audio recordings, ECGs, or EEGs, are often corrupted by noise—unwanted signals that interfere with the desired signal. Noise can originate from various sources, such as electronic interference, environmental conditions, or human activity. Digital filters are designed to suppress or eliminate noise from these signals while retaining the valuable information. For instance, in speech enhancement, filters can be used to remove background noise from speech signals, making the speech more intelligible in noisy environments.

In biomedical applications, the need for effective noise reduction and signal enhancement is even more critical, as the accuracy of medical diagnoses can depend on the quality of the signals. Filters that operate on ECG, EEG, or EMG signals must be designed to preserve key features (such as the QRS complex in ECG) while removing unwanted noise that can distort these features.

While digital filters are potent tools, they are not without their challenges. One of the main difficulties is the trade-off between time and frequency resolution. In the case of the STFT, for example, the filter's time–frequency resolution is fixed by the window length, meaning there is a balance between how finely we can analyze the signal in time and frequency. Filters with longer windows have better frequency resolution but poorer time resolution, and vice versa. Additionally, filter stability is a crucial consideration. For IIR filters, it is important to ensure that the recursive structure of the filter does not lead to instability, where the filter might produce oscillations or grow without bound. Another challenge lies in the computational complexity of digital filters, especially when real-time processing is required. This can place a significant burden on the processor, particularly when filters with higher orders or complex designs are used. Despite these challenges, advances in adaptive filtering and machine learning have enabled the development of more sophisticated filter designs that can adjust in real-time to the signal's characteristics. These advanced techniques open up new possibilities for filtering applications, especially in dynamic environments [104, 105].

2.8.1 Low-Pass Filters

Low-pass filters (LPF) are one of the fundamental types of digital filters and analog filters used in signal processing. Their primary purpose is to allow low-frequency components of a signal to pass through while attenuating or completely blocking higher-frequency components. Low-pass filters are used in a wide variety of applications, including noise reduction, signal smoothing, and frequency separation. This chapter explores the principles, design, mathematical formulation, and practical applications of low-pass filters. A low-pass filter is a type of filter that only allows signals with frequencies lower than a specific cutoff frequency to pass through, while attenuating frequencies above the cutoff frequency. The cutoff frequency, often denoted by fc, is the point where the filter's frequency response begins to decline. The transition between the passband (frequencies that pass through) and the stopband (frequencies that are attenuated) is generally smooth, though it can vary depending on the filter's design. In the simplest terms, the low-pass filter serves as a frequency-selective filter that selectively blocks high-frequency noise or unwanted high-frequency components while preserving the integrity of the low-frequency signal [106].

Mathematical Formulation

The behavior of a low-pass filter can be described by its transfer function, H(s), which defines the relationship between the input and output signals in the Laplace domain for analog filters or in the Z-domain for digital filters.

The most straightforward low-pass filter is the first-order RC low-pass filter. The transfer function H(s) of this filter in the analog domain is given by Eq. 2.27:

$$H(s) = \frac{1}{1 + sRC} \tag{2.27}$$

where:

- R is the resistance in ohms,
- C is the capacitance in farads,
- S is the complex frequency variable in the Laplace domain ($s = \sigma + j\omega$ where σ is the fundamental part and ω is the imaginary part),
- H(s) represents the transfer function of the filter.

This transfer function determines the filter's behavior in response to different frequencies. At low frequencies (where s is small), the output is approximately equal to the input, indicating that low-frequency signals pass through the filter without significant attenuation. At high frequencies (where s is large), the output diminishes, indicating that the filter attenuates high-frequency signals. The cutoff frequency fc of a low-pass filter is defined as the frequency at which the magnitude of the transfer function falls to 12 (approximately 0.707) of its maximum value. For the RC low-pass filter, the cutoff frequency fc is given by Eq. 2.28:

$$f_c = \frac{1}{2\pi RC} \tag{2.28}$$

This frequency determines the boundary between the passband (frequencies that pass through the filter) and the stopband (frequencies that are attenuated). Signals with frequencies lower than fc are passed with little or no attenuation, while signals with frequencies higher than fc are progressively attenuated.

Frequency Response of Low-Pass Filters

The frequency response of a low-pass filter describes how the filter modifies the amplitude and phase of the input signal at different frequencies. The frequency response H(jω) of the analog low-pass filter is given by Eq. 2.29:

$$H(j\omega) = \frac{1}{1 + j\omega RC} \tag{2.29}$$

where ω is the angular frequency, $\omega = 2\pi f$, and f is the frequency in hertz.

- **At ω = 0 (DC or low frequencies)**: The transfer function magnitude is 1, meaning the filter passes the signal without any attenuation.
- **At ω = ∞ (high frequencies)**: The transfer function magnitude approaches 0, meaning the filter blocks high-frequency signals.

In practice, the transition between the passband and the stopband is not instantaneous but occurs over a range of frequencies, depending on the filter order.

Higher-Order Low-Pass Filters

While the first-order RC low-pass filter is the simplest, higher-order filters (such as second-order and third-order) are often used to achieve steeper roll-offs and better frequency separation. A second-order LC low-pass filter has the following transfer function (Eq. 2.30):

$$H(s) = \frac{\omega_0^2}{s^2 + 2\zeta\omega_0 s + \omega_0^2} \tag{2.30}$$

where:

- ω_0 is the natural frequency (in radians per second),
- ζ is the damping factor, which controls the width of the filter's transition band.

Higher-order filters provide more aggressive attenuation of high frequencies. However, their design is more complex and involves using multiple reactive components (inductors and capacitors for analog filters or more advanced algorithms for digital filters).

Low-Pass Filters in Digital Signal Processing

In DSP, low-pass filters are implemented by performing calculations on discrete signals. These filters operate in the Z-domain, and the following difference equation typically describes their behavior (Eq. 2.31):

$$y[n] = \sum_{k=0}^{M} b_k x[n-k] \tag{2.31}$$

where:

- y[n] is the output at time index n,
- x[n − k] is the input at previous time steps (indexed by k),
- bk are the filter coefficients, and
- M is the number of taps in the filter.

Digital low-pass filters are designed using methods like windowing, bilinear transformation, or frequency sampling. Standard algorithms for digital low-pass filters include the FIR filter and the IIR filter.

FIR Low-Pass Filters

An FIR low-pass filter is characterized by having a finite number of taps, meaning its impulse response is of finite duration. The design of an FIR filter involves choosing the appropriate coefficients that define its frequency response [107, 108]. For an FIR low-pass filter, the transfer function is (Eq. 2.32):

$$H(z) = \sum_{k=0}^{N} b_k z^{-k} \tag{2.32}$$

where $\mathrm{b_k}$ are the filter coefficients, and N is the filter order.

IIR Low-Pass Filters

An IIR low-pass filter has feedback, meaning that its impulse response is infinite. It is more computationally efficient than FIR filters because it requires fewer coefficients to achieve a similar frequency response. The transfer function for an IIR low-pass filter is (Eq. 2.33):

$$H(z) = \frac{b_0 + b_1 z^{-1} + \cdots + b_M z^{-M}}{1 + a_1 z^{-1} + \cdots + a_N z^{-N}} \tag{2.33}$$

where $\mathrm{b_k}$ and $\mathrm{a_k}$ are the filter coefficients, and M and N are the numbers of zeros and poles of the filter, respectively.

Low-pass filters are used in a wide range of applications across various fields of signal processing [63, 109].

1. Noise reduction in biomedical signals: In ECG and EEG signal processing, low-pass filters are employed to remove high-frequency noise, such as muscle artifacts or electrical interference, while retaining the important low-frequency components of the signal.
2. Audio processing: Low-pass filters are crucial in audio processing for removing high-frequency noise or unwanted signals. They are used in equalizers, tone control systems, and bass boost circuits to shape the audio signal.
3. Image processing: Low-pass filters in image processing can be used to smooth or blur an image, thereby removing high-frequency noise, such as pixel irregularities or sharp edges.
4. Telecommunications: In telecommunications, low-pass filters are employed to remove high-frequency interference and limit the bandwidth of transmitted signals to a desired range.

2.8.2 *High-Pass Filter*

A high-pass filter is a frequency-selective filter that allows signals with frequencies higher than a specific cutoff frequency to pass through, while attenuating frequencies below the cutoff. The cutoff frequency, denoted as fc, is the point at which the filter's frequency response begins to reduce the amplitude of the signal. Below the cutoff frequency, the filter effectively blocks the signal, allowing only high-frequency components to pass. In simple terms, high-pass filters serve to attenuate low-frequency noise or components of a signal, allowing higher-frequency components to pass through without distortion. This makes them essential tools in many signal processing applications where low-frequency interference needs to be removed while preserving high-frequency information [110].

Mathematical Formulation of High-Pass Filters

The behavior of high-pass filters can be understood through their transfer function, which describes the relationship between the input signal and the output signal in the frequency domain. This can be formulated using the Laplace transform for analog filters and the Z-transform for digital filters [44, 69]. A simple and commonly used high-pass filter is the first-order RC high-pass filter. The transfer function H(s) of this filter in the analog domain is given by Eq. 2.34:

$$H(s) = \frac{sRC}{1 + sRC} \tag{2.34}$$

where:

- R is the resistance in ohms,
- C is the capacitance in farads,
- s is the complex frequency variable in the Laplace domain ($s = \sigma + j\omega$, where σ is the fundamental part and ω is the imaginary part

This transfer function defines the relationship between the input and output signals, where:

- For low frequencies ($s \approx 0$), the transfer function approaches 0, meaning the signal is attenuated.
- For high frequencies ($s \to \infty$), the transfer function approaches 1, indicating that high-frequency components are passed through without attenuation.

The **cutoff frequency** fc for a high-pass filter is the frequency at which the magnitude of the filter's transfer function falls to approximately 0.707 of its maximum value. For the first-order RC high-pass filter, the cutoff frequency is given by Eq. 2.35:

$$f_c = \frac{1}{2\pi RC} \tag{2.35}$$

This frequency serves as the boundary between the passband (where high-frequency components pass through) and the stopband (where low-frequency components are attenuated). The **frequency response** of a high-pass filter describes how the magnitude and phase of the filter's output vary concerning different input frequencies. The frequency response $H(j\omega)$ of the analog high-pass filter is given by Eq. 2.36:

$$H(j\omega) = \frac{j\omega RC}{1 + j\omega RC} \tag{2.36}$$

where ω is the angular frequency, $\omega = 2\pi f$, and f is the frequency in hertz.

- **At low frequencies** ($\omega = 0$): The transfer function magnitude is 0, meaning low-frequency signals are blocked.
- **At high frequencies** ($\omega \to \infty$): The transfer function magnitude approaches 1, meaning high-frequency signals pass through without attenuation.

This frequency response illustrates the high-pass behavior, where the filter allows higher frequencies to pass and attenuates lower frequencies [70, 111].

Higher-Order High-Pass Filters

Higher-order high-pass filters provide more aggressive filtering of low frequencies. A second-order LC high-pass filter has the following transfer function (Eq. 2.37):

$$H(s) = \frac{s^2}{s^2 + 2\zeta\omega_0 s + \omega_0^2} \tag{2.37}$$

where:

- ω_0 is the natural frequency (in radians per second),
- ζ is the damping factor, which controls the width of the filter's transition band.

Higher-order filters are helpful in applications where it is important to sharply attenuate lower-frequency components and allow only a narrow band of higher frequencies to pass.

High-Pass Filters in Digital Signal Processing

In digital signal processing (DSP), high-pass filters are implemented using discrete-time signals. These filters operate in the Z-domain, and the following difference equation often describes their behavior (Eq. 2.38):

$$y[n] = \sum_{k=0}^{M} b_k x[n-k] - \sum_{k=1}^{N} a_k y[n-k] \tag{2.38}$$

where:

- y[n] is the output signal at the n-th time step,
- x[n − k] is the input signal at previous time steps (indexed by k),
- b_k are the filter coefficients for the feedforward part (numerator),
- a_k are the filter coefficients for the feedback part (denominator),
- M and N are the numbers of taps in the feedforward and feedback parts, respectively.

The design of high-pass filters in digital signal processing often involves choosing appropriate filter coefficients based on the desired frequency response [89]. Digital high-pass filters are typically designed using methods such as windowing, bilinear transformation, or frequency sampling.

FIR High-Pass Filters

A Finite Impulse Response (FIR) high-pass filter is characterized by a finite number of taps, meaning its impulse response has a limited duration. The transfer function for an FIR high-pass filter is given by Eq. 2.39:

$$H(z) = \sum_{k=0}^{N} b_k z^{-k} \tag{2.39}$$

where b_k are the filter coefficients, and N is the filter order.

IIR High-Pass Filters

An Infinite Impulse Response (IIR) high-pass filter is characterized by feedback, meaning that its impulse response is theoretically infinite. IIR filters are more computationally efficient than FIR filters, as they require fewer coefficients for the same level of performance. The transfer function for an IIR high-pass filter is (Eq. 2.40):

$$H(z) = \frac{b_0 + b_1 z^{-1} + \cdots + b_M z^{-M}}{1 + a_1 z^{-1} + \cdots + a_N z^{-N}} \tag{2.40}$$

where b_k and a_k are the filter coefficients, and M and N are the numbers of zeros and poles, respectively.

High-pass filters have a wide range of applications in various fields of signal processing, particularly in scenarios where low-frequency noise or drift needs to be removed.

1. Noise reduction in biomedical signals: In ECG, EEG, and other biomedical signals, high-pass filters are employed to eliminate low-frequency noise, such as baseline wander or drift, caused by movement, respiratory patterns, or equipment interference.
2. Audio processing: High-pass filters are widely used in audio systems to eliminate unwanted low-frequency noise, such as hum or rumbles, and to enhance the clarity of higher-frequency sounds. They are used in equalizers, mixing consoles, and audio effects.
3. Image processing: In image processing, high-pass filters are often used for edge detection and image sharpening, as they emphasize high-frequency details such as edges and delicate textures, while suppressing low-frequency background or smoothing effects.
4. Telecommunications: High-pass filters are used in telecommunications to remove low-frequency interference in data transmission systems. By eliminating unwanted low-frequency noise, high-pass filters help to improve the signal-to-noise ratio and the quality of the communication link.

2.8.3 Band-Pass Filter

Band-pass filters (BPF) are critical tools in signal processing that allow only a specific range of frequencies, or a band, to pass through, while attenuating frequencies outside of this range. These filters are widely used in fields such as communications, audio processing, biomedical signal processing, and instrumentation, where it is essential to isolate a particular frequency range from a broader spectrum of signals. A band-pass filter can be viewed as a combination of a low-pass filter and a high-pass filter, allowing frequencies within a specific range to pass while blocking both higher and lower frequencies [50, 112].

A band-pass filter is a frequency-selective filter that passes frequencies within a specific range and attenuates frequencies outside this range. Two important parameters characterize the filter:

- **Lower cutoff frequency (f_L)**: The frequency below which signals are attenuated.
- **Upper cutoff frequency (f_H)**: The frequency above which signals are attenuated.
- **Bandwidth (BW)**: The difference between the upper and lower cutoff frequencies, i.e., $BW = f_H - f_L$. It defines the frequency range that is allowed to pass.

In simple terms, a band-pass filter isolates a specific range of frequencies while eliminating both lower and higher frequencies. The region between the two cutoff frequencies is referred to as the passband, while frequencies outside this range are

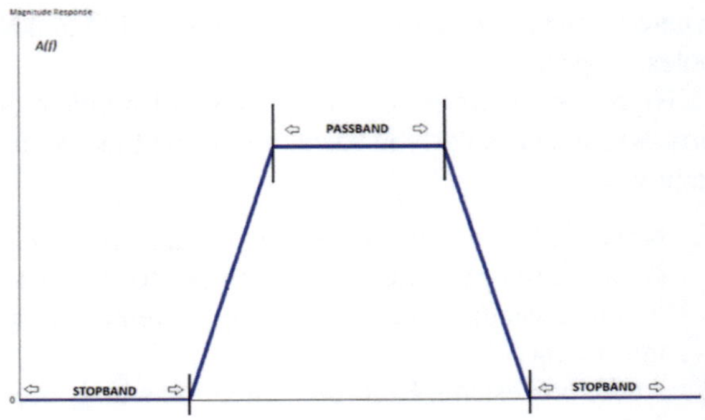

Fig. 2.13 Response of a band-pass filter (https://commons.wikimedia.org/wiki/File:Filter_Example.png)

classified as being in the stopband. The behavior of the band-pass filter is defined by the center frequency f0, which represents the midpoint of the passband (Eq. 2.41):

$$f_0 = \sqrt{f_L \cdot f_H} \tag{2.41}$$

This center frequency is essential when designing filters for specific applications, as it determines the frequency around which the filter will pass signals most effectively (Fig. 2.13).

Mathematical Formulation of Band-Pass Filters

The operation of a band-pass filter can be described using its transfer function in both the analog and digital domains. The transfer function relates the input and output signals in terms of frequency and is crucial for understanding how the filter alters the frequency content of the signal [67, 113]. For an analog band-pass filter, one standard configuration is the second-order RLC band-pass filter, whose transfer function H(s) in the analog domain is given by Eq. 2.42:

$$H(s) = \frac{s}{s^2 + \frac{\omega_0}{Q}s + \omega_0^2} \tag{2.42}$$

where:

- s is the complex frequency variable ($s = j\omega$ in the frequency domain),
- ω_0 is the **natural frequency** (in radians per second), which is related to the center frequency f_0 by $\omega_0 = 2\pi f$,
- Q is the **quality factor**, which controls the **selectivity** of the filter and determines the width of the passband.

For a band-pass filter, Q determines how sharply the filter allows frequencies around the center frequency to pass while attenuating other frequencies. A high value of Q results in a narrow passband (high selectivity), while a lower value of Q broadens the passband.

The **bandwidth** BW of a band-pass filter is related to the quality factor Q and the center frequency f_0 by the following relation (Eq. 2.43):

$$BW = \frac{f_0}{Q} \tag{2.43}$$

This equation shows how adjusting the quality factor affects the bandwidth of the filter. A higher Q-factor results in a narrower bandwidth, making the filter more selective at its center frequency. Conversely, a lower Q-factor leads to a wider passband and a less selective filter.

Transfer Function in the Digital Domain

In digital signal processing (DSP), a band-pass filter can be implemented using FIR or IIR techniques. The design of a digital band-pass filter typically involves creating a difference equation that relates the current and past samples of the input and output signals. For a digital band-pass filter, the difference equation takes the following general form (Eq. 2.44):

$$y[n] = \sum_{k=0}^{M} b_k x[n-k] - \sum_{k=1}^{N} a_k y[n-k] \tag{2.44}$$

where:

- y[n] is the output signal at the n-th time step,
- x[n − k] is the input signal at previous time steps (indexed by k),
- b_k are the filter coefficients for the feedforward part (numerator),
- a_k are the filter coefficients for the feedback part (denominator),
- M and N are the numbers of taps in the feedforward and feedback parts, respectively.

Frequency Response of Band-Pass Filters

The frequency response H(jω) of a band-pass filter describes how the filter alters the amplitude and phase of different frequency components in the input signal. It can be derived from the transfer function and gives insight into the filter's behavior in the frequency domain. For a second-order band-pass filter, the frequency response is given by Eq. 2.45:

$$H(j\omega) = \frac{j\omega}{(j\omega)^2 + \frac{\omega_0}{Q} j\omega + \omega_0^2} \tag{2.45}$$

At low frequencies (ω ≪ ω0), the filter attenuates the signal, and the transfer function approaches zero. At high frequencies (ω ≫ ω0), the filter also attenuates the signal. However, at the center frequency ω0, the filter allows the signal to pass with minimal attenuation, resulting in the maximum gain.

2.8.4 *Notch Filters*

Notch filters, also known as band-stop filters or reject filters, are signal processing tools that are designed to remove a specific range of frequencies from a signal. They work by attenuating frequencies within a narrow band while allowing the rest of the frequency spectrum to pass through relatively unaffected. Notch filters are essential in various applications, including audio processing, communication systems, and biomedical signal processing, where specific frequencies need to be eliminated without significantly altering the rest of the signal.

A notch filter is a filter that attenuates a specific frequency or a narrow range of frequencies, effectively rejecting those components from the signal. This type of filter is a specialized version of a band-stop filter, featuring a significantly narrower stopband. Notch filters are used to suppress unwanted frequencies that may interfere with the desired signal while allowing other frequencies to pass through unaffected [116, 117]. Key parameters of a notch filter include:

- **Center frequency ($\mathbf{f_0}$)**: The frequency around which the notch is applied. The notch filter will attenuate this frequency most strongly.
- **Bandwidth (BW)**: The width of the frequency range that is attenuated. This is a critical parameter in designing the filter to reject specific frequencies without affecting the overall signal quality.
- **Quality factor (Q)**: A measure of the filter's selectivity. A high Q-factor results in a narrower notch, while a low Q-factor broadens the notch.

A notch filter is designed to have a frequency response that removes the frequency components in the stopband (the range around f_0) and passes the rest of the frequencies with minimal distortion (Fig. 2.14).

Mathematical Formulation of Notch Filters

The operation of a notch filter can be described in both the analog and digital domains. The most common analog implementation of a notch filter is a second-order system that uses a parallel resonant circuit or an RLC circuit. The mathematical description of such a filter is given by its transfer function in the s-domain [67, 118]. For a simple second-order notch filter, the transfer function H(s) in the analog domain is given by Eq. 2.46:

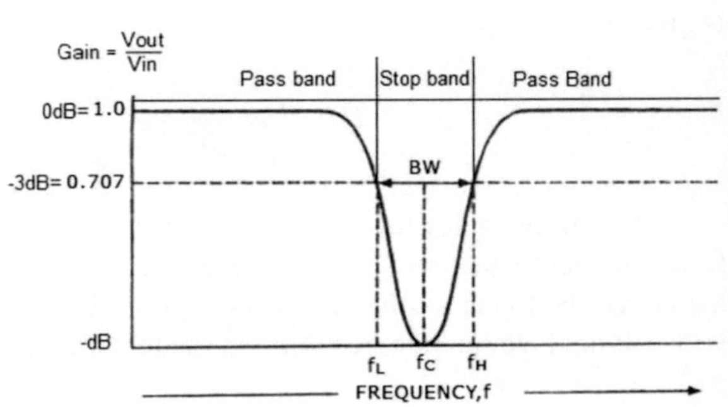

Fig. 2.14 Frequency response of a notch filter (https://commons.wikimedia.org/wiki/File:Frekvencijski_odziv_aktivnog_Twin_T_reznog_filtra.png)

$$H(s) = \frac{s^2 + \omega_0^2}{s^2 + \frac{\omega_0}{Q}s + \omega_0^2} \tag{2.46}$$

where:

- s is the complex frequency variable ($s = j\omega$ in the frequency domain),
- ω_0 is the **angular center frequency**, where $\omega_0 = 2\pi f$
- Q is the **quality factor** of the filter, which determines the sharpness of the notch.

The numerator $s_2 + \omega_0$ ensures that the filter does not attenuate any frequencies other than the center frequency. The denominator $s^2 + (\omega_0/Q)s + \omega_0^2$ introduces the attenuation at the center frequency and controls the bandwidth of the notch. The frequency response of the notch filter will show a deep attenuation at the center frequency f0, with minimal attenuation outside the stopband.

The bandwidth BW of the notch filter is related to the quality factor Q and the center frequency f0 by the following Eq. 2.47:

$$BW = \frac{f_0}{Q} \tag{2.47}$$

A higher Q-factor results in a narrower notch (more selective), meaning that only a small range of frequencies around the center frequency f_0 is rejected. Conversely, a lower Q-factor leads to a wider notch, affecting a broader range of frequencies.

Digital Notch Filter Transfer Function

In the digital domain, a notch filter is typically implemented using FIR or IIR techniques. The general difference equation for a digital notch filter can be expressed as Eq. 2.48:

$$y[n] = \sum_{k=0}^{M} b_k x[n-k] - \sum_{k=1}^{N} a_k y[n-k] \tag{2.48}$$

where:

- y[n] is the output signal at the n-th time step,
- x[n − k] is the input signal at previous time steps (indexed by k),
- b_k are the filter coefficients for the feedforward part (numerator),
- a_k are the filter coefficients for the feedback part (denominator),
- M and N are the numbers of taps in the feedforward and feedback parts, respectively.

The design of the filter coefficients b_k and a_k depends on the desired notch center frequency f0, bandwidth, and the sample rate of the signal [119, 120].

Frequency Response of Notch Filters

The frequency response H(jω) of a notch filter describes how the filter modifies the amplitude and phase of the frequency components of the input signal. For a second-order notch filter, the frequency response is given by Eq. 2.49:

$$H(j\omega) = \frac{j\omega^2 + \omega_0^2}{(j\omega)^2 + \frac{\omega_0}{Q} j\omega + \omega_0^2} \tag{2.49}$$

The frequency response exhibits a deep attenuation at the notch center frequency f_0, allowing frequencies outside the notch to pass with minimal attenuation. The filter's frequency response is sharply peaked around the center frequency f_0, and the quality factor Q determines the extent of attenuation.

2.8.5 Design and Application of Digital Filters

Band-Pass Filters

Designing a band-pass filter involves selecting the appropriate values for the cutoff frequencies, quality factor, and center frequency to meet the specific needs of the application. There are several methods used for the design of band-pass filters, including:

1. Butterworth filter design: The Butterworth filter is characterized by a flat frequency response in the passband and a smooth transition to the stopband. The Butterworth design is commonly used when a smooth, non-rippled response is required in the passband.
2. Chebyshev filter design: The Chebyshev filter offers a steeper roll-off compared to the Butterworth filter, albeit at the expense of ripples in the passband. It is often used when a sharper cutoff is required.
3. Elliptic (Cauer) filter design: The Elliptic filter provides the steepest roll-off for a given filter order, allowing ripples in both the passband and the stopband. This design is often used in applications where an extremely sharp transition is needed between the passband and stopband.

Band-pass filters are applied in a wide range of fields, including communication systems, audio processing, and biomedical signal processing. Below are some typical applications [50, 117]:

1. Audio Signal Processing

In audio signal processing, band-pass filters are used to isolate specific frequency ranges in audio signals. For example:

2. Communications Systems

- Equalization: Band-pass filters can be used in equalizers to adjust the tonal balance of an audio signal by boosting or cutting specific frequency bands.
- Noise reduction: Band-pass filters are used to remove unwanted noise from audio recordings, such as hum or static, by filtering out frequencies outside the desired range.

In radio communication systems, band-pass filters are used to select specific frequency channels for transmission and reception. A band-pass filter allows signals within the desired frequency band to pass, while blocking interference from adjacent channels.

3. Biomedical Signal Processing

Band-pass filters play an essential role in biomedical signal processing, particularly in the analysis of signals such as ECG, EEG, and EMG. For example:

- ECG: Band-pass filters are used to remove baseline wander (low-frequency noise) and high-frequency noise (such as muscle artifacts) from ECG signals while preserving the important frequency components associated with heart activity.
- EEG: In EEG analysis, band-pass filters are used to isolate specific frequency bands, such as the alpha, beta, and delta waves, for more effective analysis and interpretation.

4. Instrumentation and Measurement

In instrumentation and measurement systems, band-pass filters are used to isolate the frequencies of interest in signals measured from physical systems. This is particularly important in systems where it is necessary to filter out noise or unwanted frequencies while preserving the signal of interest.

Notch Filters

The design of a notch filter involves selecting the appropriate values for the center frequency f_0, the bandwidth BW, and the quality factor Q. The notch filter can be designed to reject a very narrow frequency band, such as eliminating power line interference at 50 or 60 Hz, or it can be adjusted for broader bands in applications requiring more aggressive filtering.

There are different methods to design notch filters, such as:

1. Butterworth Design: The Butterworth notch filter design aims to create a flat passband and a smooth, gradual roll-off in the stopband, ensuring minimal distortion.
2. Chebyshev Design: A Chebyshev notch filter offers a faster roll-off than the Butterworth design, but it introduces ripple in the passband. This design is used when a sharper cutoff is necessary.
3. Elliptic Design: The elliptic notch filter provides the steepest roll-off for a given filter order, but it introduces ripple in both the passband and stopband.

Notch filters are widely used in various domains where the elimination of a specific frequency or a narrow band of frequencies is required. Below are some prominent applications of notch filters:

1. Power Line Interference Removal in Biomedical Signals

In biomedical signal processing, particularly when working with EEG or ECG signals, notch filters are crucial for eliminating power line interference. Power lines typically introduce a 50 Hz or 60 Hz noise signal into biomedical signals, and a notch filter is applied at this frequency to eliminate the interference without affecting the signal of interest. For instance, in ECG signal analysis, power line noise (at 50 Hz or 60 Hz) can be removed using a notch filter explicitly designed to target this frequency, ensuring that the heart-related signals are preserved.

2. Audio Processing and Noise Removal

In audio signal processing, notch filters are used to eliminate hum or buzz from recordings, particularly the 50 Hz or 60 Hz hum from electrical sources. This is common in music production, film editing, and recording studio environments, where electrical interference can degrade audio quality. By applying a notch filter at the offending frequency, it is possible to remove the noise without significantly affecting the desired audio signal.

3. Communication Systems

In communications systems, notch filters are used to eliminate interference from unwanted signals, such as spurious emissions or interference from other channels. By using notch filters at specific frequencies, systems can isolate and preserve the desired signal while rejecting interfering signals at specific frequencies.

4. Instrumentation and Measurement Systems

In instrumentation and measurement systems, notch filters are used to remove interference from specific sources, such as electrical equipment. For example, if a measurement system is affected by a known frequency interference, a notch filter can be designed to reject this specific frequency.

References

1. Sarić R, Jokić D, Beganović N, Pokvić LG, Badnjević A (2020) FPGA-based real-time epileptic seizure classification using artificial neural network. Biomed Signal Process Control 62:102106
2. Merdović N, Spahić L, Hundur M, Pokvić LG, Badnjević A (2025) Advancement of post-market surveillance of medical devices leveraging artificial intelligence: infusion pumps case study. Technol Health Care 33(2):915–921
3. Wang H, Ma Y, Yang H, Jiang H, Ding Y, Xie H (2020) MEMS ultrasound transducers for endoscopic photoacoustic imaging applications. Micromachines 11(10):928
4. Qin J, An L (2021) Optical coherence tomography for ophthalmology imaging. In: optical imaging in human disease and biological research. Springer Singapore, pp 197–216

5. Meng K, Xiao X, Wei W, Chen G, Nashalian A, Shen S, Chen J (2022) Wearable pressure sensors for pulse wave monitoring. Adv Mater 34(21):2109357
6. Minervini G, Franco R, Marrapodi MM, Crimi S, Badnjević A, Cervino G, Cicciù M (2023) Correlation between temporomandibular disorders and posture evaluated through DC/TMD: a systematic review with meta-analysis. J Clin Med 12(7):2652
7. Lee J, Kim M, Park HK, Kim IY (2020) Motion artifact reduction in wearable photoplethysmography based on multi-channel sensors with multiple wavelengths. Sensors 20(5):1493
8. Kovačević Ž, Gurbeta Pokvić L, Spahić L, Badnjević A (2020) Prediction of medical device performance using machine learning techniques: infant incubator case study. Health Technol 10(1):151–155
9. Liu K, Duan T, Zhang F, Tian X, Li H, Feng M, Wang R, Jiang B, Zhang K (2024) Flexible electrode materials for emerging electronics: materials, fabrication and applications. J Mater Chem A 12(32):20606–20637
10. Kim H, Kim E, Choi C, Yeo WH (2022) Advances in soft and dry electrodes for wearable health monitoring devices. Micromachines 13(4):629
11. Yang L, Liu Q, Zhang Z, Gan L, Zhang Y, Wu J (2022) Materials for dry electrodes for the electroencephalography: advances, challenges, perspectives. Adv Mater Technol 7(3):2100612
12. Grün D, Rudolph F, Gumpfer N, Hannig J, Elsner LK, von Jeinsen B, Keller T (2021) Identifying heart failure in ECG data with artificial intelligence—a meta-analysis. Front Digit Health 2:584555
13. Badnjević A, Pokvić LG, Deumić A, Bećirović LS (2022) Post-market surveillance of medical devices: a review. Technol Health Care 30(6):1315–1329
14. Aljović A, Badnjević A, Gurbeta L (2016) Artificial neural networks in the discrimination of Alzheimer's disease using biomarker data. In: 2016 5th Mediterranean conference on embedded computing (MECO), pp 286–289
15. Hafizović L, Čaušević A, Deumić A, Bećirović LS, Pokvić LG, Badnjević A (2021) The use of artificial intelligence in diagnostic medical imaging: systematic literature review. In: 2021 IEEE 21st international conference on bioinformatics and bioengineering (BIBE), pp 1–6
16. Alić B, Gurbeta L, Badnjević A (2017) Machine learning techniques for classification of diabetes and cardiovascular diseases. In: 2017 6th Mediterranean conference on embedded computing (MECO), pp 1–4
17. Kumar S, Yadav S, Kumar A (2024) Blood pressure measurement techniques, standards, technologies, and futuristic wearable cuff-less know-how. Sens Diagn 3(2):181–202
18. Saranteas T (2009) Limitations in ultrasound imaging techniques in anesthesia: obesity and muscle atrophy. Anesth Analg 109(3):993–994
19. Bolić M (2023) Continuous monitoring of breathing. Elsevier E-Books, pp 327–368
20. Badnjević A, Avdihodžić H, Gurbeta PL (2021) Artificial intelligence in medical devices: past, present and future. Psychiatr Danub 33(Suppl 3):101–106
21. Goshvarpour A (2024) Emerging trends of biomedical signal processing in intelligent emotion recognition. Brain Sci 14(7):628
22. Cerutti S (2008) In the spotlight: biomedical signal processing. IEEE Rev Biomed Eng 1:8–11
23. Pourahmad A, Dehghani R (2019) Two-wired current modulator active electrode for ambulatory biosignal recording. IEEE Trans Biomed Circuits Syst 13(1):15–25
24. Assambo C, Burke MJ (2009) Amplifier input impedance in dry electrode ECG recording. In: 2009 annual international conference of the IEEE engineering in medicine and biology society, pp 1774–1777
25. Wu Y, Krishnan S, Ghoraani B (2022) Computational methods for physiological signal processing and data analysis. Comput Math Methods Med 2022:9861801
26. Badnjević A, Gurbeta Pokvić L, Hasičić M, Bandić L, Mašetić Z, Kovačević Ž, Kevrić J, Pecchia L (2019) Evidence-based clinical engineering: machine learning algorithms for prediction of defibrillator performance. Biomed Signal Process Control 54:101629

27. Li L, Li B (2008) Design of experiments in biomedical signal processing course. In: 2008 30th annual international conference of the IEEE engineering in medicine and biology society, pp 1568–1570
28. Cosmanescu A, Miller B, Magno T, Ahmed A, Kremenic I (2006) Design and implementation of a wireless (Bluetooth) four-channel bio-instrumentation amplifier and digital data acquisition device with user-selectable gain, frequency, and driven reference. In: 2006 international conference of the IEEE engineering in medicine and biology society, pp 2053–2056
29. Halder RS, Basumatary B, Sahani A (2024) Development of a low-cost, compact, wireless, 16-channel biopotential data acquisition, signal conditioning and arbitrary waveform stimulator. Biomed Phys Eng Express 10(2)
30. Kovačević Ž, Gurbeta Pokvić L, Spahić L, Badnjević A (2019) Prediction of medical device performance using machine learning techniques: infant incubator case study. Health Technol 10(1):151–155
31. Badnjević A, Pokvić LG, Džemić Z, Bečić F (2020) Risks of emergency use authorizations for medical products during outbreak situations: a COVID-19 case study. Biomed Eng Online 19(1)
32. Martinek R, Ladrova M, Sidikova M, Jaros R, Behbehani K, Kahankova R, Kawala-Sterniuk A (2021) Advanced bioelectrical signal processing methods: past, present and future approach—Part I: cardiac signals. Sensors 21(15):5186
33. Abbod M, Shieh JS (2019) Advanced signal processing in intelligent systems for health monitoring. Sensors 19(21):4727
34. Cerutti S, Baselli G, Bianchi A, Caiani E, Contini D, Cubeddu R, Dercole F, Rienzo L, Liberati D, Mainardi L, Ravazzani P, Rinaldi S, Signorini M, Torricelli A (2011) Biomedical signal and image processing. IEEE Pulse 2(3):41–54
35. Astaras A, Bamidis PD, Kourtidou-Papadeli C, Maglaveras N (2008) Biomedical real-time monitoring in restricted and safety-critical environments. Hippokratia 12(Suppl 1):10–14
36. Lehmann TM, Aach T, Witte H (2006) Sensor, signal, and image informatics: state of the art and current topics. Yearb Med Inform 57–67
37. Coatrieux JL (2002) Signal processing and physiological modeling—part 1: surface analysis. Crit Rev Biomed Eng 30(1–3):9–35
38. Catic A, Gurbeta L, Kurtovic-Kozaric A, Mehmedbasic S, Badnjevic A (2018) Application of neural networks for classification of Patau, Edwards, Down, Turner and Klinefelter syndrome based on first trimester maternal serum screening data, ultrasonographic findings and patient demographics. BMC Med Genomics 11(1)
39. Kumar S, Veer K, Kumar S (2023) Current trends in feature extraction and classification methodologies of biomedical signals. Curr Med Imaging. https://doi.org/10.2174/1573405619666230309103435
40. Gurbeta L, Badnjevic A, Maksimovic M, Omanovic-Miklicanin E, Sejdic E (2018) A telehealth system for automated diagnosis of asthma and COPD. J Am Med Inform Assoc 25(9):1213–1217
41. Wei Y, Zhou J, Wang Y, Liu Y, Liu Q, Luo J, Wang C, Ren F, Huang L (2020) A review of algorithm and hardware design for AI-based biomedical applications. IEEE Trans Biomed Circuits Syst 14(2):145–163
42. Sun K, Liu J, Yan F, Sun H, Zhang Y, Ren Y, Huang L, Pi Y, Wu W, Guan J (2025) A 9.68 nW 57.51 dB SNDR SAR ADC with dual bypass windows based on non-binary split capacitors for biomedical applications. IEEE Trans Biomed Circuits Syst. https://doi.org/10.1109/TBCAS.2025.3557241
43. Somappa L, Baghini MS (2021) A 400 mV 160 nW/Ch compact energy efficient ΔΣ modulator for multichannel biopotential signal acquisition system. IEEE Trans Biomed Circuits Syst 15(4):765–776
44. Peruničić Ž, Lalatović I, Spahić L, Ašić A, Pokvić LG, Badnjević A (2024) Enhancing mechanical ventilator reliability through machine learning-based predictive maintenance. Technol Health Care

45. Dekimpe R, Bol D (2022) ECG arrhythmia classification on an ultra-low-power microcontroller. IEEE Trans Biomed Circuits Syst 16(3):456–466
46. Saeed M, Wang Q, Martens O, Larras B, Frappe A, Cardiff B, John D (2021) Evaluation of level-crossing ADCs for event-driven ECG classification. IEEE Trans Biomed Circuits Syst 15(6):1129–1139
47. Li M, Huo Y, Song S, Qu W, Ye L, Zhao M, Tan Z (2025) A 62.2 dB SNDR event-driven level-crossing ADC with SAR-assisted delay compensation loop for time-sparse biomedical signal acquisition. IEEE Trans Biomed Circuits Syst 19(2):344–356
48. Jha P, Patra P, Naik J, Acharya A, Rajalakshmi P, Singh SG, Dutta A (2015) A reconfigurable medically cohesive biomedical front-end with $\Sigma\Delta$ ADC in 0.18 μm CMOS. In: 2015 37th annual international conference of the IEEE engineering in medicine and biology society (EMBC), pp 833–836
49. Badnjević A, Pokvić LG, Smajlhodžić-Deljo M, Spahić L, Bego T, Meseldžić N, Prnjavorac L, Prnjavorac B, Bedak O (2024) Application of artificial intelligence for classification of clinical outcome and therapy in patients with viral infections: the case of COVID-19. Technol Health Care 32(3):1859–1870
50. Badnjevic A, Deumic A, Smajlhodzic-Deljo M, Pokvic LG (2023) A novel method for conformity assessment testing of infusion and perfusion pumps for post-market surveillance purposes. Technol Health Care 31(1):347–355
51. Spahić L, Pokvić LG, Badnjević A (2024) Inspection and testing of infusion pumps. In: Inspection of medical devices. Springer, pp 273–290
52. Han Y, Stephany RG, Zhao L, Ahmmed P, Bozkurt A, Jia Y (2025) A wireless miniature injectable device with memory-assisted backscatter for multimodal animal physiological monitoring. IEEE Trans Biomed Eng 72(3):988–999
53. Zhao L, Stephany RG, Han Y, Ahmmed P, Huang TP, Bozkurt A, Jia Y (2024) A wireless multimodal physiological monitoring ASIC for animal health monitoring injectable devices. IEEE Trans Biomed Circuits Syst 18(5):1037–1049
54. Guerrero FN, Spinelli EM (2017) A simple encoding method for sigma-delta ADC based biopotential acquisition systems. J Med Eng Technol 41(7):546–552
55. Luu DK, Nguyen AT, Yang Z (2019) Achieving super-resolution with redundant sensing. IEEE Trans Biomed Eng 66(8):2200–2209
56. Lu CC, Huang DK (2020) A 10-bit 50 MS/s SAR ADC based on area-efficient and low-energy switching scheme. IEEE Access 8:28257–28266
57. Wang Z, Zhang J, Verma N (2015) Realizing low-energy classification systems by implementing matrix multiplication directly within an ADC. IEEE Trans Biomed Circuits Syst 9(6):825–837
58. Sadasivuni S, Bhanushali SP, Banerjee I, Sanyal A (2022) In-sensor neural network for high energy efficiency analog-to-information conversion. Sci Rep 12(1):18253
59. Valley GC (2007) Photonic analog-to-digital converters. Opt Express 15(5):1955–1982
60. Mollazadeh M, Murari K, Cauwenberghs G, Thakor N (2009) Micropower CMOS integrated low-noise amplification, filtering, and digitization of multimodal neuropotentials. IEEE Trans Biomed Circuits Syst 3(1):1–10
61. Li Y, Poon CC, Zhang YT (2010) Analog integrated circuits design for processing physiological signals. IEEE Rev Biomed Eng 3:93–105
62. Li Y, Zhao D, Serdijn WA (2013) A sub-microwatt asynchronous level-crossing ADC for biomedical applications. IEEE Trans Biomed Circuits Syst 7(2):149–157
63. Mujkić A, Baralić E, Ombašić A, Bećirović LS, Pokvić LG, Badnjević A (2022) Machine intelligence in biomedical data modeling, processing, and analysis. In: 2022 11th Mediterranean conference on embedded computing (MECO), pp 1–10
64. Martinović M, Kosović M, Spahić L, Softić A, Pokvić LG, Badnjević A (2025) Machine learning for improved medical device management: a focus on dialysis machines. Technol Health Care
65. Pokvić LG, Deumić A, Softić A, Badnjević A (2024) Application of artificial intelligence for management of maintenance and prediction of performances. In: Inspection of medical devices. Springer, pp 97–118

66. Anastasiadou MN, Christodoulakis M, Papathanasiou ES, Papacostas SS, Mitsis GD (2017) Unsupervised detection and removal of muscle artifacts from scalp EEG recordings using canonical correlation analysis, wavelets, and random forests. Clin Neurophysiol 128(9):1755–1769
67. An X, Liu Y, Zhao Y, Lu S, Stylios GK, Liu Q (2022) Adaptive motion artifact reduction in wearable ECG measurements using impedance pneumography signal. Sensors 22(15):5493
68. Hundur M, Spahić L, Bećirović F, Gurbeta Pokvić L, Badnjević A (2025) Advancement of post-market surveillance of medical devices leveraging artificial intelligence: ECG devices case study. Technol Health Care
69. Acharya UR, Fujita H, Lih OS, Hagiwara Y, Tan JH, Adam M (2016) Automated detection of arrhythmias using different intervals of tachycardia ECG segments with convolutional neural network. Inf Sci 405:81–90
70. Bećirović F, Spahić L, Merdović N, Gurbeta Pokvić L, Badnjević A (2025) Advancement of post-market surveillance of medical devices leveraging artificial intelligence: patient monitors case study. Technol Health Care 33(2):974–980
71. Merdović N, Spahić L, Hundur M, Pokvić LG, Badnjević A (2024) Advancement of post-market surveillance of medical devices leveraging artificial intelligence: infusion pumps case study. Technol Health Care 33(2):915–921
72. Mir HY, Singh O (2021) ECG denoising and feature extraction techniques: a review. J Med Eng Technol 45(8):672–684
73. Saleem S, Khandoker AH, Alkhodari M, Hadjileontiadis LJ, Jelinek HF (2022) A two-step pre-processing tool to remove Gaussian and ectopic noise for heart rate variability analysis. Sci Rep 12(1):18396
74. Wang T, Yao S, Shao LH, Zhu Y (2024) Stretchable Ag/AgCl nanowire dry electrodes for high-quality multimodal bioelectronic sensing. Sensors 24(20):6670
75. Jia Y, Pei H, Liang J, Zhou Y, Yang Y, Cui Y, Xiang M (2024) Preprocessing and denoising techniques for electrocardiography and magnetocardiography: a review. Bioeng 11(11):1109
76. Zivanovic M, Niegowski M, Lecumberri P, Gómez M (2017) A low-rank matrix factorization approach for joint harmonic and baseline noise suppression in biopotential signals. Comput Methods Programs Biomed 141:59–71
77. Hamaneh MB, Chitravas N, Kaiboriboon K, Lhatoo SD, Loparo KA (2014) Automated removal of EKG artifact from EEG data using independent component analysis and continuous wavelet transformation. IEEE Trans Biomed Eng 61(6):1634–1641
78. Acharya UR, Vinitha Sree S, Swapna G, Martis RJ, Suri JS (2013) Automated EEG analysis of epilepsy: a review. Knowl-Based Syst 45:147–165
79. Bhattacharyya A, Pachori RB (2017) A multivariate approach for patient-specific EEG seizure detection using empirical wavelet transform. IEEE Trans Biomed Eng 64(9):2003–2015
80. Jain P, Pachori RB (2015) An iterative approach for decomposition of multi-component non-stationary signals based on eigenvalue decomposition of the Hankel matrix. J Franklin Inst 352(10):4017–4044
81. Alqudah AM, Moussavi Z (2025) A review of deep learning for biomedical signals: current applications, advancements, future prospects, interpretation, and challenges. Comput Mater Continua 0(0):1–10
82. Wang LV, Yao J (2016) A practical guide to photoacoustic tomography in the life sciences. Nat Methods 13(8):627–638
83. Magjarević R, Badnjević A (2023) Inspection and testing of electrocardiographs devices. In: Badnjević A, Cifrek M, Magjarević R, Džemić Z (eds) Inspection of medical devices. Springer, pp 139–171
84. Mahato A, Bhalerao SV, Pachori RB, Gadre VM, Mahapatra DD (2025) Neurological responses to meditation with EEG analysis using novel empirical Fourier-Bessel decomposition approach. In: 2025 10th international conference on signal processing and communication (ICSC), pp 694–699
85. Tserevelakis GJ, Barmparis GD, Kokosalis N, Giosa ES, Pavlopoulos A, Tsironis GP, Zacharakis G (2023) Deep learning-assisted frequency-domain photoacoustic microscopy. Opt Lett 48(10):2720–2723

86. Bao X, Xu Y, Lam HK, Trabelsi M, Chihi I, Sidhom L, Kamavuako EN (2023) Time-frequency distributions of heart sound signals: a comparative study using convolutional neural networks. Biomed Eng Adv 5:100093
87. Chen Y, Orlandi M, Rapa PM, Benatti S, Benini L, Li Y (2025) PhysioWave: a multi-scale wavelet-transformer for physiological signal representation. arXiv preprint
88. Singh AK, Krishnan S (2023) ECG signal feature extraction: trends in methods and applications. Biomed Eng Online 22(1):22
89. Sharma RR, Pachori RB (2018) Time–frequency representation using IEVDHM–HT with application to classification of epileptic EEG signals. IET Sci Meas Technol 12(1):72–82
90. Spahić L, Sredović U, Kurpejović Z, Mrdanović E, Pokvić G, Badnjević A (2025) Machine learning for improved medical device management: a focus on infant incubators. Technol Health Care
91. Pifferi A, Contini D, Mora AD, Farina A, Spinelli L, Torricelli A (2016) New frontiers in time-domain diffuse optics: a review. J Biomed Opt 21(9):091310
92. Badnjevic A, Cifrek M, Koruga D, Osmankovic D (2015) Neuro-fuzzy classification of asthma and chronic obstructive pulmonary disease. BMC Med Inform Decis Mak 15(Suppl 3):S1
93. Wang Y, Veluvolu KC (2017) Time-frequency analysis of non-stationary biological signals with sparse linear regression based Fourier linear combiner. Sensors 17(6):1386
94. Lin YD, Tan YK, Ku T, Tian B (2023) A frequency estimation scheme based on Gaussian average filtering decomposition and Hilbert transform: with estimation of respiratory rate as an example. Sensors 23(8):3785
95. Chaddad A, Wu Y, Kateb R, Bouridane A (2023) Electroencephalography signal processing: a comprehensive review and analysis of methods and techniques. Sensors 23(14):6434
96. Daud SNSS, Sudirman R (2022) Wavelet-based filters for artifact elimination in electroencephalography signal: a review. Ann Biomed Eng 50(10):1271–1291
97. Hussein AF, Hashim SJ, Aziz AFA, Rokhani FZ, Adnan WAW (2017) Performance evaluation of time-frequency distributions for ECG signal analysis. J Med Syst 42(1):15
98. Zheng J, Liang M, Sinha S, Ge L, Yu W, Ekstrom A, Hsieh F (2022) Time-frequency analysis of scalp EEG with Hilbert-Huang transform and deep learning. IEEE J Biomed Health Inform 26(4):1549–1559
99. Yang S, Wu H, Liao X, Lin Y, Chen J (2025) Empirical mode decomposition in clinical signal analysis: a systematic review. Comput Biol Med 194:110566
100. Wachowiak MP, Wachowiak-Smolíková R, Johnson MJ, Hay DC, Power KE, Williams-Bell FM (2018) Quantitative feature analysis of continuous analytic wavelet transforms of electrocardiography and electromyography. Philos Trans A Math Phys Eng Sci 376(2126):20170250
101. Sharmila A, Aman Raj S, Shashank P, Mahalakshmi P (2018) Epileptic seizure detection using DWT-based approximate entropy, Shannon entropy and support vector machine: a case study. J Med Eng Technol 42(1):1–8
102. Wang Z, Juhasz Z (2023) GPU implementation of the improved CEEMDAN algorithm for fast and efficient EEG time–frequency analysis. Sensors 23(20):8654
103. Ahmad I, Ansari F, Dey UK (2020) Power line interference noise removal in ECG—a comparative study. Int J Comput Sci Eng 7(2):13–18
104. Hidayat R, Lestari NS, Herawati GDR, Sudarmanto FA (2021) An approach of adaptive notch filtering design for electrocardiogram noise cancellation. Indones J Electr Eng Comput Sci. 22(3):1303–1311
105. Sundararajan D (2024) Digital signal processing: an introduction. Springer
106. Nayak C, Saha SK, Kar R, Mandal D (2021) Optimal design of zero-phase digital Riesz FIR fractional-order differentiator. Soft Comput 25:4261–4282
107. Aboutabikh K, Aboukerdah N (2015) Design and implementation of a multiband digital filter using FPGA to extract the ECG signal in the presence of different interference signals. Comput Biol Med 62:1–13
108. Kolawole ES, Ali WH, Cofie P, Fuller J, Tolliver C, Obiomon P (2015) Design and implementation of low-pass, high-pass and band-pass finite impulse response filters using FPGA. Circuits Syst 6(2):30–48

109. Roy S, Chandra A (2022) A deep learning approach for the design of narrow transition-band FIR filter. Circuits Syst Signal Process 41(10):5578–5613
110. Dutta C, Kumar B, Jain BS (2001) FIR notch filter design—a review. Circuits Syst Signal Process 14(3)
111. Ferdi Y (2021) A fractional digital high-pass notch filter, pp 1–4
112. Tulyakova N, Trofymchuk O (2024) Adaptive nonlinear filtering algorithms for removal of non-stationary noise in electronystagmographic signals. Comput Biol Med 183:109229
113. Jiang L, Zhang H, Cheng S, Lv H, Li P (2020) An overview of FIR filter design in future multicarrier communication systems. Electronics 9(4):599
114. Huang F, Wang Z, Bo X (2015) Design of 50 Hz notch filter based on virtual instrument. Adv Comput Sci Res
115. Nigam S (2013) Design and implementation of digital FIR band stop filter for removal of artifact from noisy ECG signal. Int J Electr Electron Res 1(1):4–7
116. Mondal S, Chakraborty D, Kar R, Mandal D, Ghoshal SP (2012) Novel particle swarm optimization for high-pass FIR filter design
117. Singh S, Sachin, Yadav KL (2010) Performance evaluation of different adaptive filters for ECG signal processing. Int J Comput Sci Eng 2(5)
118. Paul B, Balakrishnan AA, Prasanth K (2013) Fetal ECG extraction using adaptive filters. Res J Eng Appl Sci 2(4):2320–3765
119. Azzouz A, Bengherbia B, Wira P, Alaoui N, Souahlia A, Maazouz M, Hentabeli H (2024) An efficient ECG signal denoising technique based on the combination of particle swarm optimisation and wavelet transform. Heliyon 10(5):e26171
120. Koshita S, Kumamoto Y, Abe M, Kawamata M (2013) Adaptive IIR band-pass/band-stop filtering using high-order transfer function and frequency transformation. Interdiscip Inf Sci 19(2):163–172
121. Dong J, Jiang W (2015) Design of digital filter on ECG signal processing

Chapter 3
Processing Cardiac Signals

This section examines the acquisition, preprocessing, analysis, and clinical applications of cardiac signals, particularly the electrocardiogram (ECG). It begins with the anatomy and physiology of the heart and its electrophysiological system (Chapter 3.1), followed by methods of ECG acquisition through bipolar, unipolar, and montage-based lead systems (Chapter 3.2). Preprocessing techniques (Chapter 3.3) address noise reduction, adaptive filtering, processing, and their impact on diagnostic accuracy. Feature extraction and analysis (Chapter 3.4) are presented through QRS detection, wave delineation, and the use of morphological, temporal, and transform-based features, alongside regulatory considerations. Clinical applications (Chapter 3.5) highlight ECG's role in diagnosing arrhythmias, ischemia, infarction, heart failure, and drug or electrolyte-induced alterations, as well as its use in prognosis and population screening. This section underscores ECG's enduring role as a diagnostic and monitoring tool in cardiovascular medicine.

3.1 Anatomy of the Heart and ECG

The human heart is a remarkable and intricate organ, vital to sustaining life by circulating blood throughout the body. Its primary responsibility is to deliver oxygenated blood to various tissues and organs, enabling them to perform essential metabolic functions, while also returning deoxygenated blood to the lungs for oxygenation. The heart functions as both a pump and an electrical generator, orchestrating the mechanical and electrical activities that allow it to regulate blood flow and maintain homeostasis effectively [1, 2]. To fully understand the origin of the electrical signals recorded by the electrocardiogram (ECG), it is essential to delve into the structural and functional anatomy of the heart and how these aspects coordinate to regulate the heart's rhythm, contraction, and relaxation.

© The Author(s), under exclusive license to Springer Nature Switzerland AG 2026
A. Badnjević and L. Spahić, *Biomedical Signals and Systems*, Series in BioEngineering,
https://doi.org/10.1007/978-3-032-10982-8_3

3.1.1 Location and General Structure

The heart is a muscular organ approximately the size of a fist, weighing between 250 and 350 g, and is located in the thoracic cavity, slightly to the left of the midline.

It is housed within a protective sac called the pericardium, which not only prevents friction from surrounding structures but also allows the heart to beat with minimal resistance.

The heart sits between the lungs, resting on the diaphragm, and is shielded by the rib cage. It is oriented in such a way that the apex (or tip) points toward the left, while the base of the heart, where large blood vessels emerge, is positioned more superiorly and posteriorly [3, 4].

The heart is composed of four main chambers: the right atrium, right ventricle, left atrium, and left ventricle. These chambers are functionally divided into two halves: the right side and the left side, separated by a partition called the septum. Each side of the heart serves as a pump for blood, but the nature of the blood being pumped and the destinations of each side differ.

1. Right Side of the Heart: The right atrium and right ventricle together form the right side of the heart. Their primary role is to pump deoxygenated blood to the lungs for oxygenation. Blood enters the right atrium from the superior and inferior vena cava, large veins that return deoxygenated blood from the body. From the right atrium, blood travels through the tricuspid valve into the right ventricle. When the right ventricle contracts, it pumps blood through the pulmonary valve into the pulmonary artery, which carries it to the lungs.
2. Left Side of the Heart: The left atrium and left ventricle form the left side of the heart, and their primary function is to pump oxygenated blood throughout the body. Blood returns from the lungs, rich in oxygen, via the pulmonary veins, entering the left atrium. From there, it passes through the mitral valve into the left ventricle. The left ventricle is the largest and most muscular chamber of the heart. When it contracts, it sends oxygen-rich blood through the aortic valve into the aorta, the body's largest artery, which distributes it to various organs and tissues [5].

The heart is composed of three distinct layers, each playing a crucial role in its function:

1. Endocardium: This innermost layer lines the heart's chambers, covers the heart valves, and is continuous with the endothelial lining of the blood vessels. The endocardium provides a smooth, frictionless surface that helps facilitate the efficient flow of blood through the heart. It also plays a role in maintaining the integrity of the valves and regulating blood clotting.
2. Myocardium: The myocardium is the thick, muscular middle layer of the heart, forming the majority of the heart's mass. It is composed of cardiac muscle cells (myocytes) that contract and generate the force necessary to pump blood. The myocardium is responsible for the heart's mechanical function, and its strength and endurance are critical for maintaining adequate circulation. The left ventricle,

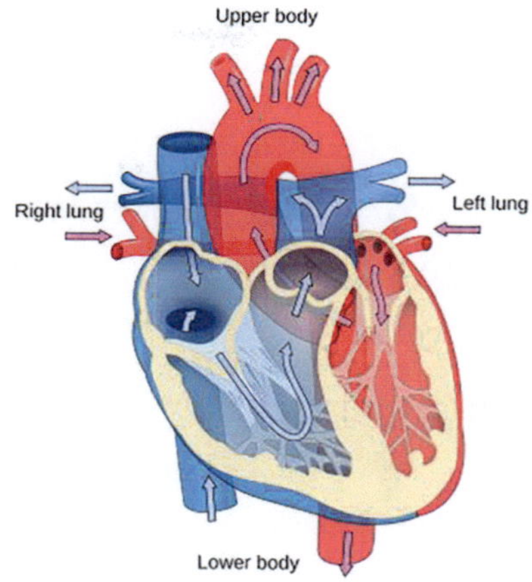

Fig. 3.1 Diagram of the heart. https://commons.wikimedia.org/wiki/File:Heart_diagram_blood_flow_en.svg

with its thicker myocardial layer, is particularly suited to generate the high pressures needed to pump blood throughout the body.

3. Epicardium: The epicardium is the outermost layer of the heart and also serves as the inner layer of the pericardium. It is a thin membrane that protects the heart, and its surface is covered by mesothelial cells that secrete a small amount of fluid to reduce friction as the heart beats. The epicardium also houses blood vessels that supply the myocardium, as well as nerves that regulate heart function.

Blood flows through the heart in a carefully coordinated sequence, ensuring that deoxygenated blood is sent to the lungs for oxygenation, while oxygen-rich blood is pumped to the body's tissues [1, 6] (Fig. 3.1).

The cycle begins when blood enters the right atrium from the superior and inferior vena cava, two large veins that bring blood from the body's upper and lower regions, respectively. The right atrium contracts, pushing blood through the tricuspid valve into the right ventricle (Fig. 3.2).

When the right ventricle contracts, it forces the deoxygenated blood through the pulmonary valve and into the pulmonary artery, which branches into smaller arteries that carry blood to the lungs. In the lungs, carbon dioxide is exchanged for oxygen, and the now-oxygenated blood returns to the heart via the pulmonary veins, entering the left atrium. From the left atrium, blood moves through the mitral valve into the left ventricle, the strongest chamber of the heart. When the left ventricle contracts, it forces oxygen-rich blood through the aortic valve into the aorta. The aorta then distributes the oxygenated blood to smaller arteries, which eventually reach the arterioles and capillaries of various tissues and organs, delivering oxygen and nutrients while picking up waste products [7]. This process constitutes the cardiac cycle, the series of events that occur during one heartbeat. The cardiac cycle is divided into two main phases:

1. Systole: The contraction phase, during which the ventricles contract to pump blood out of the heart.

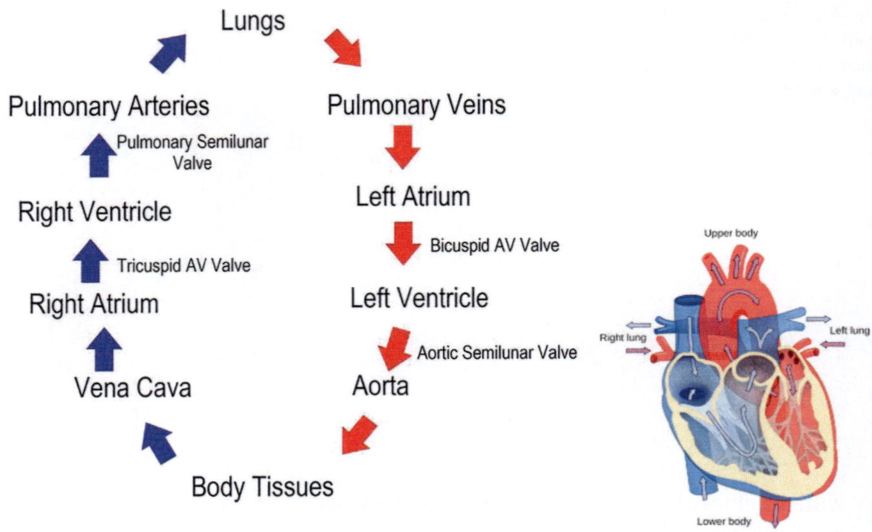

Fig. 3.2 Blood flow through the heart. https://commons.wikimedia.org/wiki/File:Circulation_of_blood_through_the_heart.png

2. Diastole: The relaxation phase, during which the heart chambers refill with blood in preparation for the next contraction.

The heart contains four vital valves that regulate the flow of blood through the heart, ensuring it moves in the correct direction. These valves act as one-way gates, preventing backflow and maintaining the unidirectional flow of blood. Each valve is composed of flaps (also known as cusps) that open and close in response to pressure differences between the chambers they separate [8] (Fig. 3.3).

1. Tricuspid valve: Located between the right atrium and the right ventricle, the tricuspid valve prevents blood from flowing backward into the atrium when the right ventricle contracts. It is called "tricuspid" because it has three flaps.
2. Pulmonary valve: Situated between the right ventricle and the pulmonary artery, the pulmonary valve prevents blood from flowing back into the right ventricle after it has been pumped into the lungs. It is a semilunar valve, meaning it has a half-moon shape.
3. Mitral valve: Positioned between the left atrium and left ventricle, the mitral valve regulates blood flow from the atrium to the ventricle. It is also known as the bicuspid valve due to its two cusps. The mitral valve ensures that blood flows in one direction—from the left atrium into the left ventricle.
4. Aortic valve: Located between the left ventricle and the aorta, the aortic valve prevents blood from flowing back into the left ventricle after it has been pumped into the systemic circulation. It is another semilunar valve.

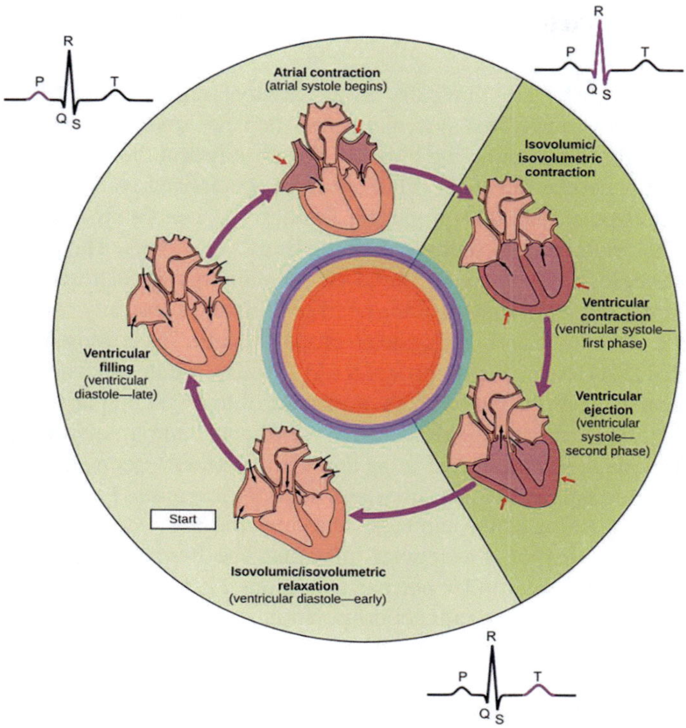

Fig. 3.3 Cardiac cycle diagram. https://commons.wikimedia.org/wiki/File:Phases_of_the_cardiac_cycle.svg

Together, these valves maintain the efficiency of blood circulation, ensuring that each chamber of the heart works in harmony with the others. The heart muscle, like any other tissue, requires oxygen and nutrients to function effectively. These are delivered by the coronary arteries, which arise from the base of the aorta. The right coronary artery supplies the right side of the heart, while the left coronary artery divides into the left anterior descending artery (LAD) and the circumflex artery, which supply the left side of the heart [9, 10]. After blood has passed through the myocardium, it returns to the heart through the cardiac veins, which drain into the right atrium via the coronary sinus.

The anatomy and physiology of the heart are fundamental to understanding how this organ works to maintain life. Its complex structure, consisting of specialized chambers, valves, and vascular supply systems, allows for the efficient circulation of blood. This ensures that oxygen and nutrients reach tissues and organs while waste products are removed, maintaining homeostasis. Understanding the heart's anatomy and function is also crucial for interpreting physiological measurements, such as the ECG, which provides valuable insights into the heart's electrical activity.

3.1.2 The Electrical System of the Heart

The heart's ability to beat rhythmically and efficiently is regulated by its intrinsic electrical conduction system. This system ensures the heart muscles contract and relax in a coordinated manner, allowing blood to flow throughout the body. The electrical impulses that drive the heartbeat originate from specialized cells within the heart, creating action potentials that propagate through the heart's structure, ultimately resulting in a synchronized contraction of the heart's chambers. The precise coordination of these electrical signals is crucial for maintaining normal heart function [11]. The electrical conduction system of the heart consists of several key components that work together to initiate and propagate electrical signals. These components include the Sinoatrial (SA) Node, Atrioventricular (AV) Node, Bundle of His, Right and Left Bundle Branches, and Purkinje Fibers. Each part of this system plays a vital role in ensuring that the electrical signals flow in a coordinated and timely manner.

The SA Node is often referred to as the heart's natural pacemaker because it is the primary site responsible for initiating electrical impulses. Located in the right atrium near the opening of the superior vena cava, the SA node generates electrical impulses spontaneously and at regular intervals. The frequency of these impulses determines the heart rate under normal conditions [12]. The cells within the SA node are specialized to generate action potentials without external stimulation, a phenomenon known as automaticity. These action potentials spread across the atria, causing the atrial muscles to contract and pump blood into the ventricles. The firing rate of the SA node determines the basic rhythm of the heart, typically at a rate of 60–100 beats per minute in a healthy adult at rest.

The SA node is influenced by the autonomic nervous system, with the sympathetic nervous system increasing the heart rate and the parasympathetic nervous system slowing it down, ensuring that the heart rate adapts to the body's needs. The AV Node is located at the junction between the atria and ventricles, in the interatrial septum near the coronary sinus. The AV node serves as a critical gatekeeper in the conduction system, slowing down electrical impulses before they pass into the ventricles [7, 13]. This delay ensures that the atria have enough time to contract and fully empty their blood into the ventricles before the ventricles contract. The AV node is essential for maintaining synchronized atrial and ventricular contractions. If the SA node fails or is damaged, the AV node can assume the role of the pacemaker of the heart. However, the AV node fires at a slower rate, typically between 40 and 60 beats per minute, resulting in a slower heart rate.

The Bundle of His is a collection of specialized fibers that serve as the continuation of the electrical signal from the AV node. Located in the interventricular septum, the Bundle of His transmits the electrical impulses from the AV node to the right and left bundle branches. The Bundle of His is crucial in ensuring that the electrical impulse travels from the atria to the ventricles, setting the stage for ventricular contraction. It acts as a bridge between the atria and ventricles, transmitting electrical signals at a rapid pace to ensure coordinated contraction of both sides of the heart (Fig. 3.4).

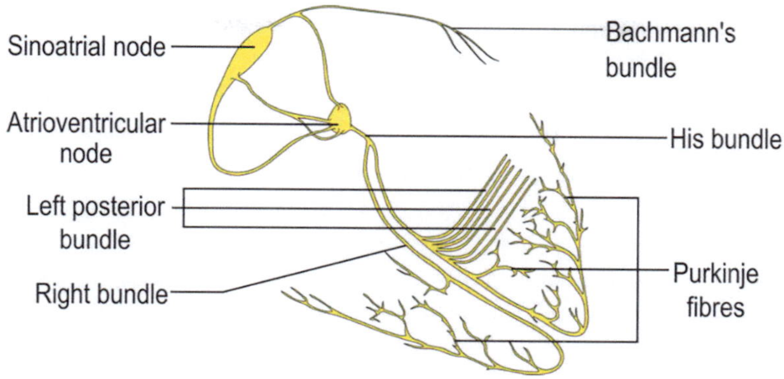

Fig. 3.4 Diagram of the Bundle of His branches. https://commons.wikimedia.org/wiki/File:ConductionsystemoftheheartwithouttheHeart.png

The electrical impulses from the Bundle of His are transmitted to the right and left bundle branches. These branches carry the electrical signals to the right and left ventricles, respectively. The right bundle branch directs the impulses to the right ventricle, while the left bundle branch sends the signals to the left ventricle [14, 15]. The left bundle branch divides into two further subdivisions: the anterior fascicle and the posterior fascicle, each responsible for transmitting the signal to different parts of the left ventricle. This dual pathway ensures that the left ventricle contracts as a whole, allowing for efficient blood pumping to the rest of the body.

The Purkinje fibers are specialized fibers that extend from the bundle branches into the myocardium of the ventricles. These fibers are responsible for rapidly conducting the electrical impulses throughout the ventricles, ensuring that both ventricles contract nearly simultaneously. The Purkinje fibers are crucial for ensuring the synchronized contraction of the ventricles, thereby facilitating efficient blood pumping [7].

The rapid spread of electrical impulses through the Purkinje fibers allows for the ventricles to contract from the apex upward, ensuring that blood is ejected efficiently into the pulmonary artery and aorta [16, 17].

The electrical impulses travel through the heart in a highly coordinated manner, ensuring that the atria contract first, followed by the ventricles. The timing of these impulses is crucial for maintaining the heart's pumping efficiency. The atria contract during the P wave of the ECG, and the ventricles contract during the QRS complex, while the T wave represents ventricular relaxation [4].

The precise sequence of electrical conduction ensures that the heart functions as an effective pump. Any disruption in this system can lead to arrhythmias, which may impair the heart's ability to circulate blood efficiently. From the SA node, which initiates the electrical impulses, to the Purkinje fibers that ensure rapid ventricular contraction, each component plays a crucial role in maintaining heart function. Understanding this electrical system is crucial for accurately interpreting the ECG, a diagnostic tool that measures the heart's electrical activity and provides valuable insights into heart health.

3.1.3 The Electrocardiogram (ECG)

The ECG is one of the most crucial diagnostic tools in modern medicine, providing a non-invasive method to monitor the heart's electrical activity. As a diagnostic method, it serves a dual purpose: offering insights into the heart's electrical system while also reflecting the physiological state of the heart muscle. From a clinical perspective, the ECG is instrumental in diagnosing arrhythmias, ischemic heart disease, and other cardiac conditions [18, 19]. The ECG graphically represents the electrical impulses that trigger heart muscle contractions, allowing clinicians to assess heart rhythm, detect abnormalities, and determine the timing of electrical events that lead to contraction. The electrical signals that lead to the formation of the ECG waveform originate from the heart's specialized conduction system, which is a highly coordinated sequence of events orchestrated by the SA node, AV node, Bundle of His, bundle branches, and Purkinje fibers [3, 20].

The heart's electrical activity forms the basis of the ECG. To understand the signals recorded by the ECG, one must first comprehend the heart's electrical conduction system. The ECG waveform reflects the sequence of electrical events occurring during each heartbeat, including the depolarization and repolarization of the heart muscle cells. The heart's electrical signals are generated by specialized myocardial cells that are capable of initiating action potentials on their own. This intrinsic ability to generate electrical impulses is what drives the heart's contraction and relaxation cycle. The electrical impulse begins at the SA node (the heart's natural pacemaker), travels through the atria, is delayed at the AV node, and then rushes through the bundle branches and Purkinje fibers to the ventricles.

The ECG is essentially a graphical representation of the changes in electrical potential between two points on the body surface, reflecting the depolarization and repolarization of the heart's muscle fibers. By detecting small voltage changes on the skin, the ECG records the combined effects of these electrical impulses across the entire heart (Fig. 3.5).

The PQRST Cycle

The ECG waveform comprises several distinct components that correspond to specific electrical events within the heart. These components include the P wave, QRS complex, and T wave, each representing different phases of myocardial electrical activity. Let us examine each of these components in detail, starting with their physiological relevance.

P Wave: Atrial Depolarization

The P wave represents the depolarization of the atria, which is the first step in the heart's contraction cycle. During this phase, the electrical impulse originates in the

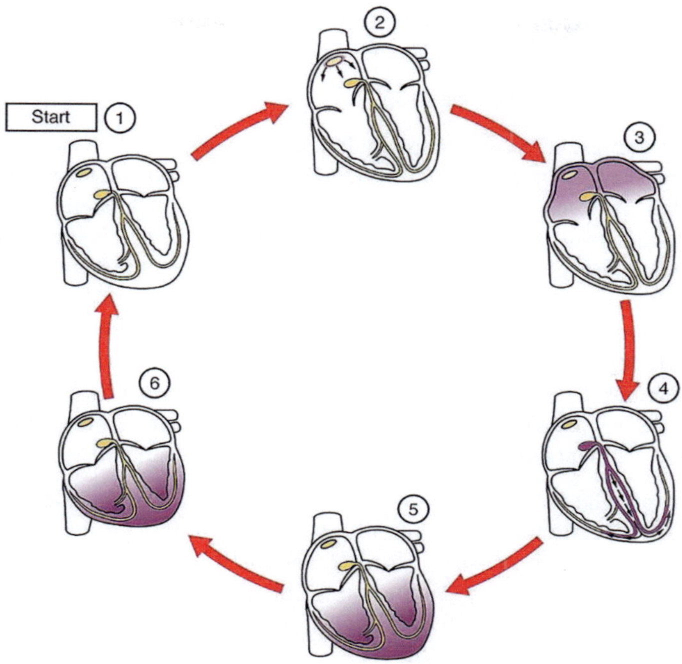

Fig. 3.5 The heart's electrical conduction system. https://commons.wikimedia.org/wiki/File:2019_Cardiac_ConductionN.jpg

SA node and travels across the atria, causing the atrial muscles to contract. The electrical activity from this depolarization is recorded as the P wave on the ECG. The depolarization of the atria is a small, localized event compared to the ventricular depolarization that occurs later, which is why the P wave is typically a low-voltage deflection. The duration of the P wave reflects the time it takes for the electrical impulse to travel through the atria, and its shape can reveal various types of atrial abnormalities, such as atrial enlargement.

QRS Complex: Ventricular Depolarization

The QRS complex represents the depolarization of the ventricles, which is the electrical event that triggers ventricular contraction. The ventricles are much larger and thicker than the atria, so the electrical activity associated with their depolarization is much more substantial, resulting in a larger and sharper deflection on the ECG. The QRS complex typically consists of three phases: the Q wave, R wave, and S wave. The Q wave is the first negative deflection after the P wave, indicating the initial depolarization of the interventricular septum. The R wave is the first positive deflection, representing the depolarization of the ventricles [21, 22]. Finally, the S wave represents the completion of ventricular depolarization, marked by a downward deflection after the R wave. The duration of the QRS complex is typically short, lasting about 0.06–0.10 s, as the ventricular conduction system (including the Purkinje fibers) allows for the rapid spread of the electrical impulse.

T Wave: Ventricular Repolarization

The T wave represents ventricular repolarization, a process that occurs after the ventricles have contracted and relaxed. During repolarization, the myocardial cells return to their resting state as potassium ions move out of the cells, restoring the negative charge inside the cell relative to the outside. The T wave is typically positive in most leads, although its morphology can vary depending on factors such as heart rate, electrolyte imbalances, or other pathological conditions. The T wave is typically broader and slower than the QRS complex because repolarization occurs more slowly than depolarization [23].

Understanding the timing and intervals between each wave is crucial for interpreting the ECG. In addition to the P wave, QRS complex, and T wave, some specific intervals and segments represent the timing of electrical events [24–26]:

- PR Interval: The time between the start of the P wave and the start of the QRS complex, reflecting the time it takes for the electrical impulse to travel from the SA node, through the atria, the AV node, and the Bundle of His.
- QT Interval: The time between the start of the QRS complex and the end of the T wave, representing the time from the beginning of ventricular depolarization to the end of ventricular repolarization.
- ST Segment: The flat section of the ECG between the end of the QRS complex and the beginning of the T wave. This segment represents the period when the ventricles are fully depolarized and the heart muscle is preparing for repolarization (Fig. 3.6).

The electrical events that generate the ECG correspond closely to mechanical events in the heart [27]. Each phase of the ECG waveform is tied to a specific step in the contraction cycle:

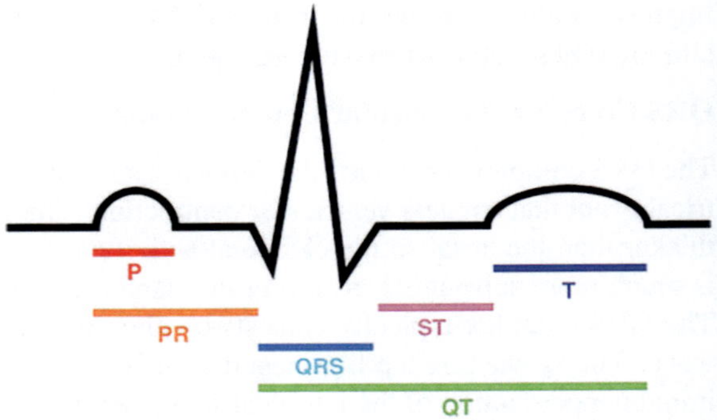

Fig. 3.6 ECG Intervals. https://commons.wikimedia.org/wiki/File:ECG_Intervals.svg

1. Atrial Contraction (P wave): The depolarization recorded in the P wave causes the atria to contract, pushing blood into the ventricles. This is the final step in the diastolic phase of the heart cycle, just before ventricular contraction begins.
2. Ventricular Contraction (QRS complex): The depolarization seen in the QRS complex triggers ventricular contraction, causing the AV valves (mitral and tricuspid) to close, and the semilunar valves (pulmonary and aortic) to open. Blood is pumped from the ventricles to the lungs and the rest of the body during this phase.
3. Ventricular Relaxation (T wave): The repolarization seen in the T wave corresponds to the relaxation of the ventricles, marking the end of systole and the start of diastole, when the heart refills with blood.

Additional ECG Waveforms and Variants

While the P wave, QRS complex, and T wave are the primary components of an ECG, several other waveforms and anomalies can be observed depending on the heart's physiological and electrical conditions [28, 29]. These include the U wave, J point, and abnormal waveforms associated with various clinical scenarios.

U Wave

The U wave is a slight deflection that follows the T wave. It is not always present, but when it does appear, it is typically seen after the T wave and before the next P wave. The exact cause of the U wave is still debated, but it is thought to represent the repolarization of the Purkinje fibers or the final stages of ventricular repolarization. The U wave is more prominent in specific leads (like lead V2) and can be seen more clearly in slower heart rates. In certain clinical conditions, such as hypokalemia (low potassium levels) or bradycardia, the U wave may become more pronounced. Conversely, in conditions like hyperkalemia, the U wave may be absent or inverted [30, 31].

J Point

The J point is the point at which the QRS complex transitions into the ST segment. It is considered the junction between ventricular depolarization and repolarization. In a normal ECG, the J point is located at the end of the QRS complex, right before the ST segment. Any abnormalities in the J point, such as elevation or depression, can be indicative of conditions like myocardial infarction, pericarditis, or ischemia [32, 33]. A depressed J point is seen in ischemic heart disease and is often followed by a down-sloping ST segment, which may be diagnostic of a subendocardial infarction or angina. Conversely, an elevated J point can be associated with early repolarization or acute pericarditis.

Abnormal Waveforms and Clinical Variants

ECG waveforms can sometimes show abnormal shapes and patterns, which can be indicative of specific diseases or pathological conditions [34, 35]. For example:

- Inverted T waves: Inverted T waves are often seen in myocardial ischemia, acute coronary syndromes, or left ventricular hypertrophy (LVH).
- Prolonged QT interval: A prolonged QT interval can be indicative of an increased risk of arrhythmias, including torsades de pointes. It is associated with specific drug effects (such as antiarrhythmic drugs), electrolyte imbalances, or congenital syndromes (like Long QT syndrome).
- Shortened QT interval: A shortened QT interval may be seen in hypercalcemia or digitalis toxicity.
- Abnormal Q waves: The presence of pathological Q waves often suggests prior myocardial infarction, particularly when they are broader and deeper than usual.

3.2 ECG Signal Acquisition

The process of signal acquisition in ECG is a multi-step procedure that involves the detection, amplification, and display of electrical signals generated by the heart. The electrodes used in this procedure detect minute electrical changes that result from the heart's contraction and relaxation. The signals are then transmitted to a recording system that produces an electrocardiogram, a graphical representation of the heart's electrical activity over time [18].

The electrodes used in an ECG are typically made of materials that can easily detect electrical impulses. These include metals like silver chloride or copper because they allow the electrical signal to be captured and transmitted to the ECG machine with minimal interference. The electrodes are placed on various parts of the body to form the lead systems.

1. Limb electrodes: These are placed on the right arm (RA), left arm (LA), and left leg (LL). These electrodes form the base for the bipolar and unipolar leads.
2. Chest electrodes (Precordial Electrodes): These are placed on the chest to capture the heart's electrical activity along the horizontal plane.
3. Reference electrodes: The reference electrode is typically placed on the right leg (RL). It acts as a neutral reference, helping to complete the electrical circuit for accurate measurement.

These electrodes pick up the action potentials that originate from the sinoatrial (SA) node and propagate through the heart's conduction system. The heart's electrical activity is typically small—on the order of millivolts—so amplification is necessary to visualize the signal clearly [36, 37] (Figs. 3.7 and 3.8).

3.2.1 Bipolar leads: Einthoven's Triangle

Einthoven's Triangle is a theoretical construct that forms the foundation for understanding the bipolar lead system. It consists of an equilateral triangle whose three

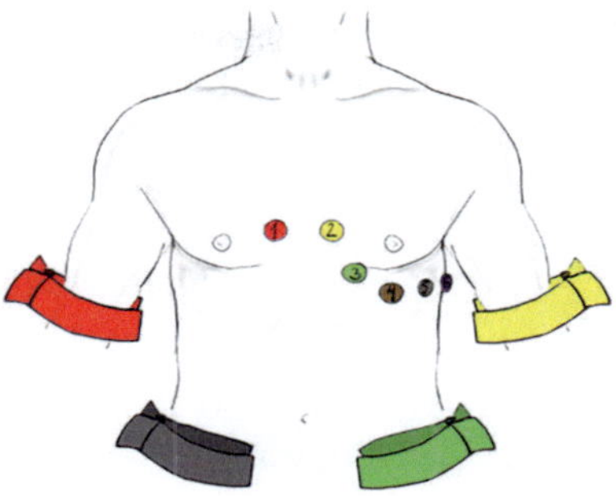

Fig. 3.7 ECG Electrode Placement. https://commons.wikimedia.org/wiki/File:ECG_limb_and_chest_electrodes_placement.png

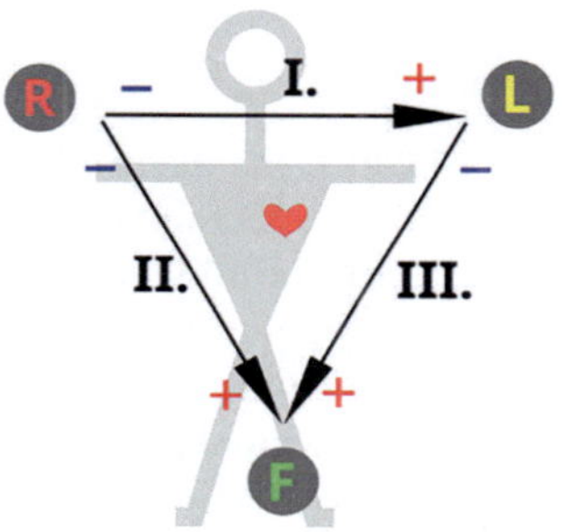

Fig. 3.8 Einthoven's Triangle. https://commons.wikimedia.org/wiki/File:ECG-Einthoven-triangle.svg

corners correspond to the three main limb electrodes: the right arm, left arm, and left leg. The triangle helps visualize the electrical vectors recorded by the three bipolar leads: Lead I, Lead II, and Lead III.

The electrical activity of the heart creates a flow of positive charges that can be measured as a potential difference between the electrodes. The bipolar leads are positioned to capture this electrical activity in a vectorial manner. Each lead measures the difference in voltage between two electrodes, with the vector pointing in the direction of the net electrical activity at that moment [7, 38].

1. Lead I: Measures the difference between the RA and LA, essentially creating a horizontal axis across the heart. The heart's electrical activity will be measured as it moves between the right and left sides, representing right-to-left conduction. This lead is particularly sensitive to abnormalities in the left side of the heart, such as left atrial enlargement or left ventricular hypertrophy. The electrical activity captured by Lead I helps assess the left-to-right axis of the heart.
2. Lead II: Measures the electrical potential between the RA and LL. This lead offers a diagonal view, effectively capturing the vertical vector of electrical activity and providing a more comprehensive view of the heart's rhythm and electrical impulse propagation. Lead II is especially useful for monitoring the SA node, as it provides a clear and continuous depiction of the heart's pacemaker activity. Lead II is often the lead of choice in clinical settings for observing heart rhythms, as it is well-aligned with the normal heart's electrical conduction.

3. Lead III: Measures the difference between the LA and LL. This leads to the capture of electrical activity moving along the left side of the heart, contributing to the view of the inferior portion of the heart. Lead III provides valuable information for assessing the left ventricle and its electrical axis. It also helps define the heart's overall electrical axis, providing additional data to identify cardiac abnormalities, such as myocardial infarction or ischemic heart disease.

The primary concept behind the bipolar leads is the vectorial view of the electrical potential differences. As the heart depolarizes, electrical impulses spread from the SA node across the atria, down the atrioventricular (AV) node, and through the His-Purkinje system, leading to ventricular depolarization. This wave of electrical activity can be visualized as a vector, with positive deflections indicating movement toward a positive electrode and negative deflections indicating movement toward a negative electrode [7, 39].

- In Lead I, the electrical vector primarily moves horizontally from right to left. Positive deflections occur when the left side of the heart is depolarizing. A negative deflection would indicate that the heart's electrical activity is moving away from the left side.
- In Lead II, the vector typically moves downward and to the left, and is aligned with the normal conduction pathway of the heart. When the heart depolarizes, the positive deflection represents the heart's electrical activity flowing in the direction of the left ventricle.
- In Lead III, the electrical activity flows along the left lateral side of the heart, with positive deflections occurring when electrical activity propagates toward the left ventricle.

Each of the bipolar leads provides information about the direction and magnitude of the heart's electrical activity. However, when considered together, they form a comprehensive picture of the heart's overall electrical axis. The electrical axis of the heart refers to the direction in which the most significant electrical activity is recorded during ventricular depolarization. In a normal heart, this axis tends to fall within a range of -30° to $+$ 90° in the frontal plane, with the vector pointing from the right shoulder (negative electrode) to the left leg (positive electrode). The overall axis is determined by looking at the relationship between the electrical deflections in Leads I, II, and III [7, 40].

For example:

- If Lead I shows a positive deflection and Lead II also shows a positive deflection, it suggests that the electrical axis is directed downward and leftward (normal axis).
- If the deflections in Lead I and Lead II are both negative, it indicates a rightward axis deviation, possibly associated with right ventricular hypertrophy or conduction abnormalities.
- If the deflection in Lead I is positive and Lead III is negative, it can suggest a left axis deviation, often associated with left ventricular hypertrophy or left bundle branch block.

By analyzing these leads together, physicians can determine whether the heart's electrical axis is normal, rightward, or leftward, which provides crucial insights into cardiac conditions such as ventricular hypertrophy, conduction abnormalities, and myocardial infarctions. The bipolar leads, particularly Lead II, are widely used in clinical monitoring because of their ability to provide a consistent view of the heart's rhythm and pacemaker activity. Lead II is often used in continuous cardiac monitoring, especially in hospital settings, due to its alignment with the heart's natural electrical conduction pathway [7, 41].

- Lead I is essential in evaluating left-sided pathologies, such as left atrial enlargement or left ventricular hypertrophy, and is often used in diagnostic ECGs.
- Lead III provides additional insight into inferior wall myocardial infarctions and is often helpful in evaluating the left ventricle.

The bipolar lead system, although older, remains indispensable in the initial diagnostic workup of many cardiac conditions. By integrating data from Lead I, Lead II, and Lead III, a comprehensive understanding of the heart's electrical activity is formed, enabling clinicians to diagnose, monitor, and manage various cardiac conditions.

To further visualize how these leads work in practice, consider a simplified version of Einthoven's Triangle:

- Lead I corresponds to the horizontal axis connecting the right arm and left arm.
- Lead II represents the diagonal axis, from the right arm to the left leg, aligning with the natural electrical flow.
- Lead III provides insight into the leftward axis between the left arm and left leg.

Together, these leads provide a comprehensive view of the heart's electrical behavior across the frontal plane, offering critical insights into heart rhythm, electrical axis, and the presence of various cardiac pathologies.

The Einthoven Triangle in Practice

The electrical activity of the heart creates a flow of positive charges that can be measured as a potential difference between the electrodes. The bipolar leads are positioned to capture this electrical activity in a vectorial manner. Each lead measures the difference in voltage between two electrodes, with the vector pointing in the direction of the net electrical activity at that moment [42, 43].

Lead I: Measures the difference between the RA and LA, essentially creating a horizontal axis across the heart. The heart's electrical activity will be measured as it moves between the right and left sides, representing right-to-left conduction. This lead is particularly sensitive to abnormalities in the left side of the heart, such as left atrial enlargement or left ventricular hypertrophy. The electrical activity captured by Lead I helps assess the left-to-right axis of the heart.

Lead II: Measures the electrical potential between the right arm (RA) and left leg (LL). This lead offers a diagonal view, effectively capturing the vertical vector of electrical activity and providing a more comprehensive view of the heart's rhythm and electrical impulse propagation. Lead II is especially useful for monitoring the

SA node, as it provides a clear and continuous depiction of the heart's pacemaker activity. Lead II is often the lead of choice in clinical settings for observing heart rhythms, as it is well-aligned with the normal electrical conduction of the heart.

Lead III: Measures the difference between the LA and LL. This leads to the capture of electrical activity moving along the left side of the heart, contributing to the view of the inferior portion of the heart. Lead III provides valuable information for assessing the left ventricle and its electrical axis. It also helps define the heart's overall electrical axis, providing additional data to identify cardiac abnormalities, such as myocardial infarction or ischemic heart disease.

The primary concept behind the bipolar leads is the vectorial view of the electrical potential differences. As the heart depolarizes, electrical impulses spread from the SA node across the atria, down the AV node, and through the His-Purkinje system, leading to ventricular depolarization. This wave of electrical activity can be visualized as a vector, with positive deflections indicating movement toward a positive electrode and negative deflections indicating movement toward a negative electrode [44–46].

- In Lead I, the electrical vector primarily moves horizontally from right to left. Positive deflections occur when the left side of the heart is depolarizing. A negative deflection would indicate that the heart's electrical activity is moving away from the left side.
- In Lead II, the vector typically moves downward and to the left, and is aligned with the normal conduction pathway of the heart. When the heart depolarizes, the positive deflection represents the heart's electrical activity flowing in the direction of the left ventricle.
- In Lead III, the electrical activity flows along the left lateral side of the heart, with positive deflections occurring when electrical activity propagates toward the left ventricle.

Each of the bipolar leads provides information about the direction and magnitude of the heart's electrical activity. However, when considered together, they form a comprehensive picture of the heart's overall electrical axis. The electrical axis of the heart refers to the direction in which the most significant electrical activity is recorded during ventricular depolarization. In a normal heart, this axis typically falls within a range of $-30\,°$ to $+\,90\,°$ in the frontal plane, with the vector pointing from the right shoulder (negative electrode) to the left leg (positive electrode). The overall axis is determined by looking at the relationship between the electrical deflections in Leads I, II, and III.

For example:

- If Lead I shows a positive deflection and Lead II also shows a positive deflection, it suggests that the electrical axis is directed downward and leftward (normal axis).
- If the deflections in Lead I and Lead II are both negative, it indicates a rightward axis deviation, possibly associated with right ventricular hypertrophy or conduction abnormalities.

- If the deflection in Lead I is positive and Lead III is negative, it can suggest a left axis deviation, often associated with left ventricular hypertrophy or left bundle branch block.

3.2.2 Goldberger Unipolar Lead System

Goldberger's modification of the standard 12-lead electrocardiogram (ECG) system is pivotal in expanding our understanding of the heart's electrical activity by introducing augmented unipolar leads, providing a broader and more comprehensive view of the heart's electrical behavior. This innovation enhances traditional bipolar limb leads by providing a more comprehensive three-dimensional representation of the heart's electrical field [47, 48] (Fig. 3.9).

Goldberger's unipolar leads—denoted as aVR, aVL, and aVF—are positioned to provide an augmented perspective of the heart's activity along the frontal plane. These leads do not measure the difference between two limb electrodes, as seen in bipolar leads, but rather capture the electrical potential of the heart relative to a central reference point, often the midpoint between the limbs. By focusing on this central point, Goldberger's leads provide a more direct and focused view of electrical impulses originating from different angles of the heart [12].

1. aVR (Augmented Voltage Right Arm) is designed to observe the electrical activity of the heart from the right shoulder. This lead typically provides a negative deflection because the electrical impulses usually flow away from the right side of the heart. It is particularly useful for identifying abnormalities in the right atrium or right ventricle and is often employed to detect conditions such as right atrial enlargement. Although less commonly used in routine diagnostic procedures, aVR offers valuable insights, especially when interpreting unusual findings in the context of other leads.
2. aVL (Augmented Voltage Left Arm) monitors the heart's electrical potential from the left shoulder. As it records the voltage difference between the left arm and

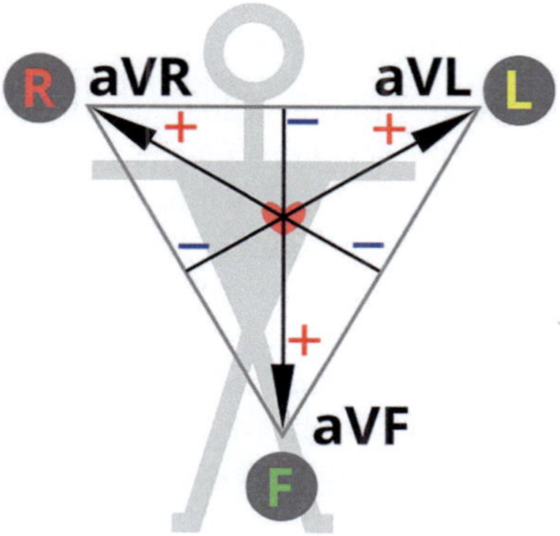

Fig. 3.9 Goldberger system. https://commons.wikimedia.org/wiki/File:ECG-Goldberger.svg

the central reference point, aVL captures electrical activity moving toward the left side of the heart. Positive deflections indicate activity from the left atrium and left ventricle, providing valuable data on left-sided heart conditions such as left ventricular hypertrophy. It is essential in diagnosing conditions affecting the lateral wall of the left ventricle, particularly when the deflection patterns diverge from expected norms.

3. aVF (Augmented Voltage Foot) focuses on the inferior aspect of the heart, measuring the electrical potential between the left leg and the central reference point. This lead captures the downward electrical flow of activity, providing insight into the heart's inferior wall. Positive deflections in aVF occur when the electrical activity moves toward the inferior heart regions, and it is especially useful for diagnosing inferior myocardial infarctions. Abnormalities in aVF can indicate significant changes in the inferior wall's electrical activity, contributing to the overall assessment of myocardial health.

Together, the augmented leads contribute to a more comprehensive picture of the heart's electrical axis, especially when combined with the traditional bipolar leads (I, II, and III). Goldberger's augmented leads offer distinct advantages by enhancing the diagnostic capability for conditions like myocardial infarction, hypertrophy, and arrhythmias. They allow for precise analysis of the heart's electrical orientation and direction, particularly in identifying deviations in the electrical axis.

In practice, the augmented leads aid clinicians by providing a better understanding of the electrical flow, supporting the diagnosis of various cardiac pathologies [49, 50]. For example:

- In aVR, a positive deflection is rare. However, it can signal a shift in the normal electrical flow, indicating right-sided heart issues such as right ventricular hypertrophy or acute right heart strain.
- In aVL, abnormal deflections can reveal the presence of left-sided pathologies like left atrial enlargement, particularly when changes in the amplitude or direction of the deflection occur.
- In aVF, significant deflection shifts are crucial for detecting inferior wall infarctions or ischemic damage to the heart.

Precordial Leads

There are six precordial leads (V1-V6) in the Goldberger system, each positioned at specific locations on the chest to capture the electrical potential from different perspectives. Lead V1 is positioned at the fourth intercostal space, right sternal border, and provides a view of the heart's right atrium and right ventricle. It helps detect abnormalities in these areas, such as right atrial enlargement or right ventricular hypertrophy. The electrical vector captured by Lead V1 reflects the rightward and posterior aspect of the heart [8].

Lead V2 is placed at the fourth intercostal space, left sternal border, providing a view of the right ventricle and septum. This lead is significant for detecting ventricular arrhythmias or ischemic changes that affect the septal region of the heart. Lead V2 captures a similar vector to Lead V1, but it is slightly shifted, offering more insight into anterior and septal electrical activity. The vector of Lead V2 is slightly angled downward and leftward, giving more detail about the septum and anterior portions of the heart.

Lead V3 is placed at the fifth intercostal space, between the midline and left side of the chest, providing a view of the anterior wall of the left ventricle. This lead helps monitor electrical disturbances that may occur in the left ventricle, such as those seen in acute myocardial infarctions or left ventricular hypertrophy.

Lead V4 is placed at the fifth intercostal space, midclavicular line, and captures the electrical activity of the anterior left ventricle, offering critical information for diagnosing anterior wall infarctions. This lead helps determine ischemic changes and monitor the left ventricle's electrical axis, often complementing findings from Leads V1-V3.

Lead V5 is positioned at the fifth intercostal space, along the anterior axillary line, providing a lateral view of the heart. This lead is valuable in assessing the lateral wall of the left ventricle and detecting conditions such as ventricular hypertrophy or lateral myocardial infarctions.

Lead V6 is placed at the fifth intercostal space, midaxillary line, and is the most lateral precordial lead. It provides a view of the left ventricle and lateral left heart wall, assisting in the diagnosis of lateral myocardial infarctions and ventricular arrhythmias. Lead V6 captures the lateral electrical activity of the heart, moving laterally from the left ventricle.

The electrical activity of the heart propagates from the SA node, down the AV node, and through the His-Purkinje system to the ventricles. The electrical impulses spread throughout the heart's muscle fibers in a vectorial manner, which can be captured and analyzed through the various leads. The positive and negative deflections in the ECG trace indicate the direction of this electrical propagation [51].

- Positive deflection indicates that the electrical activity is moving toward the positive electrode of the lead.
- Negative deflection indicates that the electrical activity is moving away from the positive electrode.

The overall electrical axis of the heart refers to the direction in which the most significant electrical activity is recorded during ventricular depolarization. This axis is crucial for diagnosing a variety of cardiac conditions, such as hypertrophy, conduction abnormalities, and myocardial infarction [52, 53].

- Normal Axis: In a healthy heart, the electrical axis falls within a range of $-30°$ to $+90°$ in the frontal plane.
- Right Axis Deviation: When both Lead I and Lead II show negative deflections, it can indicate right ventricular hypertrophy or issues with the conduction system.
- Left Axis Deviation: Positive deflections in Lead I combined with negative deflections in Lead III may suggest left ventricular hypertrophy or left bundle branch block.

Integration of Goldberger's Leads and Einthoven's Triangle:

While Einthoven's Triangle helps establish a foundational understanding of the bipolar leads in the frontal plane, the introduction of Goldberger's unipolar leads enables a comprehensive three-dimensional view of the heart's electrical behavior. The six precordial leads (V1-V6) complement the bipolar leads by capturing electrical activity in the horizontal plane, further enhancing diagnostic capabilities. Together, these leads provide critical insights into heart rhythm, electrical axis, and the detection of cardiac pathologies [54, 55].

Summary of Lead Functions:

- Lead I: Horizontal, assessing right-to-left conduction.
- Lead II: Diagonal, providing insights into pacemaker activity.
- Lead III: Inferior, contributing to the left ventricle's assessment.
- V1-V6 (Goldberger's Leads): Precordial leads providing insights into the right ventricle, septal region, anterior wall, and lateral left ventricle.

By integrating the augmented leads into the standard ECG setup, clinicians gain a more precise and more detailed understanding of the heart's electrical activity from multiple perspectives. The augmented unipolar leads refine the diagnostic process, enhancing the ability to detect and evaluate cardiac conditions that may not be fully represented by the standard bipolar leads alone [56, 57]. The incorporation of Goldberger's leads into the ECG system remains a standard practice in clinical settings, facilitating continuous monitoring and providing critical insights into the heart's electrical behavior. These augmented leads have become integral in both inpatient and outpatient settings, where they assist in monitoring arrhythmias, heart failure, and other conditions requiring long-term observation. This significantly advanced our understanding of the heart's electrical activity, allowing for more accurate diagnoses of cardiac conditions. The ability to assess the heart from various angles—right, left, and inferior—gives a more nuanced picture of its electrical health, guiding clinicians toward better treatment outcomes for patients with heart disease. By synthesizing data from the bipolar and augmented leads, healthcare providers can generate a comprehensive view of the heart's function, leading to more effective and timely interventions [58, 59].

3.2.3 ECG Montages

ECG montages refer to specific lead configurations that enhance the detection of particular cardiac conditions. While the standard 12-lead ECG provides a comprehensive view of the heart's electrical activity, specific alternative lead placements are more effective in localizing ischemia, arrhythmias, or conduction abnormalities. The following table summarizes the key ECG montages, their clinical applications, and the corresponding lead placements (Table 3.1).

Table 3.1 ECG Montages and Their Clinical Applications

Montage	Purpose	Lead placement	Clinical use
Right-sided leads (V3R–V6R)	Detects right ventricular infarction	Leads are placed on the right chest, and mirror left-sided precordial leads	Suspected **inferior MI**, RV involvement
Posterior leads (V7–V9)	Evaluates posterior infarction	Leads placed over posterior chest (V7, V8, V9)	Suspected **posterior MI**, ST depression in V1-V3
Lewis lead (Modified lead I)	Enhances P-wave visibility	RA electrode at 2nd right intercostal space, LA electrode at 4th right intercostal space	Diagnosing **atrial arrhythmias**, AV dissociation
High lateral leads (V4–V6 repositioned higher)	Detects high lateral ischemia	Leads moved superiorly on the chest	Suspected **high lateral STEMI**
Esophageal lead	Provides near-field atrial activity	An electrode is inserted into the esophagus behind the heart	Detects **atrial fibrillation/flutter**, AV nodal conduction
Lead II with lower limb elevation	Augments the QRS complex	Lower limb leads placed higher on the legs	Enhances **low-voltage QRS** detection

Clinical Scenarios

Right-Sided ECG Leads (V3R–V6R): Essential for Detecting Right Ventricular Infarction

A patient presents with inferior ST-elevation myocardial infarction (STEMI), but standard ECG does not confirm right ventricular (RV) involvement. Since right ventricular infarcts can lead to hypotension and hemodynamic instability, early detection is crucial for effective management and treatment.

Why Use This Montage?

- Standard precordial leads (V1–V6) provide a left-sided view of the heart and may not accurately capture right ventricular ischemia.
- Right-sided leads mirror the standard precordial leads on the right chest and detect ST-segment elevation in the right ventricle, which is highly suggestive of RV infarction.
- Confirmation of RV involvement guides treatment decisions, such as avoiding nitrates, which can exacerbate hypotension in RV infarcts.

Posterior Leads (V7–V9): Unmasking Posterior Myocardial Infarction

A patient presents with chest pain, and the standard ECG shows ST-segment depression in leads V1–V3. This pattern may indicate either anterior ischemia or a posterior infarction, which can be challenging to differentiate.

Why Use This Montage?

- ST-segment depression in V1–V3 may be a reciprocal change from an infarction occurring in the posterior wall of the left ventricle.
- Posterior leads (V7, V8, V9) placed along the left posterior thorax can reveal actual ST-segment elevation, confirming a posterior STEMI.
- Identifying posterior infarction ensures timely revascularization, improving patient outcomes.

Lewis Lead: Enhancing P-Wave Visibility in Atrial Arrhythmias

A patient presents with palpitations and irregular heartbeats; however, the standard ECG does not differentiate between atrial fibrillation (AF), atrial flutter, or an ectopic atrial rhythm.

Why Use This Montage?

- In the standard limb lead configuration, P-waves may be too small to differentiate subtle atrial activity.
- The Lewis Lead repositions the right arm (RA) and left arm (LA) electrodes to enhance atrial activity, improving the visibility of P-waves and atrial conduction patterns.
- This technique is particularly useful for distinguishing between atrial tachycardia, atrial flutter, and AV nodal reentrant tachycardia (AVNRT).

High Lateral Leads: Improving Diagnosis of High Lateral Myocardial Infarction

A patient has chest pain, but the standard ECG does not show clear ischemic changes. However, suspicion remains for a high lateral infarction, which may be missed on standard lead placements.

Why Use This Montage?

- Leads V4–V6 can be repositioned slightly higher on the chest, improving sensitivity for high lateral ischemia.
- This technique can reveal ST-elevation or subtle ischemic changes that are not visible in the standard 12-lead ECG.
- Identifying high lateral ischemia facilitates early intervention, thereby preventing myocardial damage.

Esophageal Lead: Direct Atrial Activity Recording

A patient presents with suspected atrial fibrillation (AF) or AV conduction abnormalities, but a standard surface ECG is inconclusive due to poor P-wave visibility.

Why Use This Montage?

- An esophageal lead is placed in the esophagus, which lies directly posterior to the left atrium, allowing a near-field recording of atrial activity.

- This technique is particularly useful for identifying atrial flutter waves, concealed accessory pathways, and AV conduction defects.

Lead II with Lower Limb Elevation: Enhancing Low-Voltage QRS Detection

A patient with suspected low-voltage QRS complexes has a standard ECG that does not adequately visualize ventricular depolarization.

Why Use This Montage?

- Elevating the lower limb electrodes reduces interference and enhances QRS amplitude, particularly in conditions like obesity, pericardial effusion, or myxedema.
- This can help improve the diagnosis of low-voltage QRS and electrical alternans in pericardial effusion.

3.3 ECG Signal Preprocessing

Electrocardiography remains one of the most widely used and informative non-invasive tools for cardiac assessment. The raw ECG signal, however, is frequently contaminated by a variety of noise sources, artifacts, and signal distortions that can severely hinder clinical interpretation and automated analysis. ECG signal preprocessing refers to the initial stage in the signal analysis pipeline where the goal is to enhance signal quality by removing or minimizing these disturbances.

Accurate ECG preprocessing is essential in both real-time monitoring (e.g., in ICU settings or wearable biosensors) and offline diagnostic evaluation (e.g., Holter monitors, exercise stress tests). Properly filtered and conditioned signals allow for reliable feature extraction, classification of cardiac events, and even long-term trend analysis using artificial intelligence (AI) models [18]. Preprocessing must strike a balance between eliminating noise and preserving physiologically significant information. For instance, aggressive filtering might suppress noise but at the cost of distorting QRS complex morphology or T-wave amplitude, both critical for diagnosing arrhythmias or ischemia [7]. The goal, therefore, is selective enhancement—employing adaptive, context-aware filters that account for signal dynamics and device limitations.

3.3.1 Sources of Noise in ECG Signals

Understanding the types and sources of noise that affect ECG signals is crucial for designing effective preprocessing strategies. The ECG signal, typically ranging from 0.05 to 150 Hz, is highly susceptible to contamination due to its low amplitude and the complex bioelectrical environment in which it is recorded. Noise not only reduces the accuracy of feature extraction algorithms but can also lead to misdiagnosis, especially in automated systems [60].

Baseline Wander

Baseline wander is one of the most common artifacts in ECG recordings. It is characterized by low-frequency drift of the ECG signal's baseline, typically in the range of 0.05–0.5 Hz. This artifact is often caused by patient movement, respiration, or poor electrode contact. Baseline wander can severely distort waveforms, particularly the ST segment, leading to false elevation or depression that mimics ischemic changes [4]. Simple high-pass filtering (cutoff around 0.5 Hz) is often employed to mitigate baseline drift. More advanced techniques include spline interpolation and polynomial fitting to remove slow-varying components without distorting clinical features. In conformity testing environments, high-fidelity baseline correction is emphasized to preserve ST-segment integrity during algorithm validation [61].

Powerline Interference

Powerline interference arises from electromagnetic coupling with surrounding electrical systems and is typically centered at 50 or 60 Hz, depending on the region. This sinusoidal noise can mask important components like the P-wave or T-wave, particularly in low-amplitude recordings [62]. Notch filters centered at 50/60 Hz are commonly used to suppress this interference. However, notch filtering can introduce phase distortion. As a result, modern systems may employ adaptive noise cancellation using a reference signal from the mains power line or adaptive filtering techniques that preserve signal morphology [63].

Muscle Artifacts (EMG Noise)

Muscle artifacts, or electromyographic (EMG) noise, are introduced by skeletal muscle contractions and manifest as high-frequency bursts (20–200 Hz) overlapping with ECG waveforms. These artifacts are especially problematic in stress testing and ambulatory ECG monitoring. Filtering muscle noise requires a trade-off between noise suppression and retention of critical high-frequency components of the QRS complex. Techniques such as wavelet denoising and adaptive band-pass filtering are often applied. Wearable devices tested under post-market surveillance protocols have shown variable sensitivity to muscle noise, highlighting the need for robust preprocessing pipelines [59].

Electrode Motion Artifacts

Electrode motion artifacts are transient distortions caused by changes in skin–electrode impedance during movement. These typically appear as large-amplitude, low-frequency deviations and can mimic arrhythmias or obscure actual cardiac events.

Motion artifacts are particularly concerning in long-term monitoring scenarios, including Holter ECG and wearable biosensors. Signal quality indices (SQIs) are increasingly used to detect and flag such segments. Advanced filtering methods also use accelerometer data to implement motion-compensated signal recovery [64].

Other noise sources include environmental electromagnetic interference and internal instrumentation noise. These are often minimized through shielded cables, proper grounding, and analog front-end design. Nonetheless, preprocessing software must account for such noise, particularly in low-cost or portable ECG systems.

Filtering Techniques

Filtering is the cornerstone of ECG signal preprocessing. The goal is to suppress noise and artifact components without compromising the clinical information embedded in the signal. A well-designed filter removes baseline drift, suppresses powerline interference, attenuates high-frequency muscle noise, and maintains the morphology of the ECG waveform.

Linear Filtering

Linear filters—such as Finite Impulse Response (FIR) and Infinite Impulse Response (IIR) filters—are the most commonly used in ECG preprocessing. FIR filters are preferred for their inherent stability and linear phase response, which helps preserve waveform shape, especially the QRS complex and ST segment. IIR filters, on the other hand, are more computationally efficient but may introduce phase distortion [65]. For baseline correction, a high-pass filter with a cutoff frequency of around 0.5 Hz is often applied. For powerline interference, a narrow-band notch filter at 50 or 60 Hz is standard. However, a notch filter can distort the ECG signal, particularly when harmonics overlap with the signal's spectral content. Therefore, hybrid FIR-notch filters or adaptive alternatives are increasingly used [66].

Nonlinear and Adaptive Filtering

Nonlinear filtering strategies have emerged as superior alternatives in situations where noise characteristics are nonstationary or context-dependent. Median filtering, for example, is used to suppress impulsive noise and motion artifacts. Adaptive filters—such as Least Mean Squares (LMS) and Recursive Least Squares (RLS)—dynamically adjust filter coefficients based on a reference signal or error minimization strategy. In wearable ECG systems, LMS-based adaptive filters have been used to remove powerline and motion artifacts using accelerometer data as a reference [7]. Such techniques are ideal for real-time applications where environmental and physiological conditions vary.

Wavelet-Based Filtering

Wavelet-based filtering decomposes the ECG signal into different frequency bands, allowing selective noise suppression without distorting important components. Studies have shown that discrete wavelet transform (DWT) filtering outperforms traditional bandpass filters in preserving QRS morphology while reducing muscle noise.

Filtering in Regulatory and Surveillance Frameworks

Filters must be validated not only for their noise reduction capability but also for their ability to preserve diagnostic parameters such as ST elevation, QT interval, and

QRS duration. Advanced filtering strategies were found to be critical in maintaining conformity during high-stress testing conditions, such as emergency defibrillator scenarios or neonatal ECG acquisition. In these cases, dynamic filtering—where cutoff frequencies adjust in response to signal content—ensured both compliance and diagnostic reliability [67].

3.3.2 Adaptive Filtering and Multirate Signal Processing

As ECG signal acquisition becomes increasingly mobile and personalized, static filtering techniques often fall short in dynamic environments. Adaptive filtering and multirate signal processing have emerged as critical tools in modern ECG preprocessing pipelines, especially in ambulatory and wearable devices. These techniques are designed to adjust in real-time to changes in the signal or the noise environment, thereby improving performance in unpredictable conditions.

Adaptive filters differ from fixed-parameter filters by updating their internal parameters continuously during signal acquisition. This allows them to track and suppress nonstationary noise sources such as motion artifacts or powerline interference with variable amplitude. The most widely used adaptive filtering algorithms in ECG processing are LMS and RLS. In LMS-based filtering, the filter adjusts its coefficients to minimize the error between the ECG signal and a reference noise signal. In wearable ECG systems, accelerometers or impedance sensors often provide this reference, enabling the filter to suppress motion-induced noise in real time [7]. RLS filters, while more computationally intensive, provide faster convergence and are better suited for applications requiring high precision, such as neonatal ECG monitoring.

Adaptive filters are handy in mobile health devices where the patient is constantly in motion. These systems rely on noise cancellation techniques that operate in real-time and must maintain signal integrity across variable recording conditions. The use of real-time adaptive preprocessing was shown to significantly enhance ECG data reliability in post-market evaluations of wearable cardiac monitors. Advanced systems combine adaptive filtering with context-awareness, using onboard sensors to trigger parameter changes in the filter design. For instance, if the system detects physical activity via inertial sensors, it can switch to a higher-order adaptive filter with a faster convergence rate to cope with increased motion artifacts [18].

Multirate processing is another advanced technique that manipulates the sampling rate of a signal to improve computational efficiency or resolution. Downsampling (decimation) is often applied to reduce data volume in long-term monitoring scenarios, while upsampling (interpolation) may be used to refine waveform features during diagnostic analysis. In ECG signal processing, multirate filtering enables selective enhancement of waveform components. For example, a decimated version of the ECG may be used for heart rate monitoring, while the full-rate signal is preserved for ST segment or QRS morphology analysis. This dual-resolution strategy supports both real-time monitoring and high-fidelity offline diagnostics [68].

Wavelet decomposition is frequently used in conjunction with multirate systems, allowing simultaneous examination of ECG features at multiple scales. Such hybrid approaches have been validated in clinical applications and conformity testing [10], particularly in complex environments like emergency response units and neonatal ICUs.

Despite their advantages, adaptive and multirate systems require careful calibration and validation. Parameters such as adaptation step size, reference signal fidelity, and resampling algorithms must be optimized to avoid signal distortion. Filtering anomalies can lead to clinically significant errors in waveform interpretation, impacting device certification and patient safety. Therefore, these advanced preprocessing methods must be incorporated with robust quality assurance protocols. In regulatory frameworks, especially for post-market surveillance, these methods are increasingly being mandated to meet compliance standards and maintain device integrity over long-term use.

3.3.3 Real-Time and Offline Preprocessing Pipelines

ECG signal preprocessing can be implemented in either real-time or offline settings, each requiring different design considerations and optimizations. The increasing availability of wearable and mobile ECG monitoring devices has placed significant emphasis on low-latency, resource-efficient preprocessing solutions suitable for embedded platforms. Conversely, offline processing enables more intensive computation and retrospective analysis.

Real-time ECG preprocessing is crucial in intensive care units (ICUs), telemetry systems, and wearable health monitors where immediate feedback is required. In such systems, preprocessing must operate under strict constraints on processing time, power consumption, and memory usage. Techniques used in real-time preprocessing include:

- Efficient IIR filtering for baseline and powerline interference
- Adaptive LMS filters driven by real-time reference sensors
- Low-order wavelet decomposition for fast denoising
- Signal Quality Indices (SQIs) for immediate noise assessment

In practical implementation, real-time systems often rely on digital signal processors (DSPs) or low-power microcontrollers. These must be carefully calibrated to ensure that filtering and feature extraction do not introduce latency that compromises clinical responsiveness [69, 70]. Post-market evaluations of wearable ECG systems revealed that devices with dedicated preprocessing hardware maintained superior signal quality under ambulatory conditions. These systems are adapted in real time to detect noise, reinitialize filters, and buffer stable signal segments for downstream analysis.

Offline ECG preprocessing is typically performed in diagnostic systems such as Holter monitors, exercise stress test analyzers, or research datasets. Here, the focus is on accuracy, flexibility, and robustness rather than computational efficiency. Offline analysis enables:

- Retrospective parameter tuning (e.g., filter cutoffs, wavelet thresholds)
- Use of computationally expensive algorithms (e.g., RLS, Empirical Mode Decomposition)
- Visual inspection and manual correction of preprocessing errors
- Batch processing of large datasets

High-resolution signal restoration and advanced noise modeling techniques are often applied in offline settings to reconstruct signals affected by severe artifacts. For example, neonatal ECG signals processed offline with high-order wavelet filtering and template subtraction achieved improved QRS detection sensitivity in neonatal device assessments.

Offline preprocessing is also essential in algorithm development and clinical research, where new preprocessing pipelines can be iteratively tested and optimized using annotated benchmark datasets.

Modern ECG platforms are increasingly adopting hybrid models that combine real-time and offline processing. A wearable device may conduct initial filtering and beat detection onboard, then transmit compressed segments for cloud-based offline refinement and decision support. This two-stage model balances efficiency and accuracy, making it ideal for telemedicine and long-term remote monitoring applications. Edge computing has further advanced this concept. Smart ECG devices now include on-device AI chips that can perform both preprocessing and classification tasks in real time. These systems rely on optimized time-domain features and adaptive filters discussed in previous sections [8].

Whether deployed in real-time or offline settings, ECG preprocessing pipelines must be validated against regulatory performance standards. Conformity testing frameworks require reproducibility of signal enhancement under various noise scenarios, including movement, electrical interference, and electrode degradation. These frameworks emphasize:

- Reproducibility of noise suppression algorithms
- Fidelity of signal morphology post-filtering
- Preservation of diagnostic features (e.g., ST elevation, QT interval)

In both contexts, rigorous benchmarking and real-world testing ensure that preprocessing components support clinical safety, efficacy, and long-term device reliability.

3.3.4 Impact of Preprocessing on Diagnostic Accuracy

The clinical utility of an ECG recording is directly tied to the fidelity with which it represents the electrical activity of the heart. Consequently, the effectiveness of preprocessing significantly influences diagnostic performance. Poor signal conditioning can obscure critical waveform features, lead to misinterpretation of cardiac events, and result in false-positive or false-negative diagnoses.

Preprocessing algorithms impact core diagnostic parameters such as QRS detection, QT interval measurement, and ST segment evaluation. Baseline wander, if inadequately corrected, can distort the ST segment and lead to misdiagnosis of ischemia. Likewise, poor suppression of EMG noise may result in missed ventricular ectopic beats or misclassification of atrial fibrillation episodes [71]. Advanced preprocessing strategies—such as wavelet filtering or template matching—have demonstrated higher sensitivity in QRS detection and better reproducibility of interval measurements in clinical trials [72]. Devices that incorporated robust adaptive filtering showed higher diagnostic specificity during post-market testing, especially in pediatric and neonatal applications.

Modern ECG systems increasingly rely on automated algorithms for rhythm interpretation and risk stratification. These algorithms are sensitive to input signal quality. Preprocessing affects feature extraction accuracy, waveform classification robustness, and even machine learning performance [73]. Noise artifacts that persist after preprocessing may cause false alarms in wearable systems or fail to trigger alerts in emergency monitors. AI-based ECG classifiers trained on clean datasets may underperform in noisy environments unless preprocessing modules generalize across noise profiles. This underscores the necessity of integrating preprocessing robustness as part of algorithm validation protocols [18].

For instance, in neonatal ECG monitors, effective suppression of motion and electrode artifacts improved detection rates of bradycardia and apnea episodes. In defibrillator systems, accurate baseline restoration ensured proper rhythm classification prior to shock delivery. Benchmarks used to evaluate preprocessing quality typically include:

- Signal-to-noise ratio (SNR)
- Percentage of correctly detected QRS complexes
- Mean error in interval measurements (e.g., QTc, PR)
- Visual scoring by clinical cardiologists

Standardization of these metrics has become essential in regulatory conformity assessment. Consistent evaluation protocols improve device comparability and ensure clinical safety across product generations.

Emerging trends include dynamic quality estimation, where real-time feedback from the classification module influences preprocessing settings. For example, if an ECG rhythm analysis engine reports low confidence in its output, the system may trigger reprocessing of that segment with adjusted filter parameters or alternative denoising strategies. Another promising direction is AI-guided preprocessing optimization, where machine learning models predict optimal filter configurations based on signal characteristics. These systems can tailor preprocessing to individual patients, electrode placements, and noise environments, ultimately improving diagnostic outcomes [12].

3.3.5 Case Studies in ECG Device Testing

Case studies provide direct insight into how ECG signal preprocessing influences device performance, safety, and compliance in real-world clinical and regulatory environments. These evaluations, conducted under post-market surveillance programs, highlight the central role of preprocessing algorithms in meeting conformity assessment standards [4, 59].

In the conformity assessment of patient monitors, preprocessing algorithms were evaluated for their ability to suppress baseline wander under different physiological and motion conditions. Devices implementing high-pass filtering combined with spline-based drift correction achieved higher fidelity in ST-segment preservation, leading to improved sensitivity in ischemia detection. Devices without adaptive baseline correction showed a higher false-positive rate for ST elevation, especially during exercise ECG recordings [61, 66, 71]. The report emphasized the importance of validating filtering routines against synthetic and physiological noise scenarios in both bench and clinical environments.

Neonatal ECG presents unique preprocessing challenges due to small signal amplitudes and high susceptibility to motion artifacts. Adaptive filters using real-time accelerometer feedback significantly reduced motion-induced baseline distortion, resulting in enhanced bradycardia detection and waveform clarity. Wavelet-based filtering and signal segmentation techniques were employed to isolate motion-free segments for further analysis, demonstrating the utility of hybrid preprocessing frameworks in sensitive clinical populations.

In testing defibrillators, the preprocessing focus was on real-time QRS detection in the presence of powerline and defibrillation shock artifacts. Dynamic notch filtering, combined with real-time template matching, preserved QRS integrity even under high electromagnetic interference. These findings supported the deployment of smarter signal quality indices (SQIs) that help differentiate noise from critical clinical patterns during emergency interventions [12].

Wearable ECG devices often operate under highly variable environmental and user conditions. Post-market studies evaluated the robustness of preprocessing pipelines in mobile devices. Results showed that systems with onboard adaptive preprocessing and real-time noise tagging outperformed their counterparts in heart rate variability analysis and arrhythmia detection accuracy. These devices also used multi-rate signal processing to conserve battery life while maintaining high resolution during suspected arrhythmic events [64, 72]. Preprocessing adaptation based on user motion, activity state, and environmental context improved both signal usability and diagnostic yield.

3.3.6 Feature Extraction and Analysis in ECG

ECG feature extraction serves as the bridge between raw signal acquisition and meaningful clinical interpretation. After effective preprocessing, the ECG waveform is analyzed to identify and quantify specific features that correspond to physiological events in the cardiac cycle. These features include the P wave, QRS complex, T wave, their respective durations and amplitudes, and inter-beat intervals.

Feature extraction is foundational for both traditional diagnostic algorithms and modern AI-driven ECG analysis. Accurate delineation of ECG waves enables automated arrhythmia classification, ischemia detection, cardiac risk stratification, and heart rate variability (HRV) analysis [37]. Poor extraction can lead to propagation of errors in downstream tasks, making this step critically crucial in both clinical and regulatory contexts. Well-established cardiac electrophysiology principles guide the process of feature extraction, but it also demands advanced signal processing skills to isolate relevant patterns from noise-prone biosignals accurately. Preprocessing removes noise and artifacts, but the complexity of cardiac electrical activity and inter-patient variability necessitates intelligent algorithms that can handle diverse morphologies, rhythms, and pathological changes. This includes distinguishing between overlapping or merged waveforms, managing ectopic beats, and compensating for amplitude or axis shifts caused by posture or electrode position.

Furthermore, ECG feature extraction plays a pivotal role in regulatory testing and post-market surveillance [7]. Devices unable to reliably extract such features under clinical and stress conditions risk failing conformity assessment. Feature extraction also underpins emerging personalized medicine approaches. With the rise of wearable ECG devices, real-time feature analysis allows for early detection of critical events such as atrial fibrillation, bradyarrhythmias, and ventricular tachycardia. Integrating feature extraction algorithms with mobile health apps and cloud-based analytics platforms creates new possibilities for remote diagnostics and AI-guided triage [74].

3.3.7 Morphological Components of the ECG Signal

The primary morphological components of the ECG signal correspond to distinct phases of the cardiac electrical cycle:

- P wave: Represents atrial depolarization. It typically appears as a low-amplitude, rounded deflection preceding the QRS complex. Abnormal P wave morphology can indicate atrial enlargement or conduction delays.
- QRS complex: Denotes ventricular depolarization. It is the most prominent and shortest duration component. The width, amplitude, and configuration of the QRS complex are crucial for identifying bundle branch blocks, ventricular hypertrophy, and ectopic rhythms.
- T wave: Reflects ventricular repolarization. T wave abnormalities may suggest electrolyte imbalances, ischemia, or post-infarction changes.

Other important features include:

- RR interval: The time between successive R waves. It determines heart rate and forms the basis of HRV analysis.
- PR interval: Measured from the onset of the P wave to the beginning of the QRS complex. It reflects AV nodal conduction time. Prolongation or shortening of this interval suggests conduction system pathology.
- QT interval: Extends from the start of the QRS complex to the end of the T wave. It represents total ventricular depolarization and repolarization. Prolonged QT intervals are associated with a higher risk of torsades de pointes and sudden cardiac death.

Proper detection and quantification of these features are critical for diagnostic accuracy. Inaccurate delineation of the T wave, for example, may lead to misclassification of ST-elevation myocardial infarction (STEMI) or long QT syndrome. Algorithms must be sensitive enough to detect subtle deviations yet robust against artifacts and waveform variability [75, 76]. Special attention is required in populations such as infants or athletes, where physiological norms differ from standard reference values. Neonatal ECG systems must be tuned for smaller signal amplitudes and faster heart rates, requiring more sensitive feature extraction algorithms. Similarly, athletes often exhibit early repolarization or bradycardia that must be correctly interpreted to avoid false positives. Advanced ECG monitors incorporate continuous morphological tracking, enabling trending of PR, QRS, and QT intervals over time. This supports the early detection of progressive conditions like AV block or QT prolongation due to drug interactions.

3.3.8 QRS Detection Algorithms

QRS detection is foundational to ECG analysis because it anchors the temporal structure of subsequent waveform interpretation. Accurately identifying the R peak allows computation of heart rate, rhythm classification, beat-to-beat interval measurement, and serves as the basis for segmenting the ECG for further feature extraction.

A standard QRS detection pipeline includes several processing stages:

1. Band-pass filtering (10–25 Hz): Enhances QRS frequency content and suppresses P and T waves.
2. Differentiation: Accentuates the slope of the QRS complex.
3. Squaring or Energy Operator: Emphasizes large slope values and removes polarity.
4. Moving Window Integration: Smoothens the energy signal and highlights QRS regions.
5. Adaptive Thresholding: Applies decision logic to identify candidate R peaks.

The Pan-Tompkins algorithm remains a widely cited benchmark due to its high detection rate in clean signals. However, its performance drops in noisy or ambulatory recordings. Therefore, newer systems integrate wavelet-based multiresolution analysis to isolate QRS features across frequency bands, enhancing resilience to noise [67, 77].

Deep learning approaches, such as convolutional neural networks (CNNs), have also been proposed for QRS detection. These models, once trained, can handle a variety of ECG morphologies without manual tuning. Nevertheless, they require substantial training data and are more suitable for offline or edge-computing scenarios. In regulatory contexts, QRS detection reliability affects the classification performance of automated external defibrillators (AEDs). Devices evaluated during conformity testing were required to detect > 99% of QRS complexes within variable physiological and noise conditions. Wearable ECG devices face the additional challenge of motion artifacts, requiring robust QRS detection in dynamic environments. Systems using accelerometer-aided adaptive thresholding achieved superior beat detection rates during physical activity compared to static algorithms.

3.3.9 *Wave Delineation (P, QRS, T Waves)*

Wave delineation refers to the process of accurately identifying the onset, peak, and offset of each ECG waveform component—namely the P wave, QRS complex, and T wave. This is essential for diagnostic calculations such as PR interval, QT interval, and ST segment deviations, which have direct implications for detecting cardiac conditions like AV blocks, ischemia, and long QT syndrome [7].

P Wave Delineation

The P wave is typically the most minor and most challenging component to detect, especially in noisy recordings or ambulatory environments. It is critical for diagnosing atrial abnormalities such as left or right atrial enlargement and for determining rhythm origin (sinus vs. ectopic). Various techniques are used for P-wave delineation:

- Template matching: Uses a pre-defined P wave shape for correlation.
- Wavelet transform: Localizes the wave using time–frequency decomposition.
- Slope and curvature analysis: Finds points of maximum slope changes.

Due to its low amplitude, the P wave is particularly vulnerable to noise, and preprocessing steps must ensure a high signal-to-noise ratio (SNR) before attempting delineation. In conformity assessment testing,

QRS Complex Delineation

After R-peak detection, the delineation of QRS onset and offset involves identifying the precise beginning and end of ventricular depolarization. Accurate QRS delineation is necessary for measuring QRS duration, which is critical for diagnosing bundle branch blocks and ventricular pre-excitation.

- First derivative zero-crossing: Detects the initial slope changes.
- Cumulative energy functions: Identify inflection points where the waveform energy stabilizes.
- Multi-lead correlation: Uses multiple ECG leads to enhance the reliability of onset/offset detection.

Even minor errors in QRS duration influenced device compliance with European and FDA cardiac monitor performance standards.

T Wave Delineation

The T wave represents ventricular repolarization and is clinically significant for detecting ischemic changes, electrolyte imbalances, and drug-induced effects (e.g., QT prolongation). However, its morphology is highly variable and often affected by noise and overlapping U waves. Techniques for T wave delineation include:

- Second derivative methods: Highlight the inflection near the T wave end.
- Morphological fitting: Matches known T wave shapes using adaptive scaling.
- Wavelet-based detection: Decomposes the ECG into multiple resolutions for accurate end detection.

Clinical studies found that misidentification of the T wave end led to QT miscalculations, contributing to inappropriate therapeutic decisions. The automated T wave delineation must be validated against manually annotated databases as part of post-market surveillance [4].

Challenges in wave delineation include:

- Signal variability: Due to patient-specific cardiac anatomy or disease.
- Noise contamination: Especially EMG and motion artifacts.
- Multi-lead inconsistency: Where waveforms appear differently across leads.

Addressing these challenges requires ensemble averaging, adaptive filtering, and lead fusion algorithms that synthesize information from multiple signal sources. The clinical impact of accurate delineation is profound:

- Incorrect QT measurement: May lead to overlooked arrhythmic risk.
- Inaccurate ST-segment analysis: Can cause false alarms in ischemia detection.
- Poor feature localization: Hinders AI model performance for predictive analytics.

As remote and wearable ECG systems become more prevalent, real-time delineation accuracy becomes a critical performance metric. Regulatory guidance increasingly mandates waveform-level testing to ensure safety and clinical reliability.

3.3.10 Feature Types: Amplitudes, Durations, Slopes, Areas

ECG analysis is fundamentally quantitative. Once wave delineation is complete, a wide variety of features can be extracted to support diagnosis, classification, and trend analysis. These features are typically categorized into four types: amplitude, duration, slope, and area-based measures. Each provides distinct insight into the physiological status of the heart [78, 79].

Amplitude Features

Amplitude features refer to the vertical values (in millivolts) of waveform components. Clinically important amplitudes include:

- P wave amplitude: Indicative of atrial size. Increased amplitude may reflect right atrial enlargement.
- R wave amplitude: Can reflect left ventricular hypertrophy or electrode placement.
- T wave amplitude: Changes with electrolyte imbalances (e.g., hyperkalemia).
- ST segment elevation or depression: Used in myocardial infarction detection.

Automated ECG systems must calculate these amplitudes relative to a consistent baseline, often the TP segment. Misalignment in baseline correction can introduce amplitude errors that may mimic pathology [7].

Duration Features

Durations are measured in milliseconds and describe the temporal span of specific waveform segments:

- PR interval (120–200 ms): Prolongation may indicate first-degree AV block.
- QRS duration (<120 ms normal): Prolonged QRS is seen in bundle branch blocks or ventricular rhythms.
- QT interval: Prolonged QT is a marker of arrhythmic risk.
- RR interval: Determines instantaneous heart rate and is used in HRV analysis.

QT and QTc (corrected for heart rate) intervals are crucial in drug safety evaluation. Errors in QT detection can lead to misclassification of torsades de pointes risk. Hence, these measurements are tightly regulated.

Slope Features

Slope or derivative-based features describe the steepness or rate of change of waveform components. These are particularly useful in:

- QRS slope analysis: For distinguishing between supraventricular and ventricular complexes.
- T wave upslope or downslope: Used in ischemia detection and risk stratification.

Steep slopes in the QRS complex are associated with rapid ventricular depolarization, while flattened slopes may indicate conduction slowing. Derivative features are commonly used in AI models to enhance the discriminatory power between similar-looking waveforms [7]. Slope analysis also forms part of real-time algorithms in wearable ECGs, as the slope-to-amplitude ratio can serve as a noise-resilient QRS marker.

Area-Based Features

The integral (area under the curve) of ECG segments provides composite measures that reflect both amplitude and duration. These include:

- QRS area: Used in ventricular dyssynchrony assessment.
- T wave area: Useful in evaluating repolarization abnormalities.
- ST area deviation: Enhanced sensitivity for ischemic detection compared to point-wise ST elevation alone.

Area-based measures are instrumental in machine learning contexts where high-dimensional features improve classification robustness [18]. For example, area-under-curve metrics were used to validate the fidelity of ECG waveform reproduction post-filtering. Devices that preserved area features more consistently across noise conditions were rated higher in signal fidelity.

While the ECG can yield dozens of potential features, not all are equally informative. Feature selection algorithms such as Principal Component Analysis (PCA), Mutual Information, or Genetic Algorithms are used to retain the most predictive features and discard redundant ones. This is critical for real-time systems with limited computational resources. Wearable ECG monitors, for example, often transmit only a subset of features to remote servers for analysis.

3.3.11 Transform-Based Features and Clustering

Beyond direct waveform analysis, ECG signals can be transformed into alternative domains—such as frequency, time–frequency, and nonlinear feature spaces—to extract hidden patterns that are not easily visible in the time domain. Transform-based features often offer superior discriminative power, especially when combined with clustering techniques for pattern recognition and disease classification [80, 81].

Frequency Domain Features

Transforming ECG signals into the frequency domain using the Fourier Transform (FT) or Power Spectral Density (PSD) analysis enables the extraction of periodicity, energy distribution, and frequency peaks. Key frequency domain features include:

- Spectral energy in low- and high-frequency bands
- Dominant frequency components
- Spectral entropy

In HRV analysis, spectral components are categorized into low frequency (LF), high frequency (HF), and very low frequency (VLF) bands, which correlate with autonomic nervous system activity. Fourier-based HRV features are now standard in arrhythmia risk analysis [10].

Time–Frequency Features.

Time–frequency transformations, such as Short-Time Fourier Transform (STFT) and Wavelet Transform (WT), allow for localized analysis of signal components. They are particularly useful when waveform characteristics change over time—as is often the case with arrhythmias or ischemic episodes. Wavelet-based features include:

- Wavelet coefficients at specific scales (e.g., QRS and T wave bands)
- Wavelet entropy
- Energy distribution across scales

Wavelet-based ECG classification has shown high accuracy in detecting atrial fibrillation, premature ventricular contractions, and myocardial infarction [59]. The time–frequency preprocessing improves delineation accuracy under noisy and ambulatory conditions, thus enhancing the reliability of extracted features.

Nonlinear and Fractal Features

Nonlinear analysis treats the ECG as a complex dynamical system. This perspective provides tools for analyzing variability, complexity, and hidden deterministic structures. Standard nonlinear ECG features include:

- Sample Entropy and Approximate Entropy
- Poincaré plot descriptors
- Detrended Fluctuation Analysis (DFA)
- Fractal Dimension

These features are especially relevant in differentiating normal sinus rhythm from chaotic rhythms like atrial fibrillation. Combining nonlinear features with standard ECG metrics improved arrhythmia classification performance in wearable devices.

Once a high-dimensional feature space is constructed, clustering techniques help to organize patterns into meaningful groups. Techniques include:

- K-means and hierarchical clustering
- Self-Organizing Maps (SOM)
- t-SNE and UMAP for visualization

Dimensionality reduction algorithms like PCA or Linear Discriminant Analysis (LDA) are used to reduce redundancy and computational burden. These are frequently used in ECG-based biometric identification systems and wearable arrhythmia detectors [8]. In conformity testing, clustering of feature vectors is used to evaluate consistency across repeated ECG measurements. Devices that clustered abnormally under standard operating conditions were flagged for waveform drift or preprocessing inconsistency.

Transform-based features serve as input to AI models such as:

- Support Vector Machines (SVMs)

- Random Forests
- Convolutional Neural Networks (CNNs)
- Long Short-Term Memory (LSTM) networks

These models classify rhythms, detect disease-specific ECG signatures, or predict adverse events. For instance, in COVID-19 monitoring, AI models using transform-based features from ECGs helped stratify patients by severity [82, 83]. By incorporating a combination of direct waveform, time–frequency, and nonlinear features, ECG analysis becomes robust to noise, variability, and recording environments. This multi-modal approach is now widely adopted in next-generation wearable monitors and AI-assisted diagnostic systems.

3.3.12 Feature Validation and Regulatory Considerations

Accurate ECG feature extraction is not only a technical objective—it is a regulatory necessity. In clinical applications, especially for life-critical monitoring, misclassification or inaccurate measurement of ECG features can result in misdiagnoses, delayed treatment, or unnecessary interventions. Therefore, validating the accuracy, reproducibility, and clinical reliability of extracted features is fundamental to device certification and continued regulatory compliance [35].

Feature extraction algorithms must be evaluated using both quantitative and clinical validation metrics. Standard validation parameters include:

- Sensitivity and Specificity for detecting critical events (e.g., QRS, T wave ends)
- Mean Absolute Error (MAE) and Root Mean Square Error (RMSE) for time interval predictions
- Bland–Altman plots comparing algorithmic measurements to expert annotations
- Correlation coefficients (r) for amplitude and interval consistency

These metrics are commonly evaluated against standardized datasets such as MIT-BIH Arrhythmia, European ST-T Database, and QT Database [12].

Regulatory Guidelines and Conformity Assessment

According to IEC 60,601–2-25 and FDA guidance for diagnostic ECG devices, waveform features like QRS duration, QT interval, and ST-segment deviations must fall within a predefined error tolerance (typically $\pm$ 10 ms for time intervals). Automated systems must prove compliance through both bench testing and real-world validation. A novel method of conformity testing in which features extracted from ECG devices were compared to gold-standard values across a variety of physiological and noise conditions. Devices that failed to detect QT prolongation or inaccurately flagged arrhythmic events were deemed non-compliant under EU MDR criteria.

AI and Black-Box Validation Challenges

With the rise of machine learning (ML) and AI in ECG interpretation, validating extracted features becomes even more complex. Deep learning models may use intermediate representations that are not explicitly visible or measurable. Regulatory bodies are increasingly asking for explainable AI (XAI) methods, where key features influencing classification must be accessible and interpretable. By combining traditional ECG features (e.g., QRS duration, ST elevation) with model-attribution maps, their system satisfied both technical performance and regulatory explainability standards [84, 85]. Validation does not end with product launch. Post-market surveillance (PMS) requires continuous monitoring of ECG feature performance over time and across populations. Anomalies in extracted features—such as systematic QT shortening or PR interval drift—can indicate hardware degradation or software bugs.

To ensure robust and compliant feature extraction:

- Use multi-site clinical validation across varied demographics and conditions.
- Integrate preprocessing and feature extraction in a closed-loop system to prevent error propagation.
- Adopt standardized reporting formats for features (e.g., HL7 annotations).
- Implement built-in self-tests for signal quality and feature reliability.
- Align design with regulatory checklists from FDA, CE, and ISO 13485.

3.3.13 Case Studies and Applications in Clinical Practice

Feature extraction plays a vital role in diverse clinical workflows. From arrhythmia detection and ischemia monitoring to risk stratification and therapy guidance, accurately extracted ECG features translate directly into diagnostic and therapeutic decisions. The following case studies highlight practical implementations of ECG feature extraction across a variety of medical domains.

Arrhythmia Detection in Ambulatory Monitoring

Ambulatory ECG monitoring—such as Holter monitoring and wearable ECG patches—requires real-time detection of abnormal rhythms. Extracted features such as RR interval variability, P wave presence, and QRS morphology are central to identifying atrial fibrillation, ventricular tachycardia, and premature contractions. Wearable ECG monitors were evaluated for their ability to identify paroxysmal atrial fibrillation episodes in outpatient settings. Devices that incorporated adaptive QRS detection and HRV feature tracking achieved higher sensitivity and lower false-positive rates compared to those relying solely on RR interval regularity [86].

QT Interval Monitoring for Drug Safety

Prolonged QT intervals are associated with life-threatening arrhythmias such as torsades de pointes. ECG feature extraction systems are used in both hospital monitoring and clinical drug trials to track QT dynamics. Case reports showed that defibrillators and telemetry systems with poor QT end delineation led to systematic QT underestimation, raising safety concerns during conformity assessments. In contrast, systems incorporating dynamic T wave endpoint detection algorithms achieved consistent QT measurement under noise and artifacts [9, 10].

Neonatal and Pediatric ECG Interpretation

Neonatal ECGs differ markedly from adult ECGs in morphology, amplitude, and frequency. Feature extraction systems in neonatal intensive care units (NICUs) must be customized to detect low-amplitude, high-rate waveforms [87]. Feature extraction failures were most frequent in P wave and PR interval identification. Systems using multi-lead template matching demonstrated improved performance and regulatory compliance in these sensitive use cases.

Predictive Analytics and Risk Scoring

Combining extracted ECG features with clinical data enables the construction of predictive models for cardiac risk. For instance, features like QT dispersion, QRS fragmentation, and T wave alternans have been associated with higher arrhythmic risk. In COVID-19 patients, the use of AI-enhanced ECG feature extraction to stratify patients by likelihood of deterioration. Systems that included both standard waveform intervals and non-linear HRV features provided early warning of respiratory decline [85].

Smart Wearables in Population Screening

Smartwatches and wearable biosensors increasingly incorporate ECG functionality for population-level screening. These devices extract single-lead features to detect atrial fibrillation, bradycardia, and prolonged QT. Devices with real-time QT and PR interval tracking, coupled with onboard anomaly detection, significantly improved diagnostic yield without increasing alert fatigue [50, 79].

Integration in Electronic Health Records (EHR)

Modern ECG systems export features into EHRs for physician review, decision support, and longitudinal trend analysis. Standards like HL7 aECG ensure interoperability. Feature-based alerts (e.g., "QRS duration > 120 ms for three consecutive beats") are increasingly used in ICU and emergency department settings to trigger rapid response protocols [41, 45]. Regulatory recommendations encourage integrated validation of these decision pathways during conformity assessments.

3.4 Clinical Applications of ECG Analysis

Electrocardiography is one of the most widely used diagnostic tools in medicine. With the proliferation of digital healthcare systems, ECG analysis has evolved far beyond traditional waveform interpretation. Today, the clinical application of ECG spans acute care diagnostics, long-term disease management, preventive screening, and personalized health monitoring. This chapter presents a comprehensive review of how ECG analysis is applied in clinical practice, focusing on both conventional and emerging use cases. Feature-rich ECG data, enabled by robust signal acquisition, preprocessing, and extraction, provides clinicians with a wealth of information to guide decisions. In addition to diagnosing cardiac arrhythmias, ischemic heart disease, and conduction disorders, ECG analysis plays a growing role in predicting adverse events, guiding pharmacological treatment, and evaluating therapy outcomes [88, 89].

This chapter highlights key clinical applications, decision algorithms, patient populations, and integration with electronic health records (EHRs) and artificial intelligence (AI)-driven tools.

3.4.1 Diagnosis of Cardiac Arrhythmias

Accurate ECG interpretation remains the gold standard for diagnosing a wide range of arrhythmias. These include:

- Atrial fibrillation (AF): Characterized by absence of P waves and irregular RR intervals.
- Ventricular tachycardia (VT): Identified by vast QRS complexes and rapid heart rate.
- Supraventricular tachycardias (SVTs): Distinguished by narrow QRS and abnormal P wave placement.
- Bradyarrhythmias: Detected through prolonged RR intervals and AV dissociation.

Modern ECG systems support automated arrhythmia classification based on extracted features such as RR variability, QRS morphology, and atrial activity indicators. Studies have shown that multi-lead detection and rhythm context algorithms significantly enhance diagnostic accuracy [7, 90]. Regulatory assessments found that arrhythmia detection accuracy varied across commercial monitors. Devices that incorporated real-time adaptive feature tracking and contextual rule engines performed better in distinguishing atrial from ventricular arrhythmias under noisy or ambulatory conditions. AI-assisted algorithms have shown promise in identifying subtle rhythm patterns. For example, convolutional neural networks trained on large ECG datasets have achieved cardiologist-level performance in detecting AF, flutter, and high-grade AV blocks [91, 92].

3.4.2 Myocardial Ischemia and Infarction Detection

The electrocardiogram remains a frontline diagnostic tool for identifying myocardial ischemia and acute coronary syndromes. ST-segment changes, T wave inversion, and Q wave formation provide critical insights into the location, severity, and timing of ischemic events. These features are not only diagnostic but also prognostic, influencing triage decisions, therapy selection, and discharge planning.

ST-Segment Analysis

ST-segment elevation in contiguous leads is the hallmark of STEMI. Non-ST elevation (NSTEMI) and unstable angina present more subtly, often with ST depression or transient T wave changes. Automated ECG systems analyze these deviations in amplitude and duration relative to isoelectric baselines to determine ischemic risk. Accurate ST-segment evaluation requires high-fidelity signal processing and baseline correction, especially in patients with pre-existing conduction abnormalities or pacemaker rhythms. Conformity testing found that several ECG monitors failed to detect subtle ST elevation during stress tests, primarily due to inadequate preprocessing algorithms [4].

T Wave Morphology and Dynamic Changes

T wave abnormalities—such as peaking, flattening, or inversion—are often the earliest indicators of ischemia. In the context of evolving infarction, T wave changes may precede ST deviation or persist during recovery. Real-time tracking of T wave evolution helps clinicians distinguish acute ischemia from chronic repolarization abnormalities. AI-enhanced ECG systems that incorporate dynamic T wave modeling have shown increased sensitivity in detecting posterior and high lateral infarctions [83, 86].

Q Waves and Infarct Characterization

The development of pathological Q waves indicates transmural infarction and suggests infarct permanence. These features are vital for localizing infarct zones and estimating the extent of myocardial damage. Modern ECG platforms integrate Q wave analysis with reciprocal lead evaluation and echocardiographic data to guide reperfusion decisions. In post-market surveillance, some portable ECG systems lacked adequate resolution to distinguish small Q waves in early anterior infarctions. ECG features extracted during chest pain evaluation feed into clinical scoring systems like the TIMI risk score and GRACE score. Accurate identification of ST-segment deviations and QT changes contributes to early risk stratification and reduces time to intervention [10]. Integration with emergency department workflows is critical. Systems capable of transmitting digitized ECGs directly into cloud-based interpretation platforms have improved door-to-balloon times for STEMI patients [93]. AI-based tools have demonstrated high accuracy in detecting silent myocardial infarction using resting ECG features [94, 95]. These advancements highlight the future of automated ischemia screening in asymptomatic or atypical populations.

3.4.3 ECG in Heart Failure and Structural Heart Disease

Electrocardiography is a critical diagnostic and prognostic tool in managing patients with heart failure (HF) and structural cardiac abnormalities. While imaging modalities such as echocardiography and MRI provide detailed anatomical information, the ECG offers rapid, non-invasive insight into the electrical consequences of mechanical dysfunction.

One of the most widely studied ECG features in heart failure is QRS duration and morphology. A QRS duration greater than 120 ms, especially with left bundle branch block (LBBB) morphology, is a strong predictor of adverse outcomes in HF and is often an indication for cardiac resynchronization therapy (CRT). Amplitude-based features, such as low QRS voltage across limb or precordial leads, are also associated with ventricular remodeling, fluid overload, and cardiac amyloidosis. Clinical trials have consistently shown that ECG findings correlate with echocardiographic measures of left ventricular ejection fraction (LVEF) and diastolic dysfunction [18, 96]. Inaccuracies in delineating the start and end of QRS complexes were found to delay appropriate intervention in some conformity test cases.

Voltage criteria for left ventricular hypertrophy (LVH), including the Sokolow-Lyon and Cornell indices, remain commonly used in ECG interpretation. ST-T changes associated with LVH ("strain pattern") also have prognostic value, indicating subendocardial ischemia due to increased wall stress. These features are often tracked longitudinally in hypertensive and aortic stenosis patients. ECG-based monitoring of LVH regression following antihypertensive therapy is a cost-effective approach in outpatient management [97, 98].

Bundle branch blocks, AV blocks, and fragmented QRS complexes (fQRS) are all indicators of electrical dyssynchrony and myocardial scarring. The presence of fQRS on a standard 12-lead ECG has been associated with increased arrhythmic risk and worse outcomes in patients with ischemic and non-ischemic cardiomyopathies [12, 99].

ECG systems must reliably capture these features, even in the presence of baseline wander or low-amplitude signals.

While an ECG alone is not sufficient to diagnose structural disease, it contributes significantly to multi-modal risk scoring models. For example, the Seattle Heart Failure Model and MADIT risk score incorporate ECG parameters such as QRS duration and PR interval into mortality predictions and device therapy decisions. Artificial intelligence is now being used to enhance ECG-based predictions of structural disease. Deep learning algorithms trained on ECG data have been shown to predict reduced ejection fraction with high sensitivity—even in patients with normal QRS duration—highlighting the evolving potential of non-invasive electrical biomarkers [8, 100].

3.4.4 ECG in Monitoring Drug Effects and Electrolyte Imbalance

Electrocardiography is a frontline method for assessing the cardiac effects of medications and electrolyte disturbances. Many commonly used drugs—including antiarrhythmics, antibiotics, and psychotropics—can alter cardiac electrophysiology, leading to changes in waveform morphology or interval duration. Similarly, electrolyte imbalances such as hyperkalemia or hypocalcemia manifest distinct patterns on the ECG.

One of the most critical ECG applications in pharmacovigilance is monitoring the QT interval. Drug-induced QT prolongation is associated with an increased risk of torsades de pointes and sudden cardiac death. Medications such as sotalol, amiodarone, erythromycin, and antipsychotics (e.g., haloperidol) have been implicated in QT prolongation. Regulatory authorities, including the FDA and EMA, mandate pre- and post-market QT monitoring for many drug classes [101]. In the post-market surveillance framework, the ECG systems used in hospitals and clinical trials must consistently detect QT changes with sub-10-ms error margins. Devices unable to delineate the end of the T wave accurately were flagged as non-compliant in conformity assessments.

Electrolyte levels—particularly of potassium, calcium, and magnesium—have direct electrophysiological consequences:

- Hyperkalemia: Peaked T waves, widened QRS, and flattened P waves.
- Hypokalemia: ST depression, flattened T waves, and U wave prominence.
- Hypocalcemia: QT prolongation due to delayed ventricular repolarization.
- Hypercalcemia: QT shortening.

Real-time ECG monitoring in emergency departments and critical care units often provides the first evidence of these imbalances. Automated systems that extract and track relevant ECG features (T wave amplitude, QT interval) are essential in guiding prompt therapy [43, 102]. Advanced ECG algorithms can estimate potassium levels based solely on ECG morphology—a technique gaining traction in remote health monitoring.

In patients with complex medication regimens, ECG feature tracking helps clinicians detect dangerous interactions. For example, the combination of a QT-prolonging antibiotic and a loop diuretic (causing hypokalemia) exponentially increases arrhythmic risk. Smart ECG systems integrated with electronic medical records (EMRs) can cross-reference active medications with ECG findings.

In clinical trials and hospital settings, serial ECG monitoring is used to track the evolution of drug-induced effects over time. Extracted features such as QTc interval, Tpeak-Tend, and QRS duration are trended to determine dose-related changes.

Standardized ECG recording and annotation protocols (e.g., ICH E14) are required in cardiotoxicity studies. Devices used in these studies must undergo rigorous validation.

3.4.5 ECG-Guided Therapy, Prognosis, and Population Screening

As cardiovascular care becomes increasingly personalized, the ECG is emerging not only as a diagnostic tool but also as a guide for therapy selection, outcome prediction, and population-wide screening initiatives. ECG-guided care supports informed clinical decisions, optimizes patient management, and identifies at-risk individuals before the onset of symptoms.

Clinical guidelines frequently use ECG criteria to guide pharmacologic interventions and device therapy. For instance:

- Antiarrhythmic therapy: Class I and III agents require close ECG monitoring for PR, QRS, and QT interval effects.
- CRT eligibility:
- ICD implantation: Involves ECG-derived risk factors such as QRS duration, fQRS, and T wave alternans.

Devices evaluated in CRT clinical environments must demonstrate precise and reproducible QRS measurements to support evidence-based therapy allocation [37, 85].

Several risk stratification models integrate ECG features with clinical variables to predict outcomes:

- TIMI and GRACE scores: Use ST changes, heart rate, and QT interval to estimate mortality risk.
- Seattle Heart Failure Model: Includes QRS duration and resting heart rate.
- Sudden cardiac death predictors: Incorporate T wave alternans and QT dispersion.

Modern AI models have expanded prognostic use. A deep learning model applied to 12-lead ECGs could predict reduced left ventricular ejection fraction (LVEF) with over 90% sensitivity. Integration of ECG-based prediction tools into EHR systems is transforming how patients are triaged and monitored.

ECG-based screening is increasingly used to detect silent cardiac disease in asymptomatic populations:

- Athlete screening: Identifies hypertrophic cardiomyopathy, WPW syndrome, and prolonged QT.
- Occupational fitness exams: Evaluate cardiovascular risk in high-stress professions.
- Community-based health fairs: Provide early arrhythmia or ischemia detection.

The integration of automated feature extraction improved detection of atrial fibrillation and prolonged QT in remote and underserved populations [103, 104].

Wearable ECG systems, including smartwatches and biosensor patches, now deliver continuous monitoring outside traditional clinical settings. These systems analyze ECG features to detect deviations in baseline physiology, enabling:

- Early alerts for arrhythmic episodes

- Medication adherence tracking
- Post-discharge surveillance

The studies demonstrated that home-based QT monitoring facilitated the earlier identification of cardiac decompensation. Real-time feature streaming to mobile apps and cloud analytics platforms enables closed-loop systems, where ECG findings directly inform patient behavior, medication adjustments, and physician alerts.

References

1. Anderson RH, Sarwark A, Spicer DE, Backer CL (2014) Exercises in anatomy: the normal heart. Multimed Man Cardiothorac Surg 2014:mmu023
2. Hai JJ, Lachman N, Syed FF, Desimone CV, Asirvatham SJ (2014) The anatomic basis for ventricular arrhythmia in the normal heart: what the student of anatomy needs to know. Clin Anat 27(6):885–893
3. Xiong G, Sun P, Zhou H, Ha S, Hartaigh BO, Truong QA, Min JK (2017) Comprehensive modeling and visualization of cardiac anatomy and physiology from CT imaging and computer simulations. IEEE Trans Vis Comput Graph 23(2):1014–1028
4. Hundur M, Spahić L, Bećirović F, Gurbeta Pokvić L, Badnjević A (2025) Advancement of post-market surveillance of medical devices leveraging artificial intelligence: ECG devices case study. Technol Health Care
5. Hakacova N (2016) The ability of mitral papillary muscle positions to explain QRS complex characteristics in humans. J Electrocardiol 49(5):680–685
6. Triposkiadis F, Xanthopoulos A, Boudoulas KD, Giamouzis G, Boudoulas H, Skoularigis J (2022) The interventricular septum: structure, function, dysfunction, and diseases. J Clin Med 11(11):3227
7. Badnjevic A, Deumic A, Ademovic A, Pokvic LG (2023) A novel method for conformity assessment testing of therapeutic ultrasounds for post-market surveillance purposes. Technol Health Care 31(1):339–346
8. Pokvić LG, Deumić A, Softić A, Badnjević A (2024) Application of artificial intelligence for management of maintenance and prediction of performances. In: Badnjević A, Cifrek M, Magjarević R, Džemić Z (eds) Inspection of Medical Devices. Springer, pp 97–118
9. Ambhore A, Teo SG, Bin Omar AR, Poh KK (2014) Importance of QT interval in clinical practice. Singapore Med J 55(12):607–612
10. Magjarević R, Badnjević A (2023) Inspection and testing of electrocardiographs (ECG) devices. In: Badnjević A, Cifrek M, Magjarević R, Džemić Z, eds. Inspection of Medical Devices: For Regulatory Purposes. Springer, pp 139–171
11. Bagliani G, Leonelli FM, De Ponti R, Mesolella E, Padeletti L (2018) Atrioventricular nodal conduction disease. Card Electrophysiol Clin. 10(2):197–209
12. Spahić L, Jeremić L, Lalatović I, Muratović T, Džuho A, Gurbeta Pokvić L, Badnjević A (2024) Machine learning for improved medical device management: a focus on defibrillator performance. Technol Health Care
13. Cai X, Huang L (2019) Cardiac sympathetic innervation and arrhythmogenesis. J Physiol 597(17):4445–4446
14. Chakrabarti S, Stuart AG (2005) Understanding cardiac arrhythmias. Arch Dis Child 90(10):1086–1090
15. Clayton RH, Panfilov AV (2008) A guide to modelling cardiac electrical activity in anatomically detailed ventricles. Prog Biophys Mol Biol 96(1–3):19–43
16. Corrado D, Basso C, Judge DP (2017) Arrhythmogenic cardiomyopathy. Circ Res 121(7):784–802

17. Yanagawa N, Nakajima Y (2009) A simple dissection method for the conduction system of the human heart. Anat Sci Educ 2(2):78–80
18. Bećirović F, Spahić L, Merdović N, Gurbeta Pokvić L, Badnjević A (2025) Advancement of post-market surveillance of medical devices leveraging artificial intelligence: Patient monitors case study. Technol Health Care 33(2):974–980
19. Medjedovic E, Stanojevic M, Jonuzovic-Prosic S, Ribic E, Begic Z, Cerovac A, Badnjevic A (2024) Artificial intelligence as a new answer to old challenges in maternal-fetal medicine and obstetrics. Technol Health Care 32(3):1273–1287
20. Gonska BD (2008) Standard-EKG. Herzschrittmacherther Elektrophysiol 19(3):89–97
21. Singh DK, Peter CT (2016) Use of the surface electrocardiogram to define the nature of challenging arrhythmias. Card Electrophysiol Clin. 8(1):1–24
22. Saremi F, Krishnan S (2007) Cardiac conduction system: Anatomic landmarks relevant to interventional electrophysiologic techniques demonstrated with 64-detector CT. Radiographics 27(6):1539–1567
23. Badnjević A, Pokvić LG, Džemić Z, Bečić F (2020) Risks of emergency use authorizations for medical products during outbreak situations: A COVID-19 case study. Biomed Eng Online 19(1):75
24. Roberts A (2002) The role of anatomy and physiology in interpreting ECGs. Nurs Times 98(20):34–36
25. Ho SY (2009) Anatomic insights for catheter ablation of ventricular tachycardia. Heart Rhythm 6(8 Suppl):S77–S80
26. Karki R, Raina A, Ezzeddine FM, Bois MC, Asirvatham SJ (2023) Anatomy and pathology of the cardiac conduction system. Cardiol Clin 41(3):277–292
27. Nan J, Sugrue A, Ladas TP, Mehra N, Asirvatham SJ (2019) Anatomic considerations relevant to atrial and ventricular arrhythmias. Card Electrophysiol Clin. 11(3):421–432
28. Moak JH, Muck AE, Brady WJ (2024) ST-segment elevation myocardial infarction mimics: The differential diagnosis of nonacute coronary syndrome causes of ST-segment/T-wave abnormalities in the chest pain patient. Turk J Emerg Med. 24(4):206–217
29. Ng J, Christensen M (2024) Registered nurses' knowledge and interpretation of ECG rhythms: a cross-sectional study. Nurs Crit Care 29(5):1032–1039
30. Katsouras CS, Greakas GF, Goudevenos JA, Michalis LK, Kolettis T, Economides C, Argyri U, Pappas S, Sideris DA (2004) Localization of accessory pathways by the electrocardiogram: agreement of three algorithms. Pacing Clin Electrophysiol 27(2):189–193
31. Kennedy A, Finlay DD, Guldenring D, Bond R, Moran K, McLaughlin J (2016) The cardiac conduction system: generation and conduction of the cardiac impulse. Crit Care Nurs Clin North Am 28(3):269–279
32. Oh IY, Cha MJ, Lee TH, Seo JW, Oh S (2018) Unsolved questions on the anatomy of the ventricular conduction system. Korean Circ J. 48(11):1015–1026
33. Mirzoyev S, McLeod CJ, Asirvatham SJ (2010) Embryology of the conduction system for the electrophysiologist. Indian Pacing Electrophysiol J. 10(8):329–338
34. Platonov PG (2012) P-wave morphology: underlying mechanisms and clinical implications. Ann Noninvasive Electrocardiol 17(3):161–169
35. Liu P, Wang Q, Sun H, Qin X, Zheng Q (2021) Left bundle branch pacing: current knowledge and future prospects. Front Cardiovasc Med. 8:630399
36. Finlay DD, Nugent CD, Donnelly MP, Lux RL (2010) Eigenleads: ECG leads for maximizing information capture and improving SNR. IEEE Trans Inf Technol Biomed 14(1):69–78
37. Merdović N, Spahić L, Hundur M, Pokvić LG, Badnjević A (2024) Advancement of post-market surveillance of medical devices leveraging artificial intelligence: infusion pumps case study. Technol Health Care 33(2):915–921
38. Smith GH, Van den Heever DJ, Swart W (2021) The reconstruction of a 12-lead electrocardiogram from a reduced lead set using a focus time-delay neural network. Acta Cardiol Sin. 37(1):47–57
39. Batchvarov VN, Behr ER (2014) Clinical utility of computed electrocardiographic leads. J Electrocardiol 47(3):281–287

40. Madias JE (2013) A proposal for a 9-lead electrocardiogram recorded via the Wilson's central terminal. Ann Noninvasive Electrocardiol 18(2):103–106
41. Sampson M (2025) How to accurately record a 12-lead ECG. Nurs Stand. https://doi.org/10.7748/ns.2025.e12512
42. Peruničić Ž, Lalatović I, Spahić L, Ašić A, Pokvić LG, Badnjević A (2024) Enhancing mechanical ventilator reliability through machine learning-based predictive maintenance. Technol Health Care
43. Martinović M, Kosović M, Spahić L, Softić A, Pokvić LG, Badnjević A (2025) Machine learning for improved medical device management: a focus on dialysis machines. Technol Health Care
44. Brodnick D (2000) A method to locate electrode placement. J Electrocardiol 33(Suppl):211–218
45. Jevon P (2010) Procedure for recording a standard 12-lead electrocardiogram. Br J Nurs 19(10):649–651
46. Spahić L, Sredović U, Kurpejović Z, Mrdanović E, Pokvić G, Badnjević A (2025) Machine learning for improved medical device management: a focus on infant incubators. Technol Health Care
47. Haas NA, Driscoll DJ, Rickert-Sperling S (2024) Clinical presentation and therapy of arrhythmias. Adv Exp Med Biol 1441:1023–1031
48. Tomašić I, Trobec R (2014) Electrocardiographic systems with reduced numbers of leads–Synthesis of the 12-lead ECG. IEEE Rev Biomed Eng 7:126–142
49. Kennedy A, Finlay DD, Guldenring D, Bond RR, McEneaney DJ, Peace A, McLaughlin J (2015) Improved recording of atrial activity by modified bipolar leads derived from the 12-lead electrocardiogram. J Electrocardiol 48(6):1017–1021
50. Samol A, Bischof K, Luani B, Pascut D, Wiemer M, Kaese S (2019) Single-lead ECG recordings including Einthoven and Wilson leads by a smartwatch: a new era of patient-directed early ECG differential diagnosis of cardiac diseases? Sensors 19(20):4377
51. Shah AJ, Lim HS, Yamashita S, Zellerhoff S, Berte B, Mahida S, Hooks D, Aljefairi N, Derval N, Denis A, Sacher F, Jais P, Dubois R, Hocini M, Haissaguerre M (2014) Non-invasive ECG mapping to guide catheter ablation. J Atr Fibrillation. 7(3):1139
52. Man S, Maan AC, Schalij MJ, Swenne CA (2015) Vectorcardiographic diagnostic and prognostic information derived from the 12-lead electrocardiogram: historical review and clinical perspective. J Electrocardiol 48(4):463–475
53. Sanoski CA (2010) Overview of electrocardiographic findings and clinical presentation of common cardiac arrhythmias. Am J Health Syst Pharm 67(9 Suppl 5):S5–S10
54. Francis J (2016) ECG monitoring leads and special leads. Indian Pacing Electrophysiol J. 16(3):92–95
55. Smith S, Maisrikrod S (2025) Wearable electrocardiogram technology: Help or hindrance to the modern doctor? JMIR Cardio. 9:e62719
56. Feild DQ, Feldman CL, Horácek BM (2002) Improved EASI coefficients: Their derivation, values, and performance. J Electrocardiol 35(Suppl):23–33
57. Trobec R, Tomašić I (2011) Synthesis of the 12-lead electrocardiogram from differential leads. IEEE Trans Inf Technol Biomed 15(4):615–621
58. Su L, Borov S, Zrenner B (2013) 12-lead Holter electrocardiography: Review of the literature and clinical application update. Herzschrittmacherther Elektrophysiol 24(2):92–96
59. Badnjević A, Pokvić LG, Deumić A, Bećirović LS (2022) Post-market surveillance of medical devices: a review. Technol Health Care 30(6):1315–1329
60. Kumar A, Tomar H, Mehla VK, Komaragiri R, Kumar M (2021) Stationary wavelet transform-based ECG signal denoising method. ISA Trans 114:251–262
61. Tobon DP, Falk TH (2018) Adaptive spectro-temporal filtering for electrocardiogram signal enhancement. IEEE J Biomed Health Inform 22(4):1177–1185
62. Lin WH, Wong MYM, Pu LN, Zhang YT (2010) Comparison of median filter and discrete dyadic wavelet transform for noise cancellation in electrocardiogram. Proc IEEE EMBS 6218–6221

63. Azzouz A, Bengherbia B, Wira P, Alaoui N, Souahlia A, Maazouz M, Hentabeli H (2024) An efficient ECG signal denoising technique based on the combination of particle swarm optimisation and wavelet transform. Heliyon. 10(5):e26171
64. An X, Liu Y, Zhao Y, Lu S, Stylios GK, Liu Q (2022) Adaptive motion artifact reduction in wearable ECG measurements using impedance pneumography signal. Sensors 22(15):5493
65. Singh O, Sunkaria RK (2015) Powerline interference reduction in ECG signals using empirical wavelet transform and adaptive filtering. J Med Eng Technol 39(1):60–68
66. Jia Y, Pei H, Liang J, Zhou Y, Yang Y, Cui Y, Xiang M (2024) Preprocessing and denoising techniques for electrocardiography and magnetocardiography: a review. Bioeng 11(11):1109
67. Mir HY, Singh O (2021) ECG denoising and feature extraction techniques: a review. J Med Eng Technol 45(8):672–684
68. Xu X, Liang Y, He P, Yang J (2019) Adaptive motion artifact reduction based on empirical wavelet transform and wavelet thresholding for the non-contact ECG monitoring systems. Sensors 19(13):2874
69. Han G, Xu Z (2016) Electrocardiogram signal denoising based on a new improved wavelet thresholding. Rev Sci Instrum 87(8):084303
70. Singh O, Sunkaria RK (2017) ECG signal denoising via empirical wavelet transform. Fungal Biol 121(2):125–132
71. Qaisar SM (2020) Baseline wander and power-line interference elimination of ECG signals using efficient signal-piloted filtering. Healthc Technol Lett. 7(5):132–136
72. Saleem S, Khandoker AH, Alkhodari M, Hadjileontiadis LJ, Jelinek HF (2022) A two-step pre-processing tool to remove Gaussian and ectopic noise for heart rate variability analysis. Sci Rep 12(1):18345
73. Hafizović L, Čaušević A, Deumić A, Bećirović LS, Gurbeta L, Badnjević A (2021) The use of artificial intelligence in diagnostic medical imaging: systematic literature review. In: Proceeding IEEE 21st international conference Bioinformatics Bioeng (BIBE), pp 1–6
74. Singh AK, Krishnan S (2023) ECG signal feature extraction: trends in methods and applications. Biomed Eng Online
75. Hong S, Zhou Y, Shang J, Xiao C, Sun J (2020) Opportunities and challenges of deep learning methods for electrocardiogram data: a systematic review. Comput Biol Med
76. Darmawahyuni A, Nurmaini S, Tutuko B, Rachmatullah MN, Firdaus F, Sapitri AI, Perwira MI (2024) An improved electrocardiogram arrhythmia classification performance with feature optimization. BMC Med Inform Decis Mak
77. Joung C, Kim M, Paik T, Kong SH, Oh SY, Jeon WK, van Koert O (2024) Deep learning-based ECG segmentation for delineation of diverse arrhythmias. PLoS One
78. Wu W, Huang Y, Wu X (2022) A new deep learning method with self-supervised learning for delineation of the electrocardiogram. Entropy 24(12):1785
79. Wang R, Veera SCM, Asan O, Liao T (2024) A systematic review on the use of consumer-based ECG wearables on cardiac health monitoring. IEEE J Biomed Health Inform
80. Farrokhi S, Dargie W, Poellabauer C (2024) Reliable peak detection and feature extraction for wireless electrocardiograms. Comput Biol Med
81. Ding C, Yao T, Wu C, Ni J (2024) Advances in deep learning for personalized ECG diagnostics: a systematic review. Biosens Bioelectron
82. Chandra S, Sharma A, Singh GK (2018) Feature extraction of ECG signal. J Med Eng Technol
83. Sun JY, Shen H, Qu Q, Sun W, Kong XQ (2021) The application of deep learning in electrocardiogram: where we came from and where we should go? Int J Cardiol
84. Musa N, Gital AY, Aljojo N, Chiroma H, Adewole KS, Mojeed HA, Sikiru IA (2022) A systematic review and meta-data analysis on the applications of deep learning in electrocardiogram. J Ambient Intell Humaniz Comput
85. Lyon A, Mincholé A, Martínez JP, Laguna P, Rodriguez B (2018) Computational techniques for ECG analysis and interpretation in light of their contribution to medical advances. J R Soc Interface
86. Bayani A, Kargar M (2024) LDCNN: a new arrhythmia detection technique with ECG signals using a linear deep convolutional neural network. Physiol Rep

87. Savalia S, Emamian V (2018) Cardiac arrhythmia classification by multi-layer perceptron and convolution neural networks. Bioengineering 5(2):35
88. Qayyum SN, Iftikhar M, Rehan M, Khan GA, Khan M, Naeem R, Ansari RS, Ullah I, Noori S (2025) Revolutionizing electrocardiography: the role of artificial intelligence in modern cardiac diagnostics. Ann Med Surg
89. Tran HHV, Thu A, Fuertes A, Twayana AR, Mahadevaiah A, Mehta KA, James M, Basta M, Weissman S, Frishman WH, Aronow WS (2025) Electrocardiogram-based artificial intelligence for detection of low ejection fraction: a contemporary review. Cardiol Rev
90. Islam MS, Kalmady SV, Hindle A, Sandhu R, Sun W, Sepehrvand N, Greiner R, Kaul P (2024) Diagnostic and prognostic models based on electrocardiograms for rapid clinical applications. Can J Cardiol
91. Guan YX, An Y, Guo FY, Pan WB, Wang JX (2023) Intelligent electrocardiogram analysis in medicine: data, methods, and applications. Chin Med Sci J
92. Elyamani HA, Salem MA, Melgani F, Yhiea NM (2024) Deep residual 2D convolutional neural network for cardiovascular disease classification. Sci Rep
93. Nechita LC, Nechita A, Voipan AE, Voipan D, Debita M, Fulga A, Fulga I, Musat CL (2024) AI-enhanced ECG applications in cardiology: comprehensive insights from the current literature with a focus on COVID-19 and multiple cardiovascular conditions. Diagnostics
94. Schots BBS, Pizarro CS, Arends BKO, Oerlemans MIFJ, Ahmetagić D, van der Harst P, van Es R (2025) Deep learning for electrocardiogram interpretation: bench to bedside. Eur J Clin Invest
95. Kraik K, Dykiert IA, Niewiadomska J, Ziemer-Szymańska M, Mikołajczak K, Kreń M, Kukiełka P, Martuszewski A, Harych T, Poręba R, Gać P, Poręba M (2025) The most common errors in automatic ECG interpretation. Front Physiol
96. Attia ZI, Harmon DM, Behr ER, Friedman PA (2021) Application of artificial intelligence to the electrocardiogram. Eur Heart J
97. Ose B, Sattar Z, Gupta A, Toquica C, Harvey C, Noheria A (2024) Artificial intelligence interpretation of the electrocardiogram: a state-of-the-art review. Curr Cardiol Rep
98. Siontis KC, Noseworthy PA, Attia ZI, Friedman PA (2021) Artificial intelligence-enhanced electrocardiography in cardiovascular disease management. Nat Rev Cardiol
99. Chung CT, Lee S, King E, Liu T, Armoundas AA, Bazoukis G, Tse G (2022) Clinical significance, challenges and limitations in using artificial intelligence for electrocardiography-based diagnosis. Int J Arrhythmia
100. Xie L, Li Z, Zhou Y, He Y, Zhu J (2020) Computational diagnostic techniques for electrocardiogram signal analysis. Sensors
101. Rjoob K, Bond R, Finlay D, McGilligan V, Leslie SJ, Rababah A, Iftikhar A, Guldenring D, Knoery C, McShane A, Peace A, Macfarlane PW (2022) Machine learning and the electrocardiogram over two decades: time series and meta-analysis of the algorithms, evaluation metrics and applications. Artif Intell Med
102. Collecchia G (2023) Artificial intelligence applied to the electrocardiogram, or is there really a needle in a haystack? Recenti Prog Med
103. Mujkić A, Baralić E, Ombašić A, Bećirović LS, Pokvić LG, Badnjević A (2022) Machine intelligence in biomedical data modeling, processing, and analysis. In: Proceeding 11th Mediterr Conference Embedded Computing (MECO), pp 1–10
104. Vuković D, Šumar S, Dupovac S, Badnjević A (2024) Cost-effectiveness analysis of medical devices in legal metrology system. In: Badnjević A, Cifrek M, Magjarević R, Džemić Z (eds) Inspection of Medical Devices. Springer, pp 267–275

Chapter 4
Processing Respiratory Signals

4.1 Respiratory System Physiology

The human respiratory system is a highly specialized and finely tuned network that facilitates the exchange of oxygen and carbon dioxide, ensuring that cells receive the oxygen necessary for metabolic processes while expelling waste gases produced by cellular respiration. This function is fundamental to sustaining life, and any disruption can have severe consequences for overall health. Respiration is more than just breathing; it encompasses a complex interplay of mechanical, neural, and biochemical processes. The efficiency of the respiratory system depends on various factors, including lung structure, airway resistance, and neural control mechanisms that regulate breathing rate and depth. As the body's oxygen demands change—such as during exercise, sleep, or illness—the respiratory system dynamically adapts to maintain homeostasis [1, 2].

This chapter examines the fundamental physiological processes that regulate respiration, encompassing the mechanics of the breathing cycle, lung volumes and capacities, and the intricate forces that drive pulmonary ventilation. Through an integrated approach, we will examine how these processes function in health, how they can be disrupted in disease, and how they are measured in clinical settings.

The act of breathing, or pulmonary ventilation, is a continuous and rhythmic process that enables gas exchange between the external environment and the bloodstream. Each cycle consists of two distinct phases: inspiration and expiration. These movements are driven by changes in pressure within the thoracic cavity, which in turn are controlled by the coordinated actions of the diaphragm and intercostal muscles [3].

Inspiration is an active process that requires muscle contraction to increase the volume of the thoracic cavity. The diaphragm, a dome-shaped muscle located at the base of the lungs, plays the primary role in this phase. When it contracts, it flattens and moves downward, expanding the thoracic cavity and reducing intrapulmonary pressure relative to the external atmospheric pressure. Simultaneously, the external

© The Author(s), under exclusive license to Springer Nature Switzerland AG 2026
A. Badnjević and L. Spahić, *Biomedical Signals and Systems*, Series in BioEngineering,
https://doi.org/10.1007/978-3-032-10982-8_4

intercostal muscles, which are situated between the ribs, contract to lift the ribcage and further expand the chest. As the thoracic cavity enlarges, the lungs also expand due to their attachment to the pleural membranes. This expansion creates a negative pressure gradient, drawing air into the lungs through the airways. The degree of inspiratory effort depends on factors such as physical activity and metabolic demands. During quiet breathing, inspiration occurs passively with minimal muscular effort. However, during intense exercise, additional muscles, including the sternocleidomastoid and scalene muscles of the neck, contribute to increasing lung expansion and enhancing airflow [4, 5].

Under normal resting conditions, expiration is a passive process that occurs as the diaphragm and intercostal muscles relax. The elastic properties of the lungs and chest wall cause them to return to their resting shape, reducing thoracic volume and increasing intrapulmonary pressure. This pressure gradient forces air out of the lungs and into the external environment [3]. In contrast, during forced expiration, such as during vigorous exercise or coughing, additional muscles help expel air more forcefully. The internal intercostal muscles contract to pull the ribs downward, while the abdominal muscles push the diaphragm upward by compressing the abdominal organs. This active form of expiration is particularly important in conditions where rapid gas exchange is needed, such as during strenuous physical exertion or when clearing the airways of mucus or irritants [6, 7] (Fig. 4.1).

The breathing cycle is tightly regulated by neural mechanisms located in the brainstem, which adjust breathing rate and depth in response to chemical signals from the blood. Levels of carbon dioxide, oxygen, and pH play a crucial role in signaling the respiratory centers to modify ventilation accordingly [6, 8]. The volume of air that the lungs move during respiration varies depending on the body's oxygen demands. These volumes and their combinations, known as lung capacities, provide valuable insight into respiratory function. Lung volumes describe distinct phases of respiration and are measured using spirometry, a technique that records airflow and lung expansion during breathing [3, 9] (Fig. 4.2).

- Tidal volume (TV) represents the amount of air exchanged during a normal, resting breath. In a healthy adult, this value averages around 500 ml per breath.
- Inspiratory reserve volume (IRV) is the additional air that can be inhaled beyond an everyday tidal inspiration, typically around 3,000 ml.
- Expiratory reserve volume (ERV) is the extra amount of air that can be forcibly exhaled after a normal breath, averaging 1,100 ml.
- Residual volume (RV) is the air that remains in the lungs even after maximal expiration, preventing lung collapse and maintaining alveolar inflation. This volume is approximately 1,200 ml.

Lung capacities represent the sum of two or more lung volumes, providing a broader view of pulmonary function [10].

- Inspiratory capacity (IC) is the maximum volume of air that can be inspired after a normal expiration and is calculated as TV + IRV.

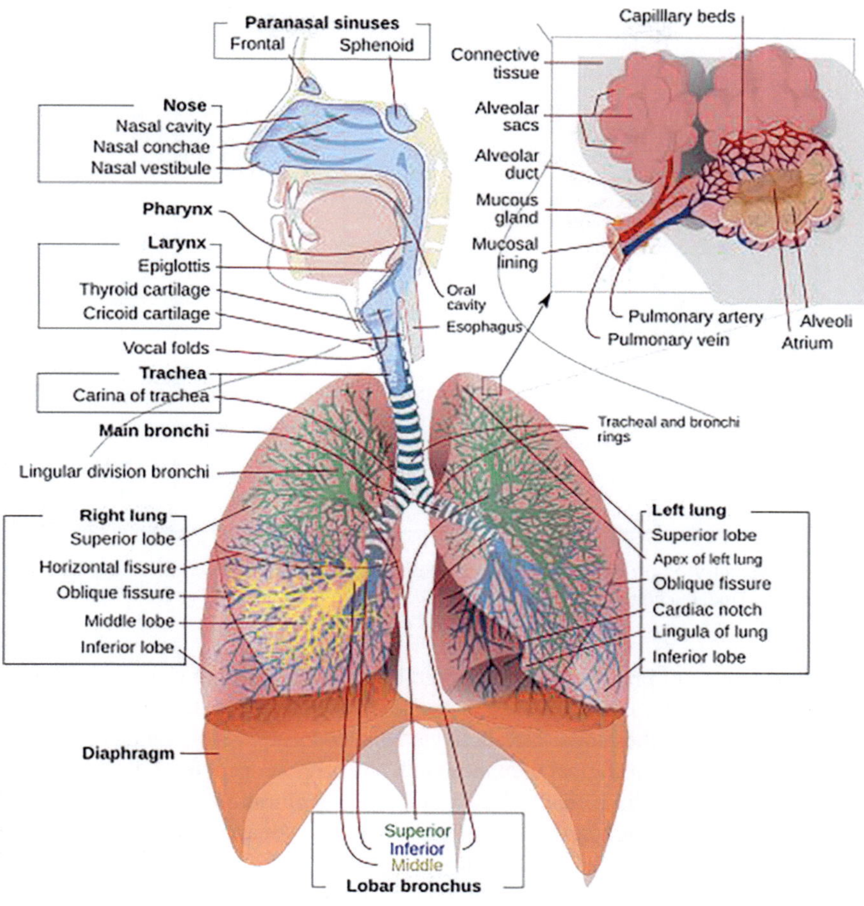

Fig. 4.1 Diagram of the respiratory system. https://commons.wikimedia.org/wiki/File:Respiratory_system_complete_en.svg

- Functional residual capacity (FRC) is the volume of air remaining in the lungs after a normal expiration, composed of ERV and RV.
- Vital capacity (VC) is the total amount of air that can be exhaled after a maximal inhalation, an essential parameter in assessing respiratory efficiency.
- Total lung capacity (TLC) is the sum of all lung volumes and represents the total air the lungs can hold.

Pulmonary diseases, such as chronic obstructive pulmonary disease (COPD) or restrictive lung disorders, significantly alter these lung volumes and capacities. Spirometry tests are commonly used in clinical settings to detect abnormalities in these values, aiding in the diagnosis and management of respiratory conditions [11, 12].

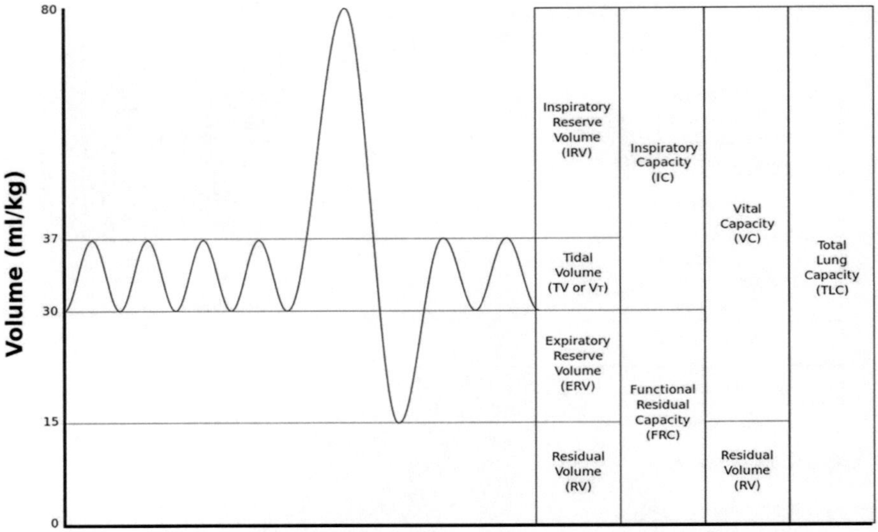

Fig. 4.2 Spirometry graph displaying lung volumes. https://commons.wikimedia.org/wiki/File:Lungvolumes_Updated.png

The mechanics of breathing are governed by the interaction of pressure differentials, lung compliance, and airway resistance, all of which contribute to the movement of air into and out of the lungs. This delicate balance ensures efficient gas exchange while adapting to physiological demands such as exercise, speech, or disease conditions. Understanding these mechanics requires an exploration of the fundamental forces at play, including pressure gradients and the dynamic changes that occur throughout the breathing cycle [13, 14].

Breathing is driven by the principle that air flows from areas of higher pressure to areas of lower pressure. This movement is dictated by Boyle's Law, which states that pressure and volume are inversely related in a closed system. When the volume of the thoracic cavity expands, the pressure within the lungs decreases, creating a vacuum that draws air into the lungs. Conversely, when the volume decreases, the pressure rises, forcing air out [15, 16].

At the core of respiratory dynamics are three key pressures that regulate airflow. Atmospheric pressure (Patm) remains relatively constant at approximately 760 mmHg at sea level, serving as the reference point against which internal pressures fluctuate. Intrapulmonary pressure (Ppul), the pressure inside the alveoli, undergoes rhythmic changes throughout the breathing cycle. During inspiration, it momentarily drops below atmospheric pressure, allowing air to rush in, while during expiration, it rises above atmospheric pressure, expelling air from the lungs [17, 18]. Surrounding the lungs is a thin fluid-filled space known as the pleural cavity, where intrapleural pressure (Pip) remains slightly negative relative to both atmospheric and intrapulmonary pressure. This negative pressure acts as a suction force, preventing lung collapse by keeping the lungs adhered to the chest wall. Typically, intrapleural pressure hovers around –4 mmHg, but as the thoracic cavity expands during inspiration, it becomes even more negative, further assisting in lung expansion.

As the diaphragm contracts and moves downward, the thoracic cavity expands, resulting in a decrease in intrapulmonary pressure. This pressure differential causes air to flow into the alveoli, a process known as inspiration, which is an active phase of breathing requiring muscle contraction. Simultaneously, the external intercostal muscles lift the ribcage, further expanding the thoracic volume. With increasing lung volume, the airways and alveoli stretch, reducing airway resistance and facilitating smooth airflow [19–21]. Expiration, in contrast, is generally a passive process during quiet breathing, driven by the natural elastic recoil of the lungs. As the diaphragm and intercostal muscles relax, the thoracic cavity contracts, reducing in size and compressing the lungs, which increases intrapulmonary pressure above atmospheric pressure. This increase forces air out, completing the breathing cycle. However, in conditions requiring forced expiration, such as vigorous exercise or coughing, additional muscles, including the internal intercostals and abdominal muscles, engage, generating greater pressure and expelling air more forcefully.

The synchronization of pressure changes, muscle movements, and airflow ensures that each breath efficiently delivers oxygen while expelling carbon dioxide. Any disruption in this balance—whether from disease, trauma, or external influences—can significantly impact respiratory function. For example, a puncture to the pleural cavity may allow air to enter, equalizing intrapleural and atmospheric pressure and leading to lung collapse (pneumothorax). Similarly, conditions that increase airway resistance, such as asthma or COPD, require greater effort to generate sufficient pressure gradients, leading to respiratory distress [3, 22, 23].

By understanding these fundamental principles, researchers and clinicians can more effectively assess respiratory health, optimize ventilation strategies, and develop targeted interventions for conditions that compromise standard breathing mechanics. The intricate regulation of pressure and airflow, as illustrated in the figure, provides a foundation for exploring further aspects of lung function, including lung compliance and airway resistance, which in turn influence respiratory efficiency.

4.2 Types of Respiratory Signals

The human respiratory system generates a variety of physiological signals that provide valuable insights into lung function, breathing patterns, and overall health. These signals, which arise from the mechanical, biochemical, and neural aspects of respiration, are essential for both clinical diagnostics and physiological research. Understanding the various types of respiratory signals is crucial for evaluating respiratory function, identifying abnormalities, and developing advanced monitoring and therapeutic technologies. Respiratory signals can be broadly categorized into mechanical, airflow-related, gas exchange, and bioelectrical signals. Mechanical signals include measurements of thoracic and abdominal movement, changes in lung volume, and airway resistance. These signals are often captured using techniques such as respiratory plethysmography or impedance pneumography, which track changes in chest expansion and contraction during breathing. The dynamics of airflow are equally important, with signals derived from spirometry and pneumotachography providing quantitative assessments of ventilation efficiency and airway function [24].

Gas exchange signals, particularly those related to oxygen and carbon dioxide levels, reflect the respiratory system's ability to maintain homeostasis. Capnography, which measures the concentration of exhaled carbon dioxide, is widely used in clinical settings to assess ventilation status and detect conditions such as respiratory depression or airway obstruction. Similarly, pulse oximetry provides continuous, non-invasive monitoring of blood oxygen saturation, offering critical information in both acute and chronic respiratory conditions. Beyond mechanical and gas exchange signals, bioelectrical activity plays a crucial role in understanding respiration. The central and peripheral nervous systems regulate breathing through a complex network of neural pathways, and this regulation is reflected in signals such as the electromyographic (EMG) activity of the diaphragm. The electrical impulses generated by respiratory muscles, captured using surface electrodes, provide insights into respiratory drive and neuromuscular function. Additionally, respiratory signals interact with cardiovascular activity, as seen in respiratory sinus arrhythmia (RSA), where heart rate fluctuates in synchrony with the breathing cycle [25].

The diversity of respiratory signals and their interdependence with other physiological systems highlight the need for advanced signal processing and integration techniques. By analyzing multiple respiratory parameters simultaneously, researchers and clinicians can obtain a more comprehensive understanding of respiratory health and disease. As technology evolves, the development of wearable sensors, artificial intelligence-driven diagnostics, and multimodal monitoring systems continues to enhance our ability to interpret these signals with greater accuracy and clinical relevance.

4.2.1 Spirometry, Plethysmography, and Respiratory Inductance

Respiratory signal acquisition is essential for understanding the intricate processes of lung function, breathing mechanics, and gas exchange. Among the most significant techniques for evaluating respiratory physiology are spirometry, plethysmography, and respiratory inductance plethysmography. These methodologies provide valuable data on pulmonary mechanics, airflow patterns, and the efficiency of gas exchange. Furthermore, they serve as critical tools for diagnosing and managing respiratory diseases such as COPD, asthma, restrictive lung disorders, and neuromuscular conditions that affect breathing. The information obtained from these assessments can be

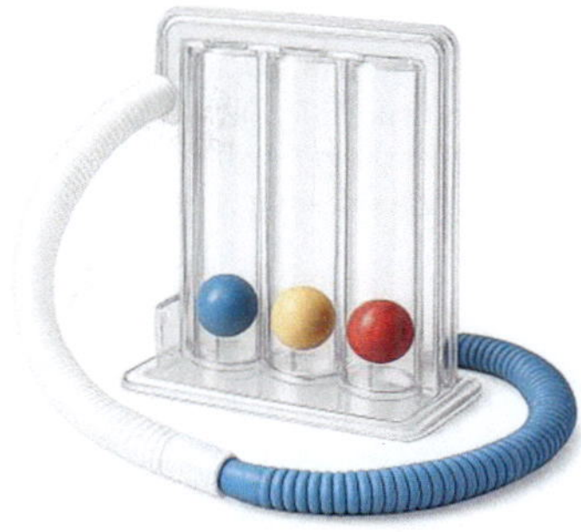

Fig. 4.3 A spirometer

linked to broader physiological signals, encompassing mechanical, airflow-related, gas exchange, and bioelectrical aspects of respiration [26, 27].

Spirometry

Spirometry is one of the most widely used and accessible pulmonary function tests, providing essential insights into airflow dynamics and the mechanical properties of the respiratory system. The technique involves measuring both the volume of air a person can inhale and exhale, as well as the speed of airflow during these maneuvers. Spirometry is conducted by instructing a patient to perform forced breathing efforts, which produce detailed flow-volume and volume-time curves. These curves serve as diagnostic markers for obstructive and restrictive lung diseases, enabling clinicians to assess the severity and progression of various pulmonary conditions [28] (Fig. 4.3).

The forced vital capacity (FVC) test, a fundamental spirometric measurement, evaluates the maximum volume of air an individual can expel after a full inhalation. The forced expiratory volume in one second (FEV1) is another critical parameter that quantifies the amount of air a person can exhale within the first second of forced expiration. Both FVC and FEV1 are essential for diagnosing obstructive diseases, such as chronic obstructive pulmonary disease (COPD), as well as restrictive diseases, including pulmonary fibrosis [29, 30].

The ratio of FEV1 to FVC is a key diagnostic marker for airflow obstruction. A reduced FEV1/FVC ratio is typically observed in obstructive conditions, where airway narrowing impairs the ability to expel air efficiently, resulting in airflow limitation. In restrictive lung diseases, both FEV1 and FVC are reduced proportionally, but the ratio remains normal or even increased. Spirometric testing is instrumental in distinguishing between these two types of respiratory disorders, aiding in targeted clinical management. Spirometry generates two main types of graphical representations: the flow-volume curve and the volume-time curve. The flow-volume curve displays the relationship between airflow and lung volume during both forced inspiration and expiration. This curve helps to identify patterns typical of obstructive and restrictive diseases. A concave, scooped-out appearance of the expiratory limb of the flow-volume curve is characteristic of obstructive conditions, whereas a more linear and reduced curve suggests restrictive lung disease (Fig. 4.4).

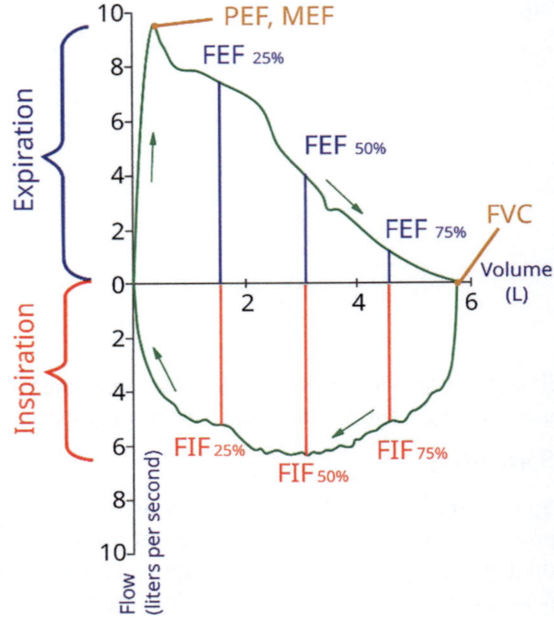

Fig. 4.4 A flow-volume curve. https://commons.wikimedia.org/wiki/File:Flow-volume-curve.svg#/media/File:Flow-volume-loop.svg

The volume-time curve, on the other hand, plots the cumulative volume of air exhaled over time, allowing for the assessment of how quickly a person can expel air. This curve is beneficial for assessing the efficiency of forced expiratory maneuvers and for verifying the completion of a complete exhalation to ensure that spirometric values are accurately recorded.

The accuracy and reliability of spirometry are heavily dependent on the sensing mechanisms employed in the equipment. Various technologies are available, each with unique advantages for assessing airflow:

1. Pneumotachographs measure airflow by detecting the pressure difference across a resistive element, such as a fine mesh screen. This pressure difference is directly proportional to the flow rate, making pneumotachographs highly effective for continuous flow measurement in dynamic respiratory assessments.
2. Turbine spirometers utilize rotating blades to estimate the velocity of airflow. As the air moves through the device, it causes a turbine to spin, with the rotation speed correlating to airflow velocity. These devices are commonly used for portable spirometry due to their simplicity and ease of use.
3. Ultrasonic spirometers use sound waves to determine airflow by measuring the time it takes for sound waves to travel through the air stream. This method provides a non-contact, highly accurate alternative, minimizing resistance and enhancing the precision of flow rate measurements.

These technologies enable real-time assessments of lung function, providing clinicians with critical information about a patient's respiratory status. Additionally,

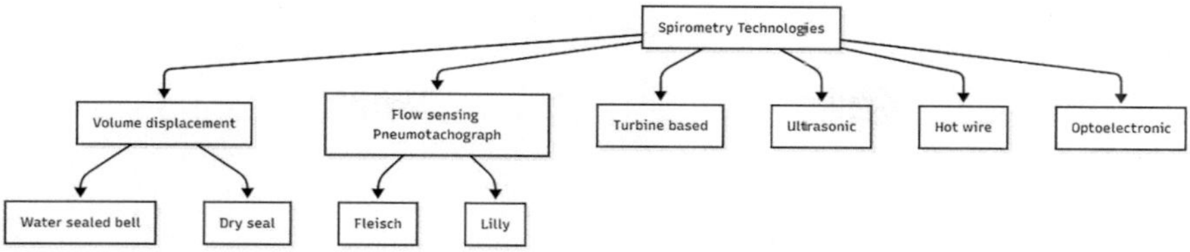

Fig. 4.5 Different spirometry technologies

modern spirometers allow for long-term monitoring of pulmonary health in both clinical and home settings, empowering patients with chronic respiratory conditions to manage their health proactively [31] (Fig. 4.5).

Body Plethysmography

While spirometry excels in capturing airflow-related parameters such as FVC and FEV1, it does not directly measure absolute lung volumes or residual volumes, which are critical for a comprehensive understanding of pulmonary function. Specifically, spirometry cannot measure lung volumes such as residual volume (RV), functional residual capacity (FRC), or TLC. These measurements are essential for diagnosing conditions related to lung hyperinflation, gas exchange abnormalities, and airway resistance. To obtain these values, body plethysmography becomes a crucial adjunct to spirometry. Body Plethysmography is a sophisticated diagnostic technique that measures lung volumes by utilizing Boyle's Law, which describes the inverse relationship between pressure and volume in a closed system. Boyle's Law is expressed mathematically as Eq. 4.1:

$$P_1V_1 = P_2V_2 \tag{4.1}$$

where:

- P_1 and P_2 are the pressures at two different points in the closed system,
- V_1 and V_2 are the corresponding volumes at those points.

In plethysmography, the patient is placed inside a sealed chamber, and as they breathe against a closed shutter (which prevents airflow), the changes in thoracic volume lead to corresponding pressure fluctuations within the chamber. These pressure changes can then be analyzed to calculate lung volumes. By applying the principles of Boyle's Law, the volume of air in the lungs at various stages of respiration can be accurately quantified, even for lung volumes that cannot be measured solely by spirometry [32, 33] (Fig. 4.6).

During the test, the patient inhales and exhales through a mouthpiece connected to a sealed chamber, while the pressure inside the chamber is continuously monitored. The closed shutter is momentarily closed at the end of a normal exhalation, causing the patient's chest wall to compress. This compression causes a measurable change in pressure, which is related to changes in the volume of the lungs [9, 34]. The ability

Fig. 4.6 A plethysmography chamber. https://commons.wikimedia.org/wiki/File:Body_Plethysmography_chamber_03.jpg

to measure these pressure fluctuations enables the system to calculate the TLC, the FRC, and the RV, which are crucial in understanding the mechanical properties of the lungs and the efficiency of gas exchange.

Total lung capacity: This is the maximum amount of air the lungs can hold, including the residual volume that remains after maximal exhalation. TLC is important for assessing conditions like emphysema, where lung hyperinflation occurs due to air trapping.

Functional residual capacity: The volume of air that remains in the lungs after normal exhalation. This measurement is particularly relevant in diagnosing airway resistance issues, as a high FRC may indicate conditions like asthma or COPD, where air trapping prevents complete exhalation.

Residual volume: The amount of air that remains in the lungs after maximal exhalation, which routine spirometry maneuvers cannot expel. Measurement of RV is essential for detecting air trapping, a hallmark of obstructive lung diseases such as emphysema and chronic bronchitis.

In addition to measuring lung volumes, plethysmography is valuable for assessing airway resistance and specific airway conductance, which are critical in diagnosing obstructive lung diseases. The resistance within the airways can be measured by analyzing the pressure–volume relationship as the patient breathes against the closed shutter. This allows the detection of minor airway dysfunctions that might not be visible through traditional spirometry. Increased airway resistance may suggest obstructive conditions, such as asthma or COPD. Specific airway conductance (SAC), which is the inverse of airway resistance, can also be quantified through plethysmography. SAC is a crucial parameter for understanding how easily air flows through the airways, and changes in SAC are commonly observed in asthma, where bronchoconstriction leads to increased airway resistance and reduced airway conductance [35, 36]. Plethysmography provides a more detailed picture of lung mechanics, especially in patients with complex or undiagnosed respiratory conditions. For instance, it is especially valuable in detecting air trapping and hyperinflation, which are often present in emphysema or severe asthma. Under these conditions, the RV increases significantly due to difficulties in fully exhaling air. Measuring RV, along with FRC and TLC, provides a deeper understanding of the extent to which lung volumes are affected, guiding both diagnosis and management. For patients with restrictive lung diseases (e.g., pulmonary fibrosis), plethysmography can help assess lung volume reduction and monitor disease progression. In restrictive conditions, the TLC may be reduced, and plethysmography is one of the most reliable methods for quantifying this decrease, which is often masked in traditional spirometry tests [37, 38].

Respiratory Inductance Plethysmography

Respiratory Inductance Plethysmography (RIP) extends the principles of plethysmography into a more portable and non-invasive modality, offering significant advantages in continuous, real-time monitoring of respiratory function. Unlike body plethysmography, which relies on a sealed chamber to measure changes in pressure and estimate lung volumes, RIP employs a more straightforward and practical approach for tracking respiratory dynamics [34, 39]. This technique does not require specialized laboratory equipment, making it suitable for ambulatory settings, sleep studies, and field research applications. RIP works by using elastic bands embedded with conductive wires, which are placed around the chest and abdomen. These bands are designed to detect changes in the thoracic (chest) and abdominal circumference during breathing. As the patient inhales and exhales, the movement of the chest and abdomen alters the circumference of the bands. The conductive wires within the bands detect these changes and provide continuous measurements of respiratory effort [40, 41] (Fig. 4.7).

By analyzing these circumferential changes, RIP can estimate important respiratory parameters, such as tidal volume (TV)—the amount of air inhaled or exhaled with each breath—and overall respiratory effort. Unlike spirometry, which primarily measures airflow and volumes in a controlled, forced breathing environment, RIP can capture respiratory patterns in more natural, resting conditions, providing more representative data of an individual's normal breathing behavior.

One of the key advantages of RIP is its ability to estimate tidal volume, which is vital for assessing ventilatory function. Tidal volume is typically measured in clinical settings using spirometry; however, RIP offers a non-invasive, continuous alternative for monitoring tidal volume, particularly when dynamic assessments of ventilation are required in real-world settings. In addition to tidal volume, RIP also quantifies respiratory effort, which provides information about the intensity of respiratory muscle activity during breathing. This is particularly useful for evaluating patients with conditions that affect respiratory muscle function or mechanics [21, 42]. RIP

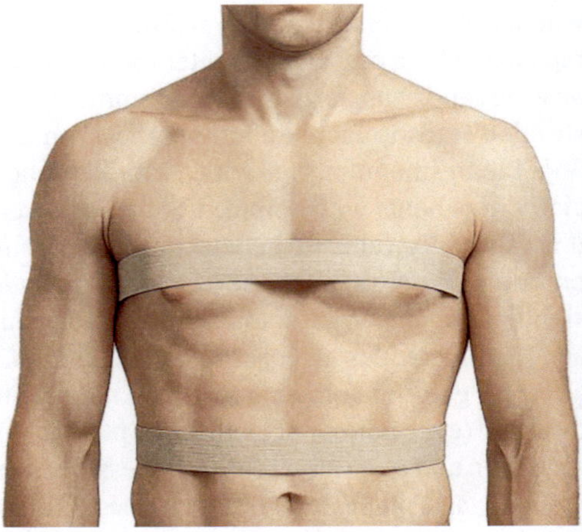

Fig. 4.7 A diagram illustrating the placement of the elastic bands around the chest and abdomen

can also analyze the phase relationships between chest and abdominal movements. During normal breathing, the chest and abdomen expand and contract in a coordinated manner. However, in patients with diaphragmatic dysfunction or respiratory muscle fatigue, this coordination may be disrupted, resulting in paradoxical breathing patterns.

In paradoxical breathing, the abdomen may move in the opposite direction of the chest during inhalation and exhalation. This is typically seen in severe respiratory conditions, such as diaphragmatic paralysis, where the diaphragm fails to contract properly, or in cases of respiratory muscle weakness due to conditions like neuromuscular disorders. RIP's ability to detect and distinguish between normal and paradoxical breathing patterns is especially valuable in clinical diagnostics. Identifying abnormal respiratory patterns early on can prompt further investigation and intervention, improving patient outcomes [43]. RIP is widely used in various clinical and research settings, thanks to its portability and non-invasive nature. Its most common applications include:

1. Sleep studies: RIP is frequently used in polysomnography to monitor breathing patterns during sleep, particularly in the diagnosis of sleep apnea and other sleep-related respiratory disorders. The ability to measure thoracic and abdominal movements while the patient sleeps provides detailed information about the respiratory phases and any interruptions in normal breathing patterns.
2. Ambulatory monitoring: Unlike laboratory-based tests, RIP allows for continuous monitoring of respiratory function in a patient's natural environment. This makes it an invaluable tool for patients with chronic respiratory conditions, such as COPD or asthma, who require regular assessment of their breathing over extended periods.
3. Research applications: RIP is also utilized in clinical research to investigate breathing patterns, respiratory control, and the impact of various treatments on lung function. Its ability to provide continuous, real-time data makes it a versatile

tool for understanding the complex dynamics of human respiration, especially in large-scale studies or studies involving interventions over prolonged periods.

The primary advantages of RIP over traditional plethysmography and spirometry are its portability, non-invasiveness, and ability to monitor respiratory patterns continuously over extended periods. Unlike body plethysmography, which requires a specialized sealed chamber and is limited to a laboratory setting, RIP can be used in a wide range of environments, from sleep studies to everyday home monitoring. This allows for greater flexibility in patient management and enables clinicians to track changes in respiratory function in a more natural setting [3, 44]. Additionally, RIP is generally more comfortable for the patient, as it does not require the patient to be confined to a chamber or perform forced breathing maneuvers. This makes it more suitable for long-term or continuous monitoring, particularly in patients who may find spirometry or body plethysmography uncomfortable or impractical (Fig. 4.8).

4.2.2 Respiratory Signals Acquisition and Respiratory Signals

The integration of respiratory signal acquisition with mechanical, airflow-related, gas exchange, and bioelectrical signals enhances our understanding of respiration as a complex physiological process. Mechanical signals, derived from plethysmography and RIP, reflect the physical expansion and contraction of the thoracic cavity, providing insights into respiratory muscle function and lung compliance. Airflow-related signals, as captured by spirometry, characterize the dynamics of ventilation and the presence of airway obstruction or restriction. Gas exchange signals, including oxygen and carbon dioxide measurements from capnography or blood gas analysis, reveal the efficiency of pulmonary oxygenation and carbon dioxide elimination. Finally, bioelectrical signals, such as diaphragmatic EMG or EEG markers of respiratory drive, offer a window into the neural control of breathing [45, 46].

Understanding these interconnected signals is essential for diagnosing and treating a range of pulmonary and neuromuscular disorders. In conditions such as COPD, airway obstruction leads to increased airway resistance, altered spirometric indices, and ventilation-perfusion mismatching, which impairs gas exchange. In restrictive lung diseases, reduced lung compliance leads to diminished lung volumes and compensatory changes in breathing effort, often detectable through RIP. In neuromuscular disorders, impaired respiratory muscle activation manifests in abnormal EMG patterns, reduced inspiratory strength, and paradoxical breathing, as detected through plethysmographic monitoring. Advances in sensor technology and computational methods continue to refine the acquisition of respiratory signals, enabling more precise and real-time monitoring of lung function. The integration of artificial intelligence and machine learning into respiratory diagnostics allows for automated detection of abnormal breathing patterns, facilitating early diagnosis and personalized treatment strategies. Portable and wearable respiratory monitoring systems

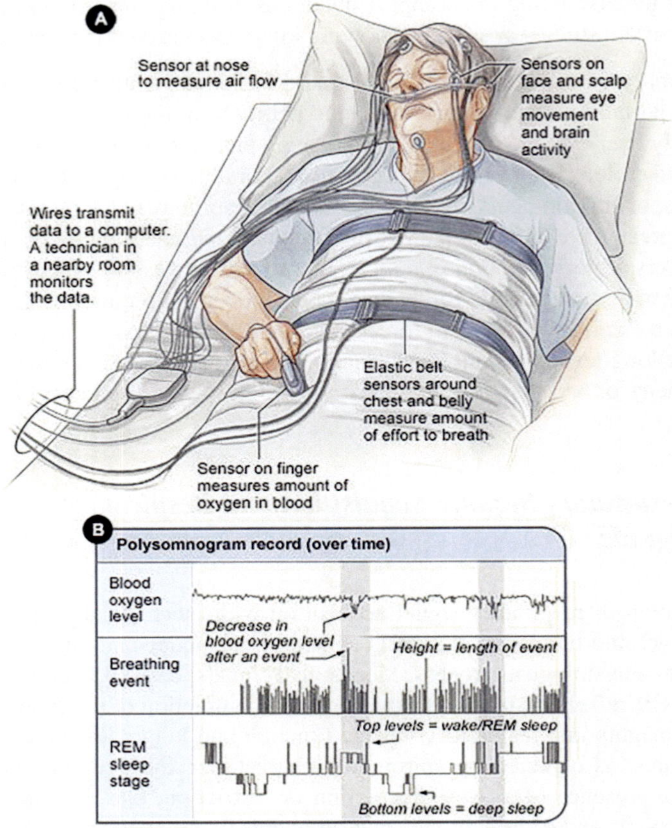

Fig. 4.8 An illustration of a sleep study setup. https://commons.wikimedia.org/wiki/File:Sleep_studies.jpg

extend these capabilities beyond clinical settings, empowering patients with chronic respiratory conditions to manage their health proactively [3, 47].

In the context of research and clinical applications, the ability to combine multiple respiratory signal modalities enhances the accuracy and depth of respiratory assessments. For instance, simultaneous spirometry and plethysmography enable the differentiation between obstructive and restrictive pathophysiology, while integrating RIP with capnography provides a comprehensive view of ventilatory efficiency and breathing effort. The synergy between mechanical, airflow, gas exchange, and bioelectrical signals forms the foundation for a holistic approach to respiratory medicine, improving both diagnostic precision and therapeutic outcomes.

The study of respiratory signal acquisition not only advances our understanding of lung physiology but also paves the way for innovative diagnostic and therapeutic approaches. Whether through conventional methods like spirometry, advanced techniques like body plethysmography, or wearable solutions like RIP, the ongoing evolution of respiratory monitoring technologies holds great promise for the future of pulmonary medicine. The accuracy and reliability of spirometry, plethysmography, and RIP depend on both technical factors and physiological considerations. Calibration is a crucial aspect of these respiratory measurement techniques, as even minor deviations in sensor precision can lead to significant errors in assessing lung function. Modern spirometers require routine calibration using standardized gas volumes to ensure consistency in airflow measurements. Similarly, plethysmographic chambers must maintain a precise pressure balance to produce valid estimations of lung volumes. In RIP, signal drift due to sensor displacement can introduce variability, necessitating proper placement and secure fixation of inductive bands to ensure accurate readings [3].

Despite these challenges, the benefits of these methodologies in both research and clinical practice are immense. Spirometry remains the primary test for assessing pulmonary function, providing real-time feedback on airflow limitation, lung volume abnormalities, and bronchodilator responsiveness. Plethysmography, particularly body plethysmography, provides a comprehensive evaluation of airway resistance and residual lung volume, parameters that are crucial for distinguishing between different restrictive and obstructive lung diseases. Respiratory inductance plethysmography, on the other hand, allows for continuous monitoring of respiratory effort and is particularly useful in sleep studies, ambulatory monitoring, and non-invasive assessments of breathing patterns [38, 48].

Understanding mechanical, airflow-related, gas exchange, and bioelectrical signals in respiration requires integrating multiple modalities of measurement. Mechanical signals, such as those derived from plethysmography and RIP, provide insight into thoracic expansion, lung compliance, and respiratory muscle engagement. These signals are directly linked to airflow-related measurements, as the movement of air through the respiratory tract corresponds to changes in intrapulmonary pressure and airway resistance [9, 14].

Airflow signals, captured primarily by spirometry, describe the velocity and volume of air entering and exiting the lungs. Peak expiratory flow rate (PEFR), FEV1, and FVC are key indicators of airway obstruction or restriction. By comparing airflow parameters to plethysmographic lung volume measurements, clinicians can differentiate between purely restrictive diseases, where lung volume is reduced, and obstructive conditions, where airflow limitation occurs despite relatively preserved lung volume. Gas exchange signals, assessed via capnography, pulse oximetry, and arterial blood gas analysis, provide direct information about the effectiveness of pulmonary ventilation. Capnography measures end-tidal CO_2 ($EtCO_2$), reflecting the balance between alveolar ventilation and metabolic CO_2 production. In contrast, pulse oximetry estimates oxygen saturation (SpO_2) by analyzing light absorption through the skin, offering a non-invasive means of monitoring blood oxygenation [49].

Beyond these physiological signals, bioelectrical signals, such as those captured by diaphragmatic EMG, EEG, and phrenic nerve stimulation, provide a unique perspective on respiratory drive and neuromuscular function. EMG recordings of the diaphragm and intercostal muscles are particularly valuable in assessing neuromuscular disorders, respiratory fatigue, and central drive abnormalities. In conditions such as amyotrophic lateral sclerosis (ALS) and muscular dystrophy, diaphragmatic EMG signals show progressive decline in amplitude, indicating respiratory muscle weakness.

4.2.3 Clinical Applications of Multi-Modal Respiratory Monitoring

The ability to integrate spirometry, plethysmography, and respiratory inductance plethysmography with mechanical, airflow-related, gas exchange, and bioelectrical signals has revolutionized the diagnosis and management of respiratory diseases. In COPD, for instance, spirometry reveals airflow limitation, while plethysmography uncovers air trapping and lung hyperinflation, key factors contributing to dyspnea and exercise intolerance. Capnography and blood gas analysis provide additional information on ventilation-perfusion mismatch and CO_2 retention, critical in assessing COPD exacerbations and respiratory failure [13].

In asthma, spirometry and plethysmography play crucial roles in assessing disease severity and monitoring the response to bronchodilators. A classic feature of asthma is variable airflow obstruction, evidenced by improvements in FEV1 following administration of bronchodilators. Plethysmography also enables the measurement of specific airway resistance, which increases during asthma exacerbations. In neuromuscular disorders, where respiratory muscle weakness can lead to hypoventilation and respiratory failure, RIP and bioelectrical signal monitoring provide essential data. Patients with conditions such as Duchenne muscular dystrophy or ALS often develop progressive diaphragmatic weakness, detectable via diaphragmatic EMG and plethysmography. Early detection of respiratory decline allows for timely intervention with non-invasive ventilation and other supportive therapies.

4.2.4 Advancements in Respiratory Signal Acquisition and Future Directions

With the evolution of wearable sensors, artificial intelligence (AI), and machine learning, respiratory monitoring is becoming increasingly automated and personalized. New spirometric devices equipped with AI algorithms can provide real-time feedback on breathing patterns, allowing for early detection of lung function decline. Portable plethysmography systems are being developed for home-based lung volume monitoring, providing continuous assessment of lung mechanics in patients with progressive respiratory diseases [18, 41].

One of the most exciting advancements is the integration of multi-modal respiratory monitoring systems, which combine spirometry, plethysmography, capnography, and bioelectrical signal analysis into a single comprehensive respiratory health platform. These multi-sensor systems enable clinicians to evaluate lung function holistically, capturing both ventilatory mechanics and gas exchange efficiency. In the future, cloud-based remote monitoring systems may enable real-time transmission of respiratory data to healthcare providers, allowing for early intervention and personalized treatment strategies.

By refining the tools used for respiratory signal acquisition, the field of pulmonary medicine is moving toward a future where diagnostic precision, early disease detection, and real-time monitoring will significantly enhance patient care. Spirometry, plethysmography, and respiratory inductance plethysmography, when combined with advanced computational analysis, are shaping the next generation of respiratory diagnostics and therapeutic interventions.

4.3 Signal Processing of Respiratory Signals

The human respiratory system generates a wide variety of physiological signals, each of which plays a significant role in assessing lung function, monitoring breathing patterns, and evaluating overall health. These signals originate from the mechanical, biochemical, and neural aspects of respiration, and their processing is crucial for both clinical diagnostics and physiological research. Effective signal processing of these respiratory signals is crucial to understanding respiratory health, detecting abnormalities, and developing advanced technologies that support patient management and therapeutic interventions. This chapter discusses the importance of signal processing techniques applied to respiratory signals, the challenges involved, and the state-of-the-art methodologies used in extracting meaningful information from respiratory data.

Signal processing in the context of respiratory signals entails transforming raw data captured from physiological measurements into actionable insights. Respiratory signals are often complex, noisy, and prone to various artifacts that can interfere with accurate analysis [50, 51]. The goal of signal processing is to enhance the quality of the raw signal, remove unwanted noise, extract key features, and ultimately enable more accurate diagnosis, prognosis, and monitoring. The primary challenges in signal processing respiratory signals stem from the variability of the signals, their susceptibility to external influences, and the inherent complexity of the physiological systems they represent. The first step in respiratory signal processing typically involves data acquisition, where sensors or devices such as spirometers, plethysmographs, and electrocardiograms capture signals from the patient. These measurements may include airflow patterns, lung volume changes, electrical activity of respiratory muscles, or gas exchange data such as oxygen and carbon dioxide levels. While these signals provide critical information about the respiratory system, they are often subject to interference from external sources, such as patient movement, electrical noise, and respiratory artifacts like coughing or swallowing. Effective signal processing techniques must therefore include robust filtering methods that mitigate these issues and ensure.

One of the primary challenges in respiratory signal processing is the removal of noise and artifacts that can distort the signal. Respiratory signals are often contaminated by various types of noise, such as electrical interference, movement artifacts, and physiological noise from other systems. Filtering methods are essential in this context to clean the raw data and isolate the respiratory information of interest.

Digital filtering techniques, including low-pass, high-pass, band-pass, and notch filters, are commonly used to address different types of noise. For example, low-pass filters are frequently employed to remove high-frequency noise, while high-pass filters can be used to remove low-frequency drift or baseline shifts. Band-pass filters can be applied to retain only the frequency range that corresponds to the respiratory signal, ensuring that irrelevant noise is excluded. In the case of gas exchange signals, for example, capnographic data can be filtered to isolate the key components of the respiratory cycle and remove fluctuations caused by body movements or other interferences [52]. Moreover, adaptive filtering techniques, such as Kalman filters or Wiener filters, are often employed to dynamically adjust the filter parameters in response to the changing characteristics of the signal. These advanced techniques can be particularly beneficial in real-time applications, where signal characteristics may vary due to factors such as breathing rate, posture, or activity level.

Once the raw respiratory signals have been filtered and cleaned, the next step in the signal processing pipeline is to extract features. This process involves identifying key characteristics of the signal that provide valuable insights into respiratory health. Feature extraction is crucial for quantifying various aspects of respiratory function, including lung volume, airflow rates, and respiratory muscle activity. For example, in airflow-related signals derived from spirometry or pneumotachography, feature extraction may involve calculating parameters such as FEV1, PEFR, and TV. These parameters are crucial for diagnosing obstructive and restrictive lung diseases, and their extraction from raw data enables efficient and accurate assessment of pulmonary function. Similarly, in the case of respiratory plethysmography, features such as chest expansion and changes in thoracic volume can be extracted to assess lung compliance and respiratory mechanics. In bioelectrical respiratory signals, such as EMG activity of the diaphragm or respiratory muscles, feature extraction might focus on identifying the frequency, amplitude, and duration of muscle contractions. These features can be used to assess the strength and coordination of respiratory muscles, particularly in patients with neuromuscular diseases. Advanced signal processing techniques, such as wavelet transforms or time–frequency analysis, are often employed to extract meaningful features from non-stationary bioelectrical signals, which exhibit variations in both time and frequency domains [53, 54].

Following feature extraction, the next step in respiratory signal processing involves analyzing the extracted features to gain insights into the patient's physiological status. This stage often requires the use of advanced signal analysis techniques to interpret the data and detect abnormalities. Various methods, including statistical analysis, spectral analysis, and machine learning, can be applied to respiratory signals to extract meaningful information and inform clinical decisions. Statistical methods are commonly used to analyze the distribution of extracted features and identify patterns or trends in respiratory behavior. For instance, analyzing the variability in respiratory rate or tidal volume over time can reveal changes in respiratory function that may signal the onset of disease or the effects of treatment. Statistical tests can also be used to compare respiratory data across different patient populations, helping to identify specific biomarkers associated with certain conditions. Spectral analysis, which involves analyzing the frequency components of a signal, is another valuable tool in respiratory signal processing. For example, in the analysis of heart rate variability and respiratory sinus arrhythmia (RSA), spectral methods can be used to assess the coupling between respiration and cardiovascular activity. Similarly, power spectral density (PSD) analysis of respiratory signals can provide insights into the frequency characteristics of the respiratory cycle, which can help identify abnormal patterns such as hyperventilation or periodic breathing. Machine learning techniques are being increasingly applied to respiratory signal analysis, as they offer the potential to classify and predict respiratory events based on large datasets automatically. Machine learning algorithms, such as decision trees, support vector machines (SVMs), and deep learning networks, can be trained to detect respiratory abnormalities, classify different types of breathing patterns, and predict disease progression. These techniques are particularly valuable in the context of wearable sensors and real-time monitoring systems, where large volumes of data are generated continuously [55].

A key challenge in respiratory signal processing is integrating signals from multiple modalities, as respiratory signals are often influenced by or interact with other physiological systems. For example, respiratory function is closely linked to cardiovascular activity, and respiratory patterns may be influenced by neural regulation. Multimodal signal integration aims to combine data from various sources, including airflow, gas exchange, and bioelectrical signals, to provide a more comprehensive understanding of respiratory health. One approach to multimodal integration is the use of data fusion techniques, which combine information from multiple sensors or devices to generate a unified estimate of the patient's respiratory status. For instance, combining capnography data with respiratory plethysmography can provide insights into both the efficiency of ventilation and the adequacy of gas exchange. Similarly, combining respiratory muscle EMG with airflow data can help assess whether respiratory muscle dysfunction is contributing to abnormal ventilation patterns. Multimodal signal integration also plays a key role in the development of advanced monitoring systems, such as wearable devices that continuously monitor multiple aspects of respiratory function. These devices often incorporate sensors for tracking airflow, oxygen saturation, and respiratory effort, among others. By integrating these signals, the system can provide real-time feedback to the patient or clinician, enabling early detection of respiratory abnormalities and facilitating timely intervention.

4.3.1 Detection of Breathing Rate and Patterns

The detection and analysis of breathing rate and breathing patterns are fundamental to understanding respiratory health and diagnosing various pulmonary conditions. Breathing rate, also known as respiratory rate (RR), refers to the number of breaths a person takes per minute, serving as a primary indicator of respiratory function. Breathing patterns refer to the specific rhythm and variability of breathing, which can provide additional insights into the underlying mechanisms of respiratory health. By analyzing these factors, healthcare professionals can identify irregularities such as tachypnea, bradypnea, apnea, dyspnea, and other forms of respiratory distress that may signal the presence of underlying diseases. Detecting breathing rate and patterns involves a variety of techniques and technologies, including traditional methods like spirometry and plethysmography, as well as modern sensor-based systems such as wearable devices, chest wall motion sensors, and capnography. Each of these technologies relies on specific signal processing approaches to quantify breathing rates and detect subtle variations in breathing patterns.

The RR is typically measured as the number of complete breathing cycles (inhalation and exhalation) per minute. Normal respiratory rates vary by age, health status, and activity level, with typical ranges from 12 to 20 breaths per minute in healthy adults at rest. In clinical contexts, detecting deviations from normal rates can serve as an early indicator of respiratory problems. The detection of breathing rate is often done by measuring airflow, chest expansion, or thoracic movements. A variety of techniques exist for this purpose, including spirometry, impedance pneumography, and RIP, among others. Each of these approaches relies on different principles to capture the rhythmic expansion and contraction of the chest during respiration [56, 57].

One of the most straightforward methods for measuring breathing rate is through impedance pneumography, where electrodes placed on the skin detect changes in the electrical impedance of the chest as it expands and contracts during breathing. The impedance signal can then be analyzed to determine the frequency of respiratory cycles, thus allowing for an accurate measurement of breathing rate. This method is non-invasive, relatively inexpensive, and suitable for continuous monitoring in both clinical and home settings.

Equation 4.2 Breathing rate calculation from impedance signal

$$RR = \frac{1}{T_{\text{cycle}}} = \frac{60}{T_{\text{cycle}}} \tag{4.2}$$

where:

- RR = Breathing rate (breaths per minute)
- T_{cycle} = Duration of one complete respiratory cycle (seconds)

For more advanced applications, spirometry is used to measure airflow during both inhalation and exhalation. Spirometers record the volume of air inhaled and exhaled over time, and the respiratory rate can be derived by counting the number of breaths within a set time interval. Additionally, pneumotachographs, which measure airflow velocity during breathing, can also be used to determine the respiratory rate with high accuracy by analyzing the timing of peak airflow during each cycle [49, 58].

4.3.2 Analysis of Breathing Patterns: Variability and Rhythmicity

In addition to breathing rate, the pattern of respiration provides critical information regarding the function of the respiratory system. Breathing patterns are often characterized by regularity, depth, and rhythmicity. Various physiological processes, including neural regulation, muscle activity, lung compliance, and airway resistance influence these factors. Understanding breathing patterns can assist in the diagnosis of conditions such as obstructive sleep apnea, COPD, and restrictive lung disease.

Various signal processing techniques are applied to detect and characterize breathing patterns. One such method is the analysis of chest wall motion using RIP. RIP measures the changes in thoracic and abdominal circumference during the breathing cycle, capturing both inspiratory and expiratory phases of respiration. The data obtained from RIP can be used to track tidal volume, respiratory effort, and breathing pattern, all of which are crucial for identifying abnormal respiratory behaviors such as rapid shallow breathing, deep breathing, or periodic breathing.

The pneumotachograph also provides valuable data for analyzing breathing patterns. By recording the flow volume loop during respiration, this device can provide detailed information about inspiratory and expiratory flow rates, as well as the shape of the flow-volume curve, which can reveal patterns of airway obstruction, restriction, or normal ventilation. The timing, flow, and volume of breaths provide insight into respiratory effort and the adequacy of ventilation, which is especially useful for monitoring patients with chronic respiratory conditions [59, 60]. Once the raw respiratory signals are acquired, several signal processing techniques are used to analyze and quantify breathing patterns. One of the primary methods for analyzing breathing patterns is time-series analysis, where respiratory signals are examined as a series of data points sampled over time [3]. By applying Fourier transforms or wavelet transforms, the frequency and temporal components of the breathing pattern can be extracted.

In the case of regular, rhythmic breathing patterns, Fourier analysis can be used to determine the dominant frequency of respiration. This is especially useful for identifying any shifts in standard breathing patterns that may indicate respiratory distress. For instance, patients with obstructive conditions such as asthma may exhibit irregular breathing with a noticeable increase in frequency and shorter inspiratory times. Similarly, patients with hypoventilation or apnea may show longer expiration times and a reduced breathing frequency.

Equation 4.3 Fourier transform of respiratory time series

$$X(f) = \int_{-\infty}^{\infty} x(t)e^{-j2\pi ft}\,dt \tag{4.3}$$

where:

- X(f) = Frequency domain representation of the respiratory signal
- x(t) = Time-domain respiratory signal
- f = Frequency in Hertz (cycles per second)

For non-rhythmic or non-stationary signals, such as those often observed in patients with respiratory distress or disease, wavelet transforms offer an effective method for analyzing the time–frequency characteristics of the signal. Wavelet analysis allows for the identification of transient events, such as irregular breathing or apnea episodes, by decomposing the signal into components at multiple scales [61, 62]. This enables a detailed examination of the local features of the breathing pattern, even when the respiratory signal is highly variable.

Equation 4.4 Continuous Wavelet Transform

$$W_{\psi}(a, b) = \int_{-\infty}^{\infty} x(t)\,\psi^{*}\left(\frac{t-b}{a}\right)dt \tag{4.4}$$

where:

- Wψ(a,b) = Wavelet transform of the signal x(t)x(t)x(t)
- a = Scale parameter
- b = Translation parameter
- ψ = Wavelet function

4.3.3 Breathing Pattern Abnormalities and Their Detection

The detection of abnormal breathing patterns is a critical aspect of respiratory diagnostics. Deviations from standard breathing patterns can indicate a range of underlying respiratory conditions, including apnea, hypoventilation, hyperventilation, and Cheyne-Stokes respiration. Advanced signal processing techniques are required to detect these irregularities and provide timely alerts to healthcare providers.

Apnea refers to the temporary cessation of breathing for at least 10 s. The detection of apnea is crucial, especially in conditions such as sleep apnea, where periodic cessation of breathing occurs during sleep. The most common method for apnea detection is through capnography, which measures the concentration of carbon dioxide (CO_2) in exhaled air. A sudden drop in CO_2 levels, accompanied by a lack of airflow, can be indicative of an apnea event. In addition to capnography, EEG and ECG are also employed in sleep studies to monitor breathing patterns during sleep. Signal processing techniques, such as thresholding, peak detection, and pattern recognition, are applied to identify apnea episodes in the time-series data of respiratory signals [63, 64].

Tachypnea refers to abnormally rapid breathing, while bradypnea refers to abnormally slow breathing. Both of these conditions can be indicative of underlying respiratory or systemic issues, such as respiratory infections, fever, heart failure, or drug overdose. The detection of tachypnea and bradypnea is achieved by analyzing breathing rate over time and identifying deviations from normal ranges. Advanced signal processing techniques are employed to detect transient changes in breathing rate. For instance, by continuously monitoring respiratory signals from wearable devices or chest-wall sensors, rapid changes in breathing rate can be flagged as abnormal. Additionally, short-term variability in breathing rate can be analyzed to distinguish between healthy variations and potentially harmful trends. This analysis is especially important in critical care settings, where real-time detection of tachypnea or bradypnea can help clinicians identify patients at risk of respiratory failure.

4.3.4 Real-Time Monitoring of Breathing Rate and Patterns

The evolution of sensor technology and signal processing techniques has enabled the development of advanced systems for real-time monitoring of breathing rate and patterns. These systems are essential in both clinical settings and home healthcare applications, where continuous assessment of respiratory function is crucial for managing patients with chronic respiratory diseases or those recovering from acute episodes [39, 65].

One of the most significant advancements in respiratory monitoring has been the development of wearable devices that track breathing rate and patterns in real-time.

These devices utilize a combination of sensors, including accelerometers, gyroscopes, and strain gauges, to capture chest wall movement, thoracic expansion, and respiratory effort. By processing the data from these sensors, algorithms can calculate breathing rate and analyze patterns of respiratory variability (Fig. 4.9).

The use of accelerometers and gyroscopes enables motion-based monitoring, which measures changes in body movement that occur during respiration. These devices track both linear acceleration and angular velocity of the chest during the breathing cycle. This method is particularly useful for detecting breathing rate and

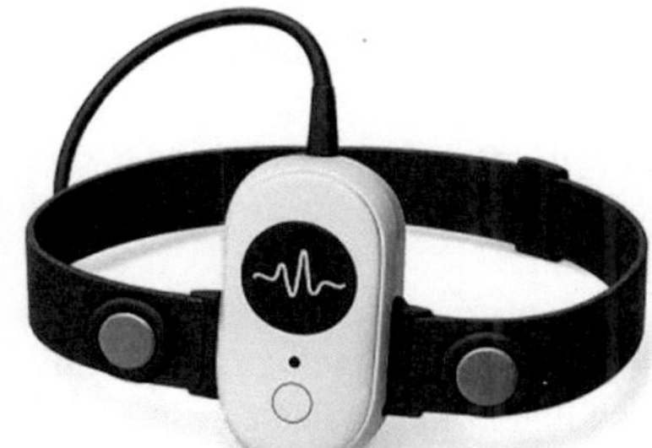

Fig. 4.9 An illustration of a wearable respiratory monitor

respiratory effort, and can be easily integrated into everyday wearables, such as smartwatches and fitness trackers. In addition to motion-based sensors, some wearable devices incorporate strain gauges or pressure sensors to monitor the expansion and contraction of the chest and abdomen during breathing. By detecting these mechanical changes in the chest wall, these sensors can calculate tidal volume, inspiratory/expiratory ratios, and respiratory rate. Advanced wearable devices can also incorporate algorithms that detect abnormal breathing patterns such as sleep apnea, tachypnea, or periodic breathing. For example, machine learning-based models can analyze patterns in breathing data to identify apnea episodes and periodic breathing events in real time. These devices send alerts to healthcare providers or caregivers if abnormal breathing patterns are detected, ensuring timely intervention [66, 67].

Telemedicine platforms are increasingly being utilized to monitor patients with chronic respiratory conditions, including asthma, COPD, and pulmonary fibrosis. These systems utilize wearable sensors in conjunction with cloud-based data analysis to continuously monitor respiratory parameters. The data from the sensors is transmitted to healthcare providers, who can use the information to adjust treatment plans or intervene if a patient's condition worsens.

For instance, some systems use a smartphone app that integrates with wearable respiratory sensors to track breathing rate and patterns. The app provides real-time feedback to patients about their respiratory health, and the data is accessible to clinicians for continuous monitoring. This approach enables the early detection of exacerbations and helps reduce hospital readmissions by facilitating proactive care. The integration of machine learning and AI into these systems enables the automatic detection of abnormal breathing events, triggering real-time alerts. AI algorithms analyze the collected data to identify patterns that may not be immediately obvious to clinicians, enhancing the accuracy of early diagnosis and intervention.

While spirometry and plethysmography measure mechanical aspects of breathing, capnography offers a unique perspective by focusing on gas exchange during respiration. Capnography measures the concentration of CO_2 in exhaled breath, providing valuable information about ventilation and metabolic function. The most common form of capnography used in clinical practice is mainstream capnography, where a sensor is placed directly in the exhaled air pathway to measure CO_2 levels during the exhalation phase [3]. Sidestream capnography, on the other hand, involves sampling the exhaled air remotely and transporting it to the sensor for analysis. By monitoring the end-tidal CO_2 ($ETCO_2$) levels, clinicians can gain insights into the patient's ventilatory status. Abnormalities in $ETCO_2$ levels, such as sudden drops or increases, can signal issues like apnea, hypoventilation, or hyperventilation, which require immediate intervention. In addition to detecting abnormal breathing patterns, capnography can also be used to assess the efficiency of ventilation and the adequacy of gas exchange.

Quantifying breathing patterns goes beyond simply measuring the rate of respiration. It involves analyzing variability, depth, and rhythm to detect subtle abnormalities that could indicate respiratory distress or disease progression. A comprehensive approach to breathing pattern analysis involves integrating multiple parameters, including tidal volume, breathing rate variability, and the inspiratory-expiratory ratio.

One novel approach for quantifying breathing patterns is the use of fractal analysis, which examines the self-similarity and complexity of respiratory signals over time. Fractal analysis can be used to characterize the irregularity of breathing patterns and identify abnormal fluctuations in respiration. For example, healthy breathing is typically characterized by a high degree of regularity and predictability. In contrast, patients with COPD or heart failure may exhibit more erratic breathing patterns.

Another promising approach is the application of chaos theory to respiratory signals. Chaotic breathing refers to highly irregular and unpredictable breathing patterns, often seen in patients with severe respiratory conditions. By applying chaotic time series analysis to respiratory data, clinicians can identify non-linear dynamics in breathing patterns that are indicative of underlying pathophysiology. These techniques can be particularly useful for early detection of respiratory decompensation in critically ill patients [3, 68]. The application of machine learning (ML) and deep learning (DL) techniques to the detection and classification of breathing patterns has opened new avenues in respiratory diagnostics. By training ML models on large datasets of respiratory signals, these algorithms can learn to differentiate between normal and abnormal breathing patterns with high accuracy. ML-based models can classify breathing patterns, including normal, obstructive, central, and mixed apnea, as well as other conditions such as tachypnea and bradypnea.

One promising technique is convolutional neural networks (CNNs), which have shown excellent performance in analyzing time-series respiratory signals. CNNs are particularly suited for extracting spatial and temporal features from sensor data such as chest wall motion or capnographic waveforms. By training CNNs on large datasets of labeled breathing signals, these models can recognize subtle patterns in the data and classify them according to clinical significance.

Equation 4.5 CNN model for breathing pattern classification

$$y = f(x, \theta) = \mathrm{CNN}(x; \theta) \tag{4.5}$$

where:

- x = Input respiratory signal (e.g., chest wall motion or capnography waveform)
- θ = Parameters of the trained CNN model
- y = Output classification (e.g., normal, apnea, tachypnea)

Machine learning models have the potential to automate the process of detecting breathing patterns, providing real-time insights into a patient's respiratory health. These systems can integrate with clinical decision support tools to alert healthcare providers of potential respiratory complications, thus improving patient outcomes.

4.3.5 Advanced Algorithms for Breathing Pattern Detection

To enhance the detection of breathing patterns and breathing rate, researchers have developed a range of advanced algorithms that utilize signal processing techniques to identify subtle variations in respiratory signals. These algorithms often employ time-domain, frequency-domain, and time–frequency analyses to extract meaningful features from respiratory data, enabling the real-time detection of various breathing patterns, including both normal and abnormal respiration [69, 70].

In time-domain analysis, the primary focus is on examining the amplitude and frequency variations in respiratory signals over time. For instance, peak detection algorithms can be applied to detect the inspiratory peaks and expiratory troughs of the respiratory signal. By counting these peaks, the breathing rate can be calculated. Additionally, breath duration can be measured by analyzing the time between successive peaks, and other features such as tidal volume and respiratory cycle duration can be extracted. The time-domain approach is often used in systems that rely on signals such as thoracic movement or strain gauge readings, where changes in the physical chest motion are tracked during inhalation and exhalation. Dynamic thresholding is another method that uses fixed or adaptive thresholds to detect inspiration and expiration events in respiratory waveforms [3].

Equation 4.6 Breathing Rate Calculation Using Time-Domain Analysis

$$BR = \frac{N}{T} \tag{4.6}$$

where:

- B = Breathing rate (breaths per minute)
- N = Number of detected respiratory cycles (inspirations)
- T = Total observation time (in minutes)

In contrast, frequency-domain analysis focuses on the spectral characteristics of the respiratory signals, typically employing techniques such as the FFT). By transforming the respiratory signal from the time domain into the frequency domain, it is possible to extract information related to the signal's periodicity and detect the dominant frequencies associated with breathing cycles. For example, a regular breathing pattern typically exhibits low-frequency components in the range of 0.1 to 0.5 Hz, which corresponds to the typical rate of normal respiration. In contrast, tachypnea (rapid breathing) typically exhibits higher-frequency components, whereas bradypnea (slow breathing) characterizes lower-frequency components. The power spectral density (PSD) of the respiratory signal can also reveal information about the regularity and rhythmicity of breathing, helping to differentiate between various abnormal breathing patterns, such as those seen in sleep apnea or COPD.

For more complex breathing patterns, particularly those that exhibit non-stationarity or irregularity, time–frequency analysis methods are employed. These methods combine the strengths of both time- and frequency-domain analysis, allowing for the detection of abnormalities in breathing patterns that vary over time. Wavelet transform and Short-Time Fourier Transform (STFT) are standard techniques for time–frequency analysis. The Wavelet Transform enables the decomposition of respiratory signals into distinct frequency bands at specific time intervals, providing a more flexible and localized representation of the signal. This is particularly useful for detecting apnea or periodic breathing events, which have highly variable frequencies over time. The wavelet transform is capable of identifying both high-frequency, rapid breathing and low-frequency, slow breathing in real-time [3, 71].

Equation 4.7 Wavelet Transform

$$\Psi\,(t,a,b)=\frac{1}{\sqrt{|a|}}\Psi\left(\frac{t-b}{a}\right) \tag{4.7}$$

where:

- Ψ(t,a,b) = Wavelet function at time t, scale a, and shift b
- a = Scale factor (determines the resolution of the wavelet)
- b = Time shift factor

4.3.6 Machine Learning Techniques for Pattern Classification

The application of ML techniques has revolutionized the field of breathing pattern detection by enabling more accurate and efficient classification of respiratory events. These techniques utilize large datasets of labeled respiratory signals to train models that can accurately identify both normal and abnormal breathing patterns. Among the most common ML approaches used in respiratory signal processing are support vector machines (SVMs), decision trees, and random forests, as well as more advanced techniques like deep learning and recurrent neural networks (RNNs).

Support Vector Machines are particularly effective for classifying respiratory signals because they are robust to noise and can handle complex, high-dimensional data. SVMs are used to classify breathing patterns into normal, obstructive, central, and mixed apneas, as well as tachypnea and bradypnea. They work by finding the optimal hyperplane that separates different classes of data points in feature space, maximizing the margin between classes [13, 72].

Equation 4.8 SVM Decision Boundary

$$f\,(x)=w^Tx+b \tag{4.8}$$

where:

- w = Weight vector
- b = Bias term
- x = Input feature vector (e.g., respiratory signal features)

More recently, deep learning methods, particularly long short-term memory (LSTM) networks, have gained popularity for analyzing temporal respiratory signals.

These recurrent neural networks are capable of capturing long-term dependencies in the data, making them well-suited for classifying sequential breathing patterns and detecting rare or intermittent events, such as apnea episodes. LSTMs can retain context from previous time steps in the sequence, allowing for a more comprehensive understanding of the respiratory signal's underlying patterns.

Equation 4.9 LSTM Update Equations

$$\begin{aligned} f_t &= \sigma\left(W_f \cdot \left[h_{t-1}, x_t\right] + b_f\right) \\ i_t &= \sigma\left(W_i \cdot \left[h_{t-1}, x_t\right] + b_i\right) \\ \hat{C}_t &= \tanh\left(W_C \cdot \left[h_{t-1}, x_t\right] + b_C\right) \\ C_t &= f_t \cdot C_{t-1} + i_t \cdot \hat{C}_t \\ o_t &= \sigma\left(W_o \cdot \left[h_{t-1}, x_t\right] + b_o\right) \\ h_t &= o_t \cdot \tanh(C_t) \end{aligned} \tag{4.9}$$

where:

- f_t = Forget gate output at time t
- i_t = Input gate output at time t
- C_t^ = Candidate cell state at time t
- C_t = Cell state at time t
- o_t = Output gate at time t
- h_t = Hidden state at time t

4.3.7 Clinical Applications and Implications

The ability to detect and analyze breathing patterns in real-time is not only beneficial for managing chronic respiratory conditions, but it also has broad applications in critical care and surgical environments. Continuous monitoring of breathing rate and patterns during intensive care unit (ICU) stays allows healthcare providers to identify early signs of respiratory failure and make timely interventions. Moreover, sleep studies have greatly benefited from advanced signal processing techniques, particularly in the detection of sleep apnea. Polysomnography, which traditionally involved multiple sensors to measure various physiological signals during sleep, can now be enhanced with machine learning algorithms that detect apnea events based on breathing patterns alone. In surgical settings, anesthesia monitoring also relies heavily on detecting breathing patterns to ensure the patient maintains adequate ventilation during procedures. Anesthesia drugs can significantly alter respiratory patterns, and detecting hypoventilation or hyperventilation in real time allows anesthesiologists to adjust their interventions appropriately. Finally, wearable devices and telemedicine platforms that integrate breathing pattern analysis enable remote monitoring of patients with respiratory conditions, reducing hospital visits and providing personalized care in the comfort of a patient's home. These advances in signal processing, combined with machine learning, will continue to evolve, enabling more precise, continuous, and non-invasive methods of monitoring respiratory health and ensuring better management of both acute and chronic respiratory conditions [44].

4.3.8 Filtering of Respiratory Signals

In respiratory signal processing, filtering plays a crucial role in improving the quality and accuracy of signals obtained from various respiratory monitoring devices. These signals are often noisy and prone to artifacts from various sources, such as motion, electrical interference, and physiological variations. Effective filtering techniques are crucial for isolating the relevant physiological components from unwanted noise, enabling precise analysis of respiratory patterns, respiratory rate, tidal volume, and other vital parameters [73].

The primary goal of filtering in respiratory signal processing is to remove unwanted noise while preserving the important features of the respiratory signals. These features include the breathing cycles, tidal volume fluctuations, and phase relationships between respiratory events. Filtering techniques are applied to various types of respiratory signals, such as airflow, thoraco-abdominal movement, gas exchange signals, and bioelectrical signals from the respiratory muscles.

Filtering serves several functions in respiratory signal analysis:

1. Noise Reduction: Respiratory signals are often contaminated by electrical interference, motion artifacts, and low-frequency drift. These noises can obscure the accurate physiological signals, making it challenging to detect relevant features. Filtering helps mitigate these effects.
2. Signal Enhancement: By removing irrelevant components from the signal, filtering enhances the ability to detect small, subtle features in the data, such as early signs of respiratory distress or abnormal patterns like apnea or hypopnea.
3. Improving Signal-to-Noise Ratio (SNR): Respiratory signals often contain low-amplitude variations that are buried in noise. Filtering enhances the SNR, facilitating the interpretation of data and the acquisition of reliable measurements.

Different types of filters are used in respiratory signal processing depending on the specific type of noise being removed and the characteristics of the signal. These filters can be broadly classified into linear filters, non-linear filters, and adaptive filters, each with distinct advantages and applications [14, 74].

Linear filters are the most commonly used type of filter in respiratory signal processing. They apply a weighted sum of the input signal values to produce the output signal, and the output is a linear function of the input. Linear filters can be further divided into low-pass, high-pass, band-pass, and band-stop filters:

1. Low-Pass Filters: These filters allow low-frequency components of the respiratory signal to pass through while attenuating high-frequency noise. In respiratory signal processing, low-pass filtering is often used to eliminate electrical interference or motion artifacts that appear at higher frequencies. For example, during

breathing cycle analysis, low-frequency respiratory rhythms are retained, while higher-frequency noise components, such as muscle tremors, are removed.
2. High-Pass Filters: These filters allow high-frequency components to pass through while attenuating low-frequency drift and baseline fluctuations. For example, high-pass filters are used to remove drift in the respiratory signal due to changes in ambient temperature or sensor placement. High-pass filtering is often applied in the analysis of respiratory muscle signals and ventilatory signals.
3. Band-Pass Filters: These filters are used to isolate a specific frequency band of interest while attenuating both low and high-frequency noise. For example, breathing frequencies typically fall within the range of 0.1 to 1 Hz, and a band-pass filter can be used to isolate the relevant respiratory signal while removing irrelevant low-frequency drift and high-frequency noise.
4. Band-Stop Filters: These filters are used to remove a specific frequency range, such as when power-line interference (50 or 60 Hz) contaminates the signal. Band-stop filters can eliminate these narrow-frequency noise components without affecting the rest of the respiratory signal.

Non-linear filters are used when non-Gaussian noise is present in the signal, or when the signal contains sharp transitions or non-stationary characteristics that linear filters cannot accurately process. For instance, median filters are often employed to eliminate spike noise without distorting the signal. These filters are handy when analyzing spontaneous breathing patterns or irregular breath sounds. The median filter operates by sorting the values in a given window and replacing the central value with the median of those values. This method preserves edges and removes impulsive noise that would otherwise distort the signal.

Equation 4.10 Median filter operation

$$y[n] = \text{median}(x[n-k], x[n-k+1], \ldots, x[n+k]) \tag{4.10}$$

where:

- x[n] = Input signal
- y[n] = Output signal after filtering
- k = Window size

Adaptive filters dynamically adjust their characteristics in response to the input signal and the noise environment. These filters are used in situations where the characteristics of the noise are not known in advance or are subject to change over time. An example is the adaptive noise cancellation (ANC) method, which uses a secondary reference signal to estimate and cancel out noise in respiratory signals, such as artifacts caused by patient movement or external interference.

In respiratory signal processing, adaptive filters can track the time-varying nature of the respiratory signal and remove noise without causing distortion. These filters are particularly effective in scenarios where respiratory signals are contaminated by environmental noise or motion artifacts.

Equation 4.11 Adaptive filter error minimization

$$e[n] = d[n] - \hat{y}[n] \tag{4.11}$$

where:

- e[n] = Error signal (difference between the desired signal and the output of the adaptive filter)
- d[n] = Desired signal (reference signal)
- y^[n] = Output of the adaptive filter

In real-time respiratory signal processing, filtering must be performed with minimal latency to provide timely feedback for clinical decision-making. For instance, continuous monitoring of respiratory rate and tidal volume during surgery, as well as the detection of apnea and hypopnea during sleep studies, requires an efficient and responsive filtering process. Real-time digital signal processors (DSPs) and hardware-based filters are often used in conjunction with software algorithms to ensure low-latency processing.

4.3.9 Filtering of Respiratory Signal Artifacts

In respiratory signal processing, artifacts are undesired distortions or interferences that can obscure or alter the accurate physiological signal. These artifacts can arise from a variety of sources, including motion, electrical interference, sensor placement, and even physiological changes unrelated to respiration, such as cardiac activity. Filtering is a critical tool used to minimize or eliminate these artifacts, allowing for accurate and reliable interpretation of respiratory signals. In this chapter, we will explore the types of respiratory signal artifacts, their sources, and the filtering techniques employed to mitigate their effects [9, 58].

Types of Respiratory Signal Artifacts

Artifacts in respiratory signals can be broadly categorized into motion artifacts, electrical interference, baseline drift, muscle activity artifacts, and cardiac-related artifacts. Each of these artifacts can distort the signal in different ways and requires specific filtering strategies to correct.

Motion artifacts occur when external forces, such as body movement or patient shifting, interfere with the respiratory signal. These artifacts are particularly common in wearable devices or ambulatory monitoring systems, where patients may move during the monitoring process. Common examples of motion artifacts include:

- Shifting of respiratory sensors: Movement of the chest or abdomen during breathing can cause mechanical strain on sensors, leading to inaccurate measurements of lung volume or airflow.
- Body posture changes: Shifting from a sitting to a lying position can alter the characteristics of respiratory signals, introducing low-frequency drifts or high-frequency fluctuations.

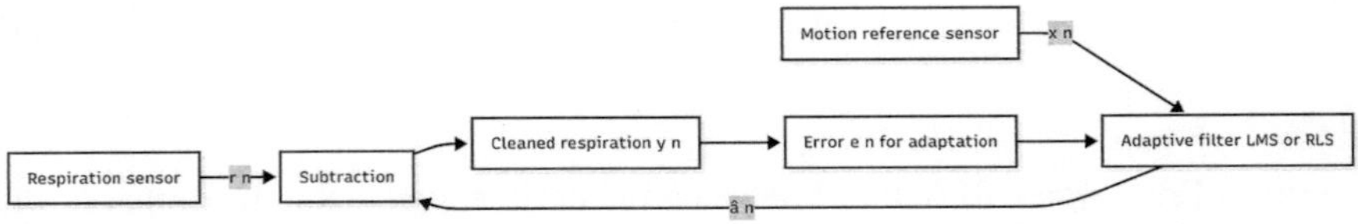

Fig. 4.10 Adaptive filtering process for removing motion artifacts in respiratory signals

To remove motion artifacts from respiratory signals, high-pass filters and adaptive filtering techniques are often used. Motion artifacts typically comprise low-frequency components that interfere with the respiratory signal. High-pass filters are effective at eliminating these unwanted frequencies without distorting the higher-frequency components of the respiratory waveform.

Equation 4.12 High-pass filter transfer function

$$H(f) = \frac{j\frac{f}{f_c}}{1 + j\frac{f}{f_c}} \tag{4.12}$$

where:

- Fc is the cutoff frequency
- f is the frequency of the signal
- j is the imaginary unit

Adaptive filters, which can dynamically adjust to changes in the signal, are also commonly employed for real-time motion artifact removal. These filters use reference signals from motion sensors to detect and subtract motion-related noise (Fig. 4.10).

Electrical interference is one of the most prevalent sources of noise in respiratory signal acquisition, especially in clinical settings. The presence of other electronic equipment, such as monitors, ECGs, and electrosurgical units, can introduce high-frequency noise into the respiratory signal. This is typically observed as spikes or sudden surges in the signal, which can obscure the underlying respiratory waveform. In addition to equipment interference, power-line interference (50 or 60 Hz) is a common artifact in signal acquisition, caused by the electrical grid. This type of interference can significantly distort the respiratory signal, particularly when measurements are made using impedance pneumography or EMG. To address electrical interference, particularly power-line noise (50 or 60 Hz), notch filters or band-stop filters are often applied. These filters specifically target the frequency of the interference, allowing the respiratory signal to pass through while attenuating the unwanted noise.

Equation 4.13 Notch filter transfer function

$$H(f) = \frac{1}{1 + j\frac{f}{f_0}} \tag{4.13}$$

where:

- f_0 is the frequency of interference (e.g., 50 or 60 Hz)

For more generalized electrical interference across a range of frequencies, a band-stop filter can be used to eliminate a wider band of noise.

Baseline drift refers to a slow and gradual shift in the respiratory signal, typically caused by sensor displacement, temperature fluctuations, or changes in electrode impedance. This artifact can cause long-term trends in respiratory signals that do not reflect true physiological changes, leading to misinterpretation of data, particularly in long-duration monitoring or chronic condition monitoring. Baseline drift typically manifests as a slow-moving offset in the signal, often appearing as a low-frequency component [18]. Baseline drift is typically handled by applying high-pass filtering techniques that remove low-frequency components. In some cases, differentiation techniques or polynomial fitting methods can be employed to track and subtract the baseline drift over time.

Equation 4.14 High-pass filter for baseline drift removal

$$H(f) = \frac{1}{1 + j\frac{f}{f_c}} \tag{4.14}$$

where:

- fc is the cutoff frequency that determines the threshold below which the signal is filtered.

In cases of long-duration monitoring, where baseline drift accumulates over time, polynomial fitting techniques can be used to model the drift and subtract it from the signal. Respiratory signals are often contaminated by EMG signals from muscles surrounding the respiratory system, such as the diaphragm, intercostals, and abdominal muscles. These EMG signals can introduce high-frequency noise that masks the true respiratory pattern. For instance, during shallow breathing or quiet breathing, muscle activity can cause high-amplitude, high-frequency spikes in the signal that are unrelated to changes in airflow or lung volume [74].

Muscle activity artifacts are typically high-frequency and can be reduced using low-pass filters or band-pass filters. The cutoff frequency of the filter is selected based on the characteristic frequency of muscle activity, which typically ranges from 50 to 150 Hz for most respiratory-related muscle signals. For more complex muscle artifact removal, wavelet transform techniques can be used. These methods decompose the signal into different frequency bands, allowing for the selective filtering of high-frequency muscle activity without compromising the respiratory signal.

Equation 4.15 Low-pass filter for muscle artifact removal

$$H(f) = \frac{1}{1 + j\frac{f}{f_c}} \tag{4.15}$$

where:

- fc is the cutoff frequency, typically set to a value above the frequency range of muscle activity.

Cardiac-related artifacts, such as cardiac oscillations or respiratory sinus arrhythmia (RSA), occur when the heart rate fluctuates in synchrony with the respiratory cycle. These artifacts can cause variations in the respiratory signal that are mistaken for changes in breathing patterns. For example, the systolic blood pressure pulse may be recorded alongside respiratory signals, producing pulsatile artifacts that complicate the analysis of breathing. Effective filtering is essential for mitigating the effects of artifacts and ensuring the accuracy of respiratory measurements. Various filtering techniques can be applied depending on the type of artifact present. The goal of filtering is to remove noise, preserve respiratory features, and maintain the integrity of the signal for further analysis. Below are some of the most commonly used filtering techniques for handling respiratory signal artifacts [66].

Cardiac-related artifacts, such as those caused by RSA or pulsatile motion, can be effectively filtered using band-pass filters that target the frequency range of the cardiac cycle (typically 0.5–2 Hz). This enables the removal of overlapping heart rate fluctuations from the respiratory signal, preserving the true respiratory pattern intact. In more advanced applications, independent component analysis (ICA) or principal component analysis (PCA) can be applied to separate the cardiac and respiratory components, further improving the accuracy of the signal.

4.4 Clinical Applications of Respiratory Signal Analysis

Respiratory signal analysis plays an increasingly vital role in clinical diagnostics, patient monitoring, therapy optimization, and prognostic evaluation. Respiratory signals—such as airflow, respiratory rate, tidal volume, end-tidal CO_2, and thoracoabdominal movements—reflect pulmonary and systemic health status. These signals are typically collected via spirometry, impedance pneumography, capnography, RIP, or nasal airflow sensors [75, 76]. Analyzing these signals enables clinicians to detect disorders like obstructive sleep apnea (OSA), COPD, asthma, hypoventilation, and acute respiratory failure. Furthermore, integration of respiratory metrics into multimodal monitoring—alongside ECG and blood oxygen saturation—enhances early warning capability.

4.4.1 Respiratory Monitoring in Critical Care and Emergency Settings

In critical care and emergency environments, respiratory signal monitoring is essential for assessing ventilation, detecting deterioration, and guiding life-saving interventions. Accurate interpretation of respiratory signals enables rapid response to hypoventilation, airway obstruction, or respiratory arrest.

In the ICU, respiratory signal analysis is used for ventilator management, weaning protocols, and sedation monitoring. Parameters like respiratory rate variability (RRV), peak inspiratory flow, and dynamic compliance guide real-time decisions.

Respiratory inductance plethysmography and capnography are widely used to evaluate ventilatory patterns and the efficacy of mechanical ventilation. Capnography, particularly $EtCO_2$, serves as a noninvasive marker for ventilation-perfusion matching and cardiac output estimation [3, 18]. Studies have shown that continuous monitoring of respiratory rate and $EtCO_2$ enhances early detection of sepsis, opioid-induced hypoventilation, and ventilator-associated complications [53, 77]. In ventilated patients, waveform analysis can reveal auto-PEEP, inspiratory flow limitation, and patient-ventilator asynchrony.

In the emergency setting, respiratory signals are vital in triage and early warning systems. Tools like impedance pneumography and nasal airflow sensors help detect obstructed breathing, apnea, or rapid shallow breathing—indicators of distress or impending respiratory failure. Integration of respiratory monitoring into Early Warning Scores (EWS) improves predictive power for cardiac arrest or ICU admission. Wearable biosensors capable of continuous respiratory tracking have been validated in EDs to monitor patients under observation protocols, especially for overdose, trauma, or COPD exacerbations [42, 78].

In mechanically ventilated patients, respiratory signal analysis aids in:

- Determining optimal positive end-expiratory pressure (PEEP)
- Identifying readiness for extubation through weaning indices
- Preventing ventilator-induced lung injury (VILI)

Smart ventilators utilize real-time respiratory waveform analytics to adjust tidal volume, flow rate, and inspiratory-expiratory ratio. Adaptive ventilation modes use feedback from airflow and pressure sensors to reduce asynchrony.

4.4.2 Respiratory Signals in Sleep Medicine and Home Monitoring

Respiratory signal analysis is essential in diagnosing and managing sleep-related breathing disorders, particularly OSA, central sleep apnea (CSA), and hypoventilation syndromes. With the rising prevalence of sleep-disordered breathing, accurate, non-invasive monitoring is crucial in both clinical sleep laboratories and at-home settings [79, 80].

Sleep Apnea Diagnosis and Classification

Polysomnography (PSG) remains the gold standard for sleep apnea diagnosis. It includes multiple respiratory channels—nasal airflow, thoracoabdominal movement, oxygen saturation, and capnography—to classify apneas and hypopneas. Obstructive events are characterized by effortful but blocked respiration, while central events lack respiratory effort. Subtle waveform reductions and desaturations identify respiratory effort-related arousals (RERAs) and hypopneas. Advanced signal processing techniques, including entropy analysis, wavelet transforms, and airflow variability analysis, have improved automated detection of apneas. AI models trained on respiratory features now offer sensitivity $> 90\%$ for OSA diagnosis [3, 81].

Home Sleep Testing and Telemonitoring

Portable sleep monitors using fewer channels (Type III devices) are increasingly used for home sleep apnea testing (HSAT). These systems typically analyze nasal airflow, respiratory effort, and oxygen saturation. Signal quality assurance and artifact rejection are critical challenges in unsupervised environments. Recent studies have validated wearable respiratory sensors (e.g., piezo belts, patch-based RIP) that transmit data to cloud servers for remote interpretation. AI-enhanced HSAT systems have demonstrated diagnostic performance approaching in-lab PSG under selected conditions [49, 82]. The evaluations emphasized that accurate digitization of respiratory waveforms—especially under low amplitude or irregular patterns—is essential for reliable classification of sleep-disordered events.

CPAP Therapy Monitoring and Adherence

Continuous positive airway pressure (CPAP) therapy is the frontline treatment for OSA. Modern CPAP machines record airflow and pressure data to monitor adherence, residual apneas, and mask leaks. Respiratory signal analysis is used to:

- Quantify apnea–hypopnea index (AHI)
- Detect expiratory pressure intolerance
- Identify central apneas emerging during treatment (complex sleep apnea)

Cloud-based CPAP platforms allow clinicians to adjust settings and assess treatment efficacy remotely.

Pediatric Sleep Monitoring

In infants and children, respiratory signals are used to diagnose congenital central hypoventilation syndrome, upper airway obstruction, and nocturnal hypoxemia associated with neuromuscular disease. Wearable devices designed for pediatric use must address motion artifacts, low signal amplitude, and variable breathing patterns. Advances in AI-based pattern recognition and signal fusion have improved reliability in this population [14, 83].

4.4.3 Respiratory Signal Use in Perioperative and Postoperative Monitoring

In the perioperative and postoperative periods, respiratory signal monitoring is critical to prevent complications such as hypoventilation, airway obstruction, and respiratory depression due to anesthesia or opioid use. These signals assist in early detection of deterioration, guide analgesia protocols, and support recovery optimization.

Preoperative evaluation of respiratory function is used to identify patients at risk for intraoperative and postoperative respiratory complications. Parameters such as baseline respiratory rate, tidal volume, and SpO_2 variability provide insight into pulmonary reserve. Patients with COPD, obesity hypoventilation syndrome, or undiagnosed OSA often exhibit subtle abnormalities in pre-op respiratory signals. Digital spirometry, portable RIP sensors, and home capnography can capture these patterns preemptively, allowing anesthesiologists to adjust ventilation strategies and airway management plans [39, 84].

During surgery, respiratory signal monitoring ensures adequate ventilation under general anesthesia. Volume-controlled and pressure-controlled ventilation modes are tailored using continuous feedback from capnography, flow sensors, and airway pressure monitors. End-tidal CO_2 monitoring is essential for confirming intubation, assessing ventilation adequacy, and detecting circuit disconnections. Sudden changes in $EtCO_2$ may indicate hypotension, embolism, or airway obstruction [53, 71]. Smart anesthesia workstations integrate respiratory signals with depth-of-anesthesia monitoring, creating adaptive feedback loops.

In the recovery phase, especially in post-anesthesia care units (PACUs), patients are vulnerable to opioid-induced respiratory depression. Continuous respiratory rate, airflow, and $EtCO_2$ monitoring can detect early signs of hypoventilation or apnea before oxygen saturation drops. Wearable devices and contactless monitoring systems are gaining traction for postoperative surveillance. AI algorithms are being used to recognize hazardous breathing patterns and differentiate between true apneas and motion artifacts [34, 85].

Respiratory signal analysis is now embedded in ERAS pathways to promote fast-track recovery. By tracking respiratory effort and oxygenation during early mobilization, clinicians can optimize analgesia and detect complications such as atelectasis or pulmonary embolism. Systems that combine ECG, pulse oximetry, and respiratory waveform monitoring allow for multidimensional assessments of cardiorespiratory status. These tools support personalized recovery planning and real-time risk stratification.

4.4.4 Predictive Analytics and Telemedicine Using Respiratory Signals

Advances in AI, remote sensing, and mobile health have transformed respiratory signal analysis into a tool not only for diagnosis but also for real-time prediction and remote patient management. Predictive analytics leverages large-scale respiratory data to anticipate clinical deterioration, optimize therapy, and streamline care delivery.

Machine learning (ML) models can be trained on respiratory features—such as rate variability, airflow shape, tidal volume dynamics, and capnographic trends—to predict adverse events like:

- Imminent respiratory failure
- Sepsis-related hypoventilation
- Need for re-intubation
- Risk of prolonged weaning from mechanical ventilation

Recent studies have shown that models incorporating respiratory rate variability and $EtCO_2$ trajectory outperform standard clinical scores in early warning detection [58, 73, 86]. Deep learning techniques, including recurrent neural networks (RNNs), can process sequential breathing waveforms to detect temporal deterioration patterns.

Wearable and ambient respiratory monitors, including chest bands, under-mattress sensors, and camera-based systems, enable long-term data collection outside clinical environments. These data streams are used for:

- Chronic disease management (e.g., COPD, CHF)
- Hospital-at-home programs
- Tele-ICU patient surveillance

Platforms like remote respiratory telemonitoring for post-COVID recovery have demonstrated improved outcomes and patient satisfaction. These systems often include AI to reduce alarm fatigue, personalize alerts, and flag high-risk trends for clinical follow-up.

AI-enabled respiratory signal analysis is also being used in patient-facing apps to provide personalized feedback. Breathing training, biofeedback for anxiety or asthma control, and rehabilitation coaching for pulmonary recovery all use signal-derived metrics. For example, guided rehabilitation for COPD patients may involve real-time visualization of respiratory depth and rhythm, encouraging adherence to breathing exercises. Machine learning algorithms adapt difficulty levels based on individual progress, increasing engagement and therapeutic benefit [3, 55, 87].

Future clinical decision-making increasingly relies on multimodal data fusion—combining respiratory signals with ECG, SpO_2, blood pressure, and activity data. Integrated analytics platforms assess these signals holistically to generate comprehensive patient risk profiles. In predictive analytics platforms, respiratory rate trends and tidal volume fluctuations contribute to composite risk scores for sepsis, cardiac arrest, or acute respiratory distress syndrome (ARDS).

References

1. Auten R, Schwarze J, Ren C, Davis S, Noah TL (2016) Pediatric pulmonology year in review 2015: Part 1. Pediatr Pulmonol 51(7):733–739
2. Birnkrant DJ, Yilmaz O, Nicolai T, Black JB, Mhanna MJ, Noah TL (2016) Pediatric pulmonology year in review 2015: Part 3. Pediatr Pulmonol 51(7):747–753
3. Badnjević A, Deumic A, Ademovic A, Pokvić LG (2023) A novel method for conformity assessment testing of therapeutic ultrasounds for post-market surveillance purposes. Technol Health Care 31(1):339–346
4. Bordoni B, Escher AR (2025) Muscles and central neural networks involved in breathing: state of the art. Cureus. 17(3):e80599
5. Brander L, Slutsky A (2005) Year in review in critical care, 2003 and 2004: Respirology and critical care. Crit Care 9(5):517–522
6. Douglas IS, Bednash JS, Fein DG, Mallampalli RK, Mansoori JN, Gershengorn HB (2019) Update in critical care and acute respiratory distress syndrome 2018. Am J Respir Crit Care Med 199(11):1335–1343
7. Zhang H, Slutsky AS (2009) Year in review 2008: Critical care – Respirology. Crit Care 13(5):225
8. Ventre KM, Wolf GK, Arnold JH (2011) Pediatric respiratory diseases: 2011 update for the Rogers' textbook of pediatric intensive care. Pediatr Crit Care Med 12(3):325–338
9. Spahić L, Jeremić L, Lalatović I, Muratović T, Džuho A, Gurbeta Pokvić L, Badnjević A (2024) Machine learning for improved medical device management: a focus on defibrillator performance. Technol Health Care
10. Fanelli V, Zhang H, Slutsky AS (2011) Year in review 2010: critical care–respirology. Crit Care 15(6):240
11. Haitsma JJ, Villar J, Slutsky AS (2006) Year in review 2005: Critical care–Respirology: mechanical ventilation, infection, monitoring, and education. Crit Care 10(3):217
12. Vasquez D, Singh JM, Ferguson ND (2007) Year in review 2006: critical care–respirology. Crit Care 11(4):224
13. Hundur M, Spahić L, Bećirović F, Gurbeta Pokvić L, Badnjević A (2025) Advancement of post-market surveillance of medical devices leveraging artificial intelligence: ECG devices case study. Technol Health Care
14. Merdović N, Spahić L, Hundur M, Pokvić LG, Badnjević A (2024) Advancement of post-market surveillance of medical devices leveraging artificial intelligence: Infusion pumps case study. Technol Health Care 33(2):915–921
15. Tufail A, Jiang Y, Cui X (2025) A review on the mucus dynamics in the human respiratory airway. Biomech Model Mechanobiol 24(1):107–123
16. Staitieh BS, Ioachimescu OC (2017) Interpretation of pulmonary function tests: beyond the basics. J Investig Med 65(2):301–310
17. Laveneziana P, Albuquerque A, Aliverti A, Babb T, Barreiro E, Dres M, Dubé BP, Fauroux B, Gea J, Guenette JA, Hudson AL, Kabitz HJ, Laghi F, Langer D, Luo YM, Neder JA, O'Donnell D, Polkey MI, Rabinovich RA, Rossi A, Verges S (2019) ERS statement on respiratory muscle testing at rest and during exercise. Eur Respir J 53(6):1801214
18. Peruničić Ž, Lalatović I, Spahić L, Ašić A, Pokvić LG, Badnjević A (2024) Enhancing mechanical ventilator reliability through machine learning-based predictive maintenance. Technol Health Care
19. Kallet RH, Volsko TA, Hess DR (2013) Respiratory care year in review 2012: Invasive mechanical ventilation, noninvasive ventilation, and cystic fibrosis. Respir Care 58(4):702–711
20. Rubin BK, Dhand R, Ruppel GL, Branson RD, Hess DR (2011) Respiratory care year in review 2010: Part 1. Asthma, COPD, pulmonary function testing, ventilator-associated pneumonia. Respir Care 56(4):488–502
21. Badnjević A, Pokvić LG, Deumić A, Bećirović LS (2022) Post-market surveillance of medical devices: a review. Technol Health Care 30(6):1315–1329

22. Karbing DS, Perchiazzi G, Rees SE, Jaffe MB (2018) Journal of clinical monitoring and computing 2017 end of year summary: respiration. J Clin Monit Comput 32(2):197–205
23. Noah TL, Auten R, Schwarze J, Davis S (2015) Pediatric pulmonology year in review 2014: Part 2. Pediatr Pulmonol 50(11):1140–1146
24. Chu M, Nguyen T, Pandey V, Zhou Y, Pham HN, Bar-Yoseph R, Radom-Aizik S, Jain R, Cooper DM, Khine M (2019) Respiration rate and volume measurements using wearable strain sensors. NPJ Digit Med. 2:8
25. Fraiwan L, Khasawneh N, Lweesy K, Elbalki M, Almarzooqi A, Abu HN (2021) Non-contact spirometry using a mobile thermal camera and AI regression. Sensors 21(22):7574
26. Glaab T, Taube C, Braun A, Mitzner W (2007) Invasive and noninvasive methods for studying pulmonary function in mice. Respir Res 8(1):63
27. Khemani RG, Flink R, Hotz J, Ross PA, Ghuman A, Newth CJ (2015) Respiratory inductance plethysmography calibration for pediatric upper airway obstruction: an animal model. Pediatr Res 77(1):75–83
28. Monaco V, Stefanini C (2021) Assessing the tidal volume through wearables: a scoping review. Sensors 21(12):4124
29. Rahman MS, Chowdhury S, Rasheduzzaman M, Doulah ABMSU (2024) Artificial intelligence-based algorithms and healthcare applications of respiratory inductance plethysmography: a systematic review. Algorithms 17(6):261
30. Toften S, Kjellstadli JT, Thu OKF, Ellingsen OJ (2022) Noncontact longitudinal respiratory rate measurements in healthy adults using radar-based sleep monitor (Somnofy): validation study. JMIR Biomed Eng 7(2):e36618
31. Vitazkova D, Foltan E, Kosnacova H, Micjan M, Donoval M, Kuzma A, Kopani M, Vavrinsky E (2024) Advances in respiratory monitoring: a comprehensive review of wearable and remote technologies. Biosensors 14(2):90
32. Ren S, Wang X, Cai M, Shi Y, Wang T, Luo Z (2025) Design of a multi-parameter fusion sensor and system for respiratory monitoring of mechanically ventilated patients in the ICU. IEEE J Biomed Health Inform 29(2):923–934
33. Hussain T, Ullah S, Fernández-García R, Gil I (2023) Wearable sensors for respiration monitoring: a review. Sensors 23(17):7518
34. Badnjević A, Pokvić LG, Smajlhodžić-Deljo M et al (2024) Application of artificial intelligence for the classification of the clinical outcome and therapy in patients with viral infections: the case of COVID-19. Technol Health Care 32(3):1859–1870
35. Andreozzi E, Centracchio J, Punzo V, Esposito D, Polley C, Gargiulo GD, Bifulco P (2021) Respiration monitoring via forcecardiography sensors. Sensors 21(12):3996
36. Pimentel MAF, Johnson AEW, Charlton PH, Birrenkott D, Watkinson PJ, Tarassenko L, Clifton DA (2017) Toward a robust estimation of respiratory rate from pulse oximeters. IEEE Trans Biomed Eng 64(8):1914–1923
37. Massaroni C, Nicolò A, Schena E, Sacchetti M (2020) Remote respiratory monitoring in the time of COVID-19. Front Physiol 11:635
38. Patel S, Park H, Bonato P, Chan L, Rodgers M (2012) A review of wearable sensors and systems with application in rehabilitation. J Neuroeng Rehabil 9:21
39. Pokvić LG, Deumić A, Softić A, Badnjević A (2024) Application of artificial intelligence for management of maintenance and prediction of performances. In: Badnjević A, Cifrek M, Magjarević R, Džemić Z (eds) Inspection of Medical Devices. Springer, Cham, pp 83–95
40. Massaroni C, Venanzi C, Silvatti AP, Lo Presti D, Saccomandi P, Formica D, Giurazza F, Caponero MA, Schena E (2018) Smart textile for respiratory monitoring and thoraco-abdominal motion pattern evaluation. J Biophotonics 11(5):e201700263
41. Wang X, Li H, Wang T, Niu X, Wang Y, Xu S, Jiang Y, Chen L, Liu H (2022) Flexible and high-performance piezoresistive strain sensors based on multi-walled carbon nanotubes@polyurethane foam. RSC Adv 12(22):14190–14196
42. Martinović M, Kosović M, Spahić L, Softić A, Pokvić LG, Badnjević A (2025) Machine learning for improved medical device management: a focus on dialysis machines. Technol Health Care

43. Alghadhban A, Ramadan RA, Alazmi M (2025) Advancing respiratory disease diagnosis: a deep learning and vision transformer-based approach with a novel X-ray dataset. Comput Biol Med 194:110501
44. Vanegas E, Igual R, Plaza I (2020) Sensing systems for respiration monitoring: a technical systematic review. Sensors 20(18):5446
45. Kalsotra S, Rice-Weimer J, Tobias JD (2023) Intraoperative electromyographic monitoring in children using a novel pediatric sensor. Saudi J Anaesth 17(3):378–382
46. Menuet C, Ben-Tal A, Linossier A et al (2025) Redefining respiratory sinus arrhythmia as respiratory heart rate variability: an international expert recommendation for terminological clarity. Nat Rev Cardiol. https://doi.org/10.1038/s41569-025-01160-z
47. Boiko A, Martínez Madrid N, Seepold R (2023) Contactless technologies, sensors, and systems for cardiac and respiratory measurement during sleep: a systematic review. Sensors 23(11):5038
48. Zhang C, Zhang L, Tian Y, Bao B, Li D (2023) A machine-learning-algorithm-assisted intelligent system for real-time wireless respiratory monitoring. Appl Sci 13(6):3885
49. Spahić L, Sredović U, Kurpejović Z, Mrdanović E, Pokvić G, Badnjević A (2025) Machine learning for improved medical device management: a focus on infant incubators. Technol Health Care
50. Fang Y, Xu J, Xiao X, Zou Y, Zhao X, Zhou Y, Chen J (2022) A deep-learning-assisted on-mask sensor network for adaptive respiratory monitoring. Adv Mater 34(17):2109718
51. Badnjević A, Pokvić LG, Džemić Z, Bečić F (2020) Risks of emergency use authorizations for medical products during outbreak situations: a COVID-19 case study. Biomed Eng Online 19(1):75
52. Keenan DB, Wilhelm FH (2005) Adaptive and wavelet filtering methods for improving accuracy of respiratory measurement. Biomed Sci Instrum 41:37–42
53. Yang L, Ding Z, Zhou J, et al (2024) Algorithmic detection of sleep-disordered breathing using respiratory signals: a systematic review. Physiol Meas 45(3). https://doi.org/10.1088/1361-6579/ad2c13.
54. Chen A, Zhang J, Zhao L et al (2021) Machine-learning enabled wireless wearable sensors to study individuality of respiratory behaviors. Biosens Bioelectron 173:112799
55. Rozo A, Moeyersons J, Morales J et al (2022) Data augmentation and transfer learning for data quality assessment in respiratory monitoring. Front Bioeng Biotechnol. 10:806761
56. Kunczik J, Hubbermann K, Mösch L, Follmann A, Czaplik M, Barbosa PC (2022) Breathing pattern monitoring by using remote sensors. Sensors 22(22):8854
57. Wang Q, Dong Z, Liu D et al (2021) Frequency-modulated continuous wave radar respiratory pattern detection technology based on multifeature. J Healthc Eng. 2021:9376662
58. Alam R, Peden DB, Lach JC (2021) Wearable respiration monitoring: Interpretable inference with context and sensor biomarkers. IEEE J Biomed Health Inform 25(6):1938–1948
59. Hao Z, Wang Y, Li F, Ding G, Gao Y (2024) MmWave-RM: a respiration monitoring and pattern classification system based on mmWave radar. Sensors 24(13):4315
60. Fekr AR, Janidarmian M, Radecka K, Zilic Z (2014) A medical cloud-based platform for respiration rate measurement and hierarchical classification of breath disorders. Sensors 14(6):11204–11224
61. Purnomo AT, Komariah KS, Lin DB, Hendria WF, Sin BK, Ahmadi N (2022) Non-contact supervision of COVID-19 breathing behaviour with FMCW radar and stacked ensemble learning model in real-time. IEEE Trans Biomed Circuits Syst 16(4):664–678
62. Kim SH, Geem ZW, Han GT (2019) A novel human respiration pattern recognition using signals of ultra-wideband radar sensor. Sensors 19(15):3340
63. Ravichandran V, Murugesan B, Balakarthikeyan V et al (2019) RespNet: a deep learning model for extraction of respiration from photoplethysmogram. Conf Proc IEEE Eng Med Biol Soc 2019:5556–5559
64. Fekr AR, Janidarmian M, Radecka K, Zilic Z (2016) Respiration disorders classification with informative features for m-health applications. IEEE J Biomed Health Inform 20(3):733–747
65. Vuković D, Šumar S, Dupovac S, Badnjević A (2024) Cost effectiveness analysis of medical devices in legal metrology system. In: Badnjević A, Cifrek M, Magjarević R, Džemić Z (eds) Inspection of Medical Devices. Springer, Cham, pp 279–289

66. Silva DF, Muders T, Reinartz S, Putensen C, Leonhardt S (2025) Adaptive cardiorespiratory separation with harmonic models and filters: the case of electrical impedance tomography. IEEE Trans Biomed Eng. https://doi.org/10.1109/TBME.2025.3566608
67. Várady P, Micsik T, Benedek S, Benyó Z (2002) A novel method for the detection of apnea and hypopnea events in respiration signals. IEEE Trans Biomed Eng 49(9):936–942
68. Yeo M, Byun H, Lee J et al (2022) Respiratory event detection during sleep using electrocardiogram and respiratory-related signals: using polysomnogram and patch-type wearable device data. IEEE J Biomed Health Inform 26(2):550–560
69. Hong JW, Kim SH, Han GT (2023) Detection of multiple respiration patterns based on 1D SNN from continuous human breathing signals and the range classification method for each respiration pattern. Sensors 23(11):5275
70. Saeed U, Shah SY, Zahid A, Ahmad J, Imran MA, Abbasi QH, Shah SA (2021) Wireless channel modelling for identifying six types of respiratory patterns with SDR sensing and deep multilayer perceptron. IEEE Sens J 21(18):20833–20840
71. McClure K, Erdreich B, Bates JHT, McGinnis RS, Masquelin A, Wshah S (2020) Classification and detection of breathing patterns with wearable sensors and deep learning. Sensors 20(22):6481
72. Dinh T, Nguyen T, Phan HP, Nguyen NT, Dao DV, Bell J (2020) Stretchable respiration sensors: advanced designs and multifunctional platforms for wearable physiological monitoring. Biosens Bioelectron 166:112460
73. Bećirović F, Spahić L, Merdović N, Gurbeta Pokvić L, Badnjević A (2025) Advancement of post-market surveillance of medical devices leveraging artificial intelligence: patient monitors case study. Technol Health Care 33(2):974–980
74. Filosa M, Massari L, Ferraro D et al (2022) A meta-learning algorithm for respiratory flow prediction from FBG-based wearables in unrestrained conditions. Artif Intell Med 130:102328
75. Dow DE, Garcia AP (2013) Detection of respiration in central venous pressure using state machine. Conf Proc IEEE Eng Med Biol Soc 2013:5299–5302
76. Selvaraj N, Lee J, Chon KH (2010) Time-varying methods for characterizing nonstationary dynamics of physiological systems. Methods Inf Med 49(5):435–441
77. Magjarević R, Badnjević A (2023) Inspection and testing of electrocardiographs (ECG) devices. In: Badnjević A, Cifrek M, Magjarević R, Džemić Z (eds) Inspection of Medical Devices: for Regulatory Purposes. Springer, Cham, pp 139–171
78. Sethi AK, Muddaloor P, Anvekar P et al (2023) Digital pulmonology practice with phonopulmography leveraging artificial intelligence. Sensors 23(13):5900
79. Theerawit P, Sutherasan Y, Ball L, Pelosi P (2017) Respiratory monitoring in adult intensive care unit. Expert Rev Respir Med 11(11):885–895
80. Simonte R, Cammarota G, Vetrugno L, De Robertis E, Longhini F, Spadaro S (2024) Advanced respiratory monitoring during extracorporeal membrane oxygenation. J Clin Med13 [Epub ahead of print]
81. McNicholas WT (2021) Getting more from the sleep recording. Sleep Med Clin 16(4):591–598
82. Mazzotti DR, Lim DC, Sutherland K et al (2018) Opportunities for utilizing polysomnography signals to characterize obstructive sleep apnea subtypes and severity. Physiol Meas 39(8):084007
83. Williams E, Dassios T (2024) What's hot in neonatal respiratory monitoring? Pediatr Res [Epub ahead of print]
84. Islam SMM (2022) Radar-based remote physiological sensing: progress, challenges, and opportunities. Front Physiol 13:1013252
85. Kraman SS, Pasterkamp H, Wodicka GR (2023) Smart devices are poised to revolutionize the usefulness of respiratory sounds. Chest 163(2):394–400
86. Lalouani W, Younis M, Emokpae RN, Emokpae LE (2022) Enabling effective breathing sound analysis for automated diagnosis of lung diseases. Smart Health 26:100311
87. Cammarota G, Simonte R, Longhini F, Spadaro S, Vetrugno L, De Robertis E (2023) Advanced point-of-care bedside monitoring for acute respiratory failure. Anesthesiology

Chapter 5
Processing EEG Signals

This section addresses the physiological basis, acquisition, preprocessing, analysis, and clinical utility of electroencephalography. It begins with the anatomy and bioelectrical properties of the brain, neuronal activity, and functional networks (Sect. 5.1). EEG acquisition methods (Sect. 5.2) are then discussed, including instrumentation, electrode placement, noise management, safety standards, and technological advances. Preprocessing techniques (Sect. 5.3) focus on artifact removal, filtering, referencing, and validation of signal quality. Feature extraction and analysis (Sect. 5.4) encompass time-, frequency-, and time–frequency-based features, as well as nonlinear methods for characterizing brain activity. Clinical applications (Sect. 5.5) highlight EEG's diagnostic role in neurology, psychiatry, intensive care, rehabilitation, and brain–computer interfaces, while also considering technical challenges and future directions. This section establishes EEG as a versatile modality for probing brain function and supporting neurological healthcare.

5.1 Anatomy and Physiology of the Brain

The brain is the central command unit of the human body and arguably the most intricate and complex biological structure known. It is responsible for controlling bodily movements, interpreting sensory information, and orchestrating thoughts, emotions, memory, and behavior. Spanning billions of neurons and trillions of synaptic connections, the brain functions through dynamic electrical and chemical processes that underlie all voluntary and involuntary activities. Its complexity is such that while significant strides have been made in understanding specific regions and pathways, a complete map of its interconnections and functions remains one of science's greatest frontiers.

© The Author(s), under exclusive license to Springer Nature Switzerland AG 2026
A. Badnjević and L. Spahić, *Biomedical Signals and Systems*, Series in BioEngineering,
https://doi.org/10.1007/978-3-032-10982-8_5

From a biomedical engineering perspective, understanding the brain's anatomy and physiology is essential for developing neuroprosthetics, diagnostic imaging technologies, neural interfaces, and brain-computer interfaces (BCIs). Furthermore, accurate knowledge of neurophysiology aids in the diagnosis and treatment of disorders such as epilepsy, Parkinson's disease, Alzheimer's disease, traumatic brain injuries, and psychiatric conditions. Contemporary neuroscience is increasingly interdisciplinary, integrating insights from biology, medicine, computer science, physics, and engineering. Recent advances in neuroimaging, computational modeling, and electrophysiology have enhanced our understanding of brain activity at molecular, cellular, and systems levels. Novel imaging modalities such as functional magnetic resonance imaging (fMRI), positron emission tomography (PET), and diffusion tensor imaging (DTI) have enabled non-invasive visualization of brain structure and function [1]. Parallel advances in genetics and molecular biology have clarified the roles of specific genes and proteins in neurodevelopment and synaptic signaling [2]. These breakthroughs are converging to form the basis of personalized medicine in neurology, allowing for tailored interventions based on individual neural profiles.

In this chapter, we explore the architecture of the human brain and the bioelectrical principles that govern neural activity. We delve into the anatomical organization of the central nervous system, the electrical dynamics of neuronal signaling, synaptic communication, and the integration of brain regions into functional networks. Emphasis is placed on the physiological mechanisms that translate structure into function and how these underlie cognition, perception, motor control, and homeostasis.

5.1.1 Historical Context and Bioelectrical Discovery

The foundations of our understanding of brain physiology are deeply rooted in the historical exploration of bioelectricity. In the late eighteenth century, Luigi Galvani, an Italian physician and physicist, conducted a pivotal experiment using frog legs. When he applied an electrical spark to the exposed nerves of a dissected frog, the muscles twitched as if alive. This observation led Galvani to propose the existence of "animal electricity," a vital force intrinsic to biological tissues. His ideas were later challenged by Alessandro Volta, who argued that the electricity originated from the contact of different metals rather than the tissues themselves. Despite their disagreement, the intellectual exchange between Galvani and Volta laid the foundation for electrophysiology and the invention of the battery.

In the nineteenth and twentieth centuries, the emergence of neurophysiology as a scientific discipline gained momentum. Hermann von Helmholtz was among the first to measure the speed of nerve impulses, debunking the notion that neural signals were instantaneous. Later, Santiago Ramón y Cajal, using refined histological techniques, provided detailed anatomical drawings of neurons and proposed the "neuron doctrine," asserting that the nervous system is composed of individual cells rather than a continuous syncytium. His work was corroborated by electron microscopy in the 1950s, establishing neurons as discrete signaling units [3].

In the 1950s, Alan Hodgkin and Andrew Huxley conducted experiments on the giant axon of the squid, elucidating the ionic mechanisms of action potentials. Their development of the Hodgkin-Huxley model, a set of nonlinear differential equations describing voltage-gated ion channels, earned them the Nobel Prize in Physiology or Medicine in 1963 and remains a cornerstone in computational neuroscience [4].

The twentieth century also saw the rise of electroencephalography (EEG) as a non-invasive technique for monitoring brain activity. Developed by Hans Berger in the 1920s, EEG measures voltage fluctuations resulting from ionic current within neurons. It is instrumental in diagnosing epilepsy, sleep disorders, and monitoring depth of anesthesia. With advancements in signal processing and machine learning, EEG remains a key tool in brain-computer interface research [5].

Today, the legacy of early electrophysiologists is evident in technologies such as deep brain stimulation (DBS), transcranial magnetic stimulation (TMS), and neural prosthetics. These innovations hinge on the ability to modulate and interpret bioelectrical signals for therapeutic and assistive purposes.

5.1.2 Structural Overview of the Brain

The brain's structural complexity is reflected in its division into distinct yet highly interconnected regions, each contributing to specific physiological and cognitive functions. At a macroscopic level, the human brain is divided into the cerebrum, cerebellum, brainstem, and diencephalon. Each component contains specialized substructures responsible for regulating different aspects of human behavior and physiology.

The Cerebrum

The cerebrum, the most significant portion of the brain, is responsible for conscious thought, voluntary movement, sensory perception, learning, memory, and language. It consists of two hemispheres—left and right—separated by the longitudinal fissure and connected via the corpus callosum, a massive bundle of myelinated axons. Each hemisphere is further divided into four lobes: frontal, parietal, temporal, and occipital. The frontal lobe manages executive functions such as decision-making, planning, social behavior, and motor control. The parietal lobe processes somatosensory information, including touch, pressure, temperature, and pain. The temporal lobe is essential for auditory processing and memory encoding, while the occipital lobe is the primary center for visual processing [6]. The cerebral cortex, a thin layer of gray matter covering the cerebrum, contains billions of neurons and is organized into layers and columns that coordinate local and long-range communication. Beneath the cortex lies the white matter, composed of axonal tracts that connect cortical areas and with subcortical structures. The arrangement of gyri (ridges) and sulci (grooves) on the cortex significantly increases surface area, facilitating complex processing.

The Cerebellum

Located beneath the cerebrum at the back of the skull, the cerebellum is primarily involved in the coordination of voluntary movement, balance, and motor learning. Though traditionally associated with motor control, recent studies also implicate the cerebellum in cognitive processes such as attention, language, and emotion regulation. The cerebellum receives input from the spinal cord, sensory systems, and cerebral cortex, and modulates motor outputs to ensure smooth, coordinated movement. It is organized into anterior, posterior, and flocculonodular lobes, with a layered cortex surrounding white matter that includes the deep cerebellar nuclei.

The Brainstem

The brainstem comprises the midbrain, pons, and medulla oblongata. It acts as a conduit between the brain and spinal cord and houses essential autonomic centers that regulate cardiovascular function, respiration, and arousal. The midbrain includes structures such as the superior and inferior colliculi (involved in visual and auditory reflexes, respectively) and the substantia nigra, which plays a key role in dopamine production and motor control [7]. The pons serves as a relay station between the cerebrum and cerebellum and helps regulate sleep and respiration. The medulla oblongata contains the cardiac, vasomotor, and respiratory centers, critical for sustaining life. Damage to the medulla is often fatal due to its control of vital autonomic functions.

The Diencephalon

Situated between the brainstem and cerebrum, the diencephalon includes the thalamus and hypothalamus. The thalamus functions as a relay station for almost all sensory inputs (except olfaction), directing signals to the appropriate cortical areas. It also plays a role in consciousness and alertness. The hypothalamus maintains homeostasis by regulating body temperature, hunger, thirst, circadian rhythms, and hormone secretion via its control of the pituitary gland. It links the nervous and endocrine systems, influencing stress responses and reproductive behaviors [8].

Ventricular System and Protective Structures

The brain contains four interconnected ventricles filled with cerebrospinal fluid (CSF), which cushions neural tissue, maintains chemical stability, and removes waste products. The choroid plexus within the ventricles produces CSF, which circulates through the ventricles, subarachnoid space, and spinal cord. Disruption in CSF flow or absorption can lead to conditions such as hydrocephalus. The meninges further protect the brain—three connective tissue layers (dura mater, arachnoid mater, and pia mater)—and the blood–brain barrier (BBB), which selectively restricts the passage of substances from the bloodstream into neural tissue. This selective permeability is essential for preventing neurotoxins and pathogens from entering the central nervous system while allowing essential nutrients and gases to pass through [9].

5.1.3 Neuronal Anatomy

At the heart of brain function lies the neuron, a highly specialized cell capable of receiving, processing, and transmitting electrical signals. Neurons form the structural and functional units of the nervous system and vary in shape, size, and connectivity depending on their roles. Despite their diversity, all neurons share key structural features: the soma (cell body), dendrites, axon, and presynaptic terminals.

The soma contains the nucleus and organelles essential for cellular metabolism and maintenance. It integrates excitatory and inhibitory inputs received by the dendrites and generates action potentials if a threshold is exceeded. The nucleus houses the cell's DNA and supports transcriptional regulation necessary for neuroplasticity and synaptic remodeling.

Dendrites are short, branched projections that receive synaptic inputs from other neurons. The surface of dendrites is studded with spines—tiny protrusions that increase synaptic contact and exhibit plasticity depending on activity levels. These spines are dynamic structures whose remodeling underlies learning and memory. The axon is a long projection specialized for transmitting action potentials to distant targets. It arises from the axon hillock, the site of action potential initiation, and often branches into axon collaterals that facilitate communication with multiple targets. Axonal terminals form synapses with other neurons, muscle fibers, or glands, where neurotransmitter release occurs.

Many axons are myelinated, a process in which glial cells (oligodendrocytes in the CNS and Schwann cells in the PNS) wrap around the axon with layers of insulating membrane. Myelin increases conduction velocity by enabling saltatory conduction, whereby action potentials leap between the nodes of Ranvier—gaps in the myelin sheath densely populated with voltage-gated sodium channels [10]. Defects in myelination can result in severe neurological disorders such as multiple sclerosis (MS), where immune-mediated demyelination leads to disrupted conduction and a variety of sensory, motor, and cognitive deficits (Fig. 5.1).

Neurons are classified based on structure and function:

- Multipolar neurons: The most common type, found in the cortex and spinal cord, with one axon and multiple dendrites.
- Bipolar neurons: Found in sensory pathways such as the retina and olfactory epithelium, with one dendrite and one axon.
- Unipolar (pseudounipolar) neurons: Common in dorsal root ganglia, these have a single process that bifurcates into peripheral and central branches.

Functionally, neurons can be categorized as sensory (afferent), motor (efferent), or interneurons. Interneurons constitute the vast majority of neurons in the CNS and mediate complex reflexes and higher-order processing. At the presynaptic terminal, synaptic vesicles containing neurotransmitters are released into the synaptic cleft in response to calcium influx triggered by the arrival of an action potential. This process of neurotransmitter release is mediated by SNARE proteins and regulated by synaptotagmin acting as a calcium sensor [11]. Once released, neurotransmitters

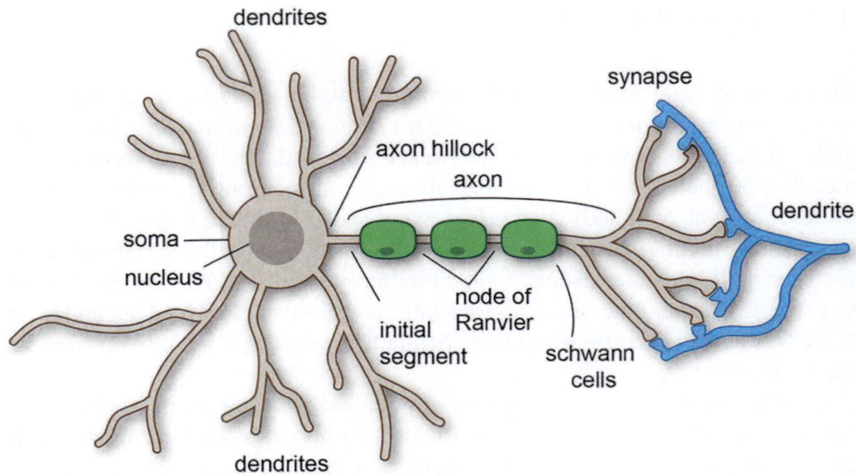

Fig. 5.1 Neuronal anatomy (https://commons.wikimedia.org/wiki/File:Anatomy_of_neuron.png)

diffuse across the cleft and bind to receptors on the postsynaptic membrane, generating either excitatory postsynaptic potentials (EPSPs) or inhibitory postsynaptic potentials (IPSPs), depending on the receptor type and ion channels involved.

5.1.4 Bioelectric Phenomena: Resting and Action Potentials

Neurons operate through rapid changes in membrane potential, a process governed by the movement of ions across the neuronal membrane. These electrical signals form the foundation for communication in the nervous system. Understanding the mechanisms of the resting membrane potential and the generation of action potentials is critical for interpreting neuronal behavior and the effects of drugs, diseases, and therapeutic interventions.

Resting Membrane Potential

The resting membrane potential is the baseline electrical charge across the neuronal membrane when the cell is not actively transmitting a signal. In most neurons, this potential is approximately -65 to -70 mV, meaning the interior of the cell is more negatively charged than the extracellular environment. This potential arises from the differential distribution of key ions—primarily sodium (Na^+), potassium (K^+), chloride (Cl^-), and negatively charged proteins—on either side of the membrane. The Na^+/K^+-ATPase pump actively transports three Na^+ ions out of the cell and two K^+ ions into the cell, maintaining the ionic gradients essential for resting potential [12]. Potassium leak channels, which allow K^+ to move out of the cell down its concentration gradient, play a significant role in setting this baseline potential. The resting membrane potential can be quantitatively estimated using the Nernst equation for individual ion species or the Goldman-Hodgkin-Katz equation when considering multiple permeable ions simultaneously. These equations highlight the importance of both concentration gradients and membrane permeability in determining electrical potential.

The Action Potential

When a neuron receives sufficient excitatory input, it depolarizes to a threshold (typically around -55 mV), triggering an action potential—a rapid, transient reversal of membrane potential that propagates along the axon. The process unfolds in several phases:

- Depolarization: Voltage-gated Na^+ channels open, allowing Na^+ to rush into the cell, causing the membrane potential to become positive.
- Repolarization: At the peak (~ + 30 mV), Na^+ channels inactivate and voltage-gated K^+ channels open, allowing K^+ to exit the cell, returning the membrane to a negative potential.
- Hyperpolarization: K^+ channels may remain open briefly, causing the membrane potential to fall below resting levels before returning to baseline.

This sequence of events, lasting just a few milliseconds, is fundamental to neuronal signaling. The action potential is all-or-none—once the threshold is reached, the complete response occurs. Its amplitude and duration are consistent, though frequency can vary with stimulus intensity [13] (Fig. 5.2).

Ion Channels and Their Regulation

Ion channels are transmembrane proteins that allow specific ions to pass in and out of the cell. They can be voltage-gated, ligand-gated, mechanically gated, or leak channels. The behavior of these channels is essential for determining neuronal excitability. Voltage-gated sodium and potassium channels are especially critical in generating and propagating action potentials. Calcium channels play a key role

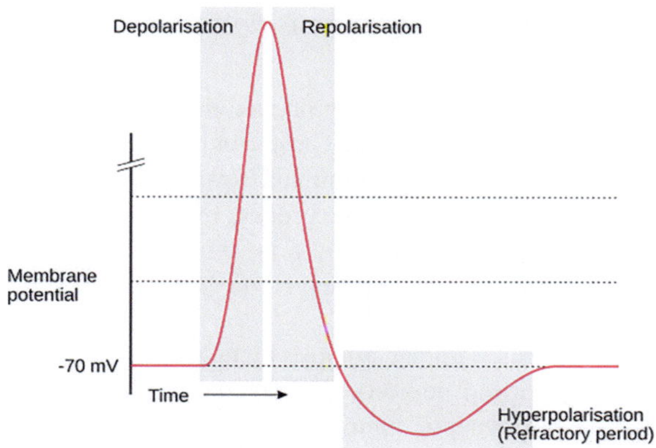

Fig. 5.2 Action potential (https://commons.wikimedia.org/wiki/File:Action_potential_schematic.svg)

in neurotransmitter release at synaptic terminals. Mutations or malfunctions in ion channels, known as channelopathies, are linked to conditions such as epilepsy, cardiac arrhythmias, and neuropathic pain [14]. Regulation of ion channel function can occur through phosphorylation, interaction with scaffolding proteins, or changes in gene expression. These mechanisms are essential for processes like synaptic plasticity and adaptation to long-term stimuli.

Propagation of the Action Potential

Action potentials propagate along the axon without decrement due to the sequential opening of voltage-gated ion channels. In myelinated axons, this propagation is accelerated by saltatory conduction, where the action potential jumps between nodes of Ranvier—regions devoid of myelin with high concentrations of sodium channels. Unmyelinated fibers conduct more slowly via continuous conduction, where depolarization spreads sequentially along the membrane. The speed of conduction depends on axon diameter, myelination, and the density and distribution of ion channels [5]. Disruption in conduction can result from demyelinating diseases such as multiple sclerosis or from local anesthetics that block sodium channels, inhibiting action potential propagation and inducing loss of sensation.

Refractory Periods and Signal Directionality

Two types of refractory periods follow an action potential:

- Absolute refractory period: During which a second action potential cannot be initiated due to inactivation of sodium channels.
- Relative refractory period: A stronger-than-normal stimulus is required to generate a new action potential due to hyperpolarization.

These periods ensure unidirectional propagation of signals along axons and prevent continuous firing, thus preserving signal fidelity and preventing synaptic overload.

5.1.5 Synaptic Transmission and Neurotransmitters

Neurons communicate with one another and with effector cells through specialized junctions called synapses. These can be either electrical or chemical, with chemical synapses being the predominant form in the mammalian central nervous system. Synaptic transmission is the process by which signaling molecules, known as neurotransmitters, are released by a presynaptic neuron, traverse the synaptic cleft, and bind to receptors on a postsynaptic neuron, thereby influencing the likelihood of action potential generation.

A typical chemical synapse includes the presynaptic terminal (axon terminal of the transmitting neuron), the synaptic cleft (a gap of approximately 20–40 nm), and the postsynaptic membrane (often on a dendritic spine). Within the presynaptic terminal are numerous synaptic vesicles filled with neurotransmitters, mitochondria, and active zones rich in docking proteins and calcium channels. When an action potential reaches the presynaptic terminal, voltage-gated calcium channels open, leading to an influx of $Ca2^+$. This triggers vesicle fusion with the membrane via SNARE complex proteins such as synaptobrevin, syntaxin, and SNAP-25, mediated by the calcium sensor synaptotagmin [15]. The neurotransmitter is then exocytosed into the synaptic cleft.

Neurotransmitters are categorized by their chemical structure and function. Major classes include:

- Amino acids: Glutamate (excitatory), GABA, and glycine (inhibitory)
- Monoamines: Dopamine, norepinephrine, serotonin, and histamine
- Acetylcholine: Crucial for neuromuscular junctions and autonomic regulation
- Neuropeptides: Substance P, endorphins, vasopressin
- Purines: ATP and adenosine.

Glutamate is the primary excitatory neurotransmitter in the CNS and plays a central role in synaptic plasticity and learning. GABA, by contrast, opens chloride channels, leading to inhibitory postsynaptic potentials (IPSPs) that hyperpolarize the neuron [16].

Neurotransmitters exert their effects through two main classes of receptors:

- Ionotropic receptors (ligand-gated ion channels): Mediate fast synaptic transmission. For example, AMPA and NMDA receptors for glutamate or GABA_A receptors.
- Metabotropic receptors (G-protein-coupled receptors): Initiate slower but prolonged effects through second messenger pathways, such as cAMP or IP_3.

Ionotropic receptors cause immediate changes in membrane potential, while metabotropic receptors modulate neuronal excitability, gene expression, and synaptic strength over longer timescales [17].

Synaptic strength is not static. Activity levels can modulate it in a phenomenon called synaptic plasticity, which includes:

- Short-term plasticity: Paired-pulse facilitation and synaptic depression, lasting milliseconds to minutes.
- Long-term potentiation (LTP): A sustained increase in synaptic strength, often studied in the hippocampus, associated with learning and memory.
- Long-term depression (LTD): A prolonged decrease in synaptic efficacy.

These forms of plasticity are regulated by intracellular calcium levels, receptor trafficking, and signaling cascades involving kinases and phosphatases. To prevent continuous stimulation, neurotransmitters must be rapidly cleared from the synaptic cleft. This is achieved through:

- Enzymatic degradation (e.g., acetylcholinesterase at cholinergic synapses)
- Reuptake into presynaptic terminals or glial cells via transporters (e.g., serotonin transporter SERT)
- Diffusion away from the synapse.

Dysregulation of neurotransmitter clearance can lead to synaptic dysfunction and is implicated in disorders such as depression, schizophrenia, and Parkinson's disease [18].

5.1.6 Supportive Cells and Structures

While neurons are often the primary focus in studies of the nervous system, glial cells and structural support systems play equally vital roles in ensuring proper brain function. Glial cells not only maintain homeostasis but also regulate neurotransmission, support metabolism, and participate in immune responses within the central nervous system (CNS). Glial cells outnumber neurons by an estimated ratio of 3:1 in the human brain [19].

Astrocytes

Astrocytes are star-shaped glial cells that serve multiple essential functions. They regulate the extracellular ionic environment, particularly by buffering potassium ions, and help maintain neurotransmitter levels by uptaking excess glutamate. They are intimately involved in the formation and maintenance of the blood–brain barrier (BBB), where they interface with endothelial cells of capillaries and pericytes to control molecular trafficking between blood and brain tissue.

Astrocytes also release gliotransmitters, including ATP, glutamate, and D-serine, which modulate synaptic transmission and plasticity. These cells respond to neural activity with calcium waves, forming a tripartite synapse along with the pre- and postsynaptic neurons. Disruptions in astrocytic function have been implicated in neurodegenerative conditions such as Alzheimer's disease and amyotrophic lateral sclerosis (ALS).

Oligodendrocytes and Myelination

Oligodendrocytes are responsible for myelinating axons in the CNS. Each oligodendrocyte can extend its processes to multiple axons, wrapping them with concentric layers of membrane to form the myelin sheath. Myelination increases the conduction velocity of action potentials via saltatory conduction and contributes to efficient neuronal signaling. The integrity of myelin is essential for coordinated neural activity. Demyelinating diseases such as multiple sclerosis are characterized by immune-mediated damage to oligodendrocytes and myelin, resulting in conduction block or slowed transmission and contributing to the neurological deficits observed in affected patients [20].

Microglia

Microglia are the resident immune cells of the CNS. Originating from mesodermal progenitors, they continually survey the neural environment using motile processes. In response to injury or infection, microglia become activated, changing morphology and releasing pro-inflammatory cytokines such as IL-1β, TNF-α, and IL-6 [21]. In addition to immune surveillance, microglia play roles in synaptic pruning during development, removing weak or unnecessary synapses to fine-tune neural circuits. Aberrant microglial activation has been linked to neuropathologies such as chronic neuroinflammation in Alzheimer's and schizophrenia.

Ependymal Cells and the Ventricular System

Ependymal cells line the walls of the brain's ventricular system and are involved in the production and circulation of cerebrospinal fluid (CSF). These ciliated cells form a barrier between the CSF and neural tissue and facilitate the distribution of nutrients and the removal of waste products. The CSF acts as a protective cushion for the brain and spinal cord, maintaining intracranial pressure and providing mechanical support. Abnormalities in CSF dynamics can lead to hydrocephalus, a condition characterized by increased ventricular volume and intracranial pressure.

The Blood–Brain Barrier (BBB)

The BBB is a selectively permeable barrier formed by tight junctions between endothelial cells lining cerebral blood vessels, supported by astrocytic endfeet and pericytes. It regulates the passage of ions, molecules, and cells between the blood and brain, maintaining the brain's microenvironment. While the BBB protects the brain from toxins and pathogens, it also poses challenges for drug delivery in neurological diseases. Researchers are investigating techniques such as focused ultrasound and nanoparticle-based carriers to transiently and safely open the BBB for therapeutic purposes [22].

5.1.7 Integration and Functional Networks

The brain's effectiveness stems not merely from its distinct regions but from the sophisticated integration of these areas into dynamic functional networks. These networks allow distributed processing and ensure rapid communication among regions responsible for sensory input, motor output, cognition, memory, emotion, and autonomic regulation.

Functional integration refers to the coordination between spatially separated brain regions to accomplish complex tasks. Unlike localized brain functions—such as the motor strip's control over movement—functional integration enables processes like attention, language, and decision-making, which recruit multiple interacting brain areas simultaneously [23]. Brain connectivity is categorized into:

- Structural connectivity: The anatomical links (white matter tracts) between brain areas, visualized using diffusion tensor imaging (DTI).
- Functional connectivity: The statistical correlation of neural activity between regions during rest or tasks, studied using fMRI or EEG.

- Effective connectivity: The influence one neural system exerts over another, assessed through causal modeling techniques such as dynamic causal modeling (DCM).

Several large-scale networks have been identified and linked to specific functions:

- Default Mode Network (DMN): Active during rest and involved in self-referential thought, memory consolidation, and mind-wandering.
- Salience Network: Detects and filters salient stimuli, facilitating the switch between the DMN and executive control network.
- Central Executive Network: Engaged during working memory, decision-making, and goal-directed behavior.
- Sensorimotor and Visual Networks: Maintain real-time interaction with the environment through sensory input and motor output.

These networks demonstrate a high degree of plasticity and can be modulated by learning, injury, and neurodegenerative diseases. Many functions emerge from interactions between cortical and subcortical structures. For instance, the basal ganglia interact with the motor cortex through direct and indirect pathways that fine-tune motor commands. Dysfunction in these loops contributes to movement disorders such as Parkinson's and Huntington's diseases. Similarly, thalamocortical loops mediate sensory transmission, attention, and consciousness. The thalamus serves as a relay and filter, modulating the flow of sensory and motor signals to and from the cortex [24].

Neural oscillations—rhythmic fluctuations in electrical activity—are another mechanism for integrating brain regions. Oscillations occur at different frequencies (e.g., delta, theta, alpha, beta, gamma), each associated with specific behavioral states and cognitive processes. Synchronized oscillations facilitate communication across distant brain areas. For example, gamma oscillations (~30–80 Hz) are implicated in perceptual binding, while theta oscillations (4–8 Hz) are crucial for memory encoding and navigation [25].

The brain exhibits remarkable plasticity, allowing for reorganization of functional networks in response to experience, learning, and injury. Neuroplastic changes include synaptic strengthening, neurogenesis, and changes in network dynamics.

After a stroke or traumatic brain injury, for example, unaffected brain regions can assume lost functions through network reconfiguration. This forms the basis of many rehabilitation strategies, including constraint-induced therapy and neurofeedback [26].

Understanding functional brain networks informs the development of neurotechnologies such as brain-computer interfaces (BCIs), deep brain stimulation (DBS), and closed-loop neuromodulation systems. These technologies rely on accurate modeling of brain network dynamics to decode intentions, restore motor function, or suppress pathological activity in disorders like epilepsy and depression.

5.2 EEG Signal Acquisition

Electroencephalography is a cornerstone of modern neuroscience and biomedical signal analysis. It involves the measurement of electrical activity produced by the firing of neurons in the brain, detected at the scalp using non-invasive electrodes. First recorded by Hans Berger in 1929, EEG has evolved from bulky analog systems to high-resolution, multichannel digital instruments used in hospitals, research labs, wearable devices, and brain-computer interfaces (BCIs) [27]. EEG provides critical insights into both normal and pathological brain function. It is widely used in the diagnosis and monitoring of epilepsy, sleep disorders, brain injuries, coma, and psychiatric conditions. It is also essential for cognitive neuroscience studies and neurorehabilitation applications, where brain activity must be monitored in real time. The importance of accurate and high-quality signal acquisition in EEG cannot be overstated. Poor signal acquisition can introduce noise, distort waveforms, and reduce diagnostic efficacy. Moreover, with the rise of AI-powered EEG analysis and portable devices, acquisition quality directly influences algorithm performance and clinical outcomes [28].

5.2.1 Neurophysiological Basis of EEG Signals

EEG signals primarily reflect the summated postsynaptic potentials generated by synchronized activity in large populations of pyramidal neurons in the cerebral cortex. These neurons are aligned perpendicularly to the cortical surface, allowing their electric fields to summate and propagate to the scalp. Frequency bands categorize the rhythmic activity observed in EEG recordings:

- Delta (0.5–4 Hz): Prominent in deep sleep
- Theta (4–8 Hz): Associated with drowsiness and meditation
- Alpha (8–13 Hz): Seen during relaxed wakefulness
- Beta (13–30 Hz): Linked to active thinking and concentration
- Gamma (>30 Hz): Associated with cognitive processes and attention.

These oscillations vary by brain state, region, and pathology. For instance, abnormal spikes or sharp waves are common biomarkers in epilepsy [29].

5.2.2 Instrumentation and Signal Path

EEG systems comprise electrodes, amplifiers, filters, analog-to-digital converters (ADCs), and data acquisition software.

Electrodes are the interface between the brain's electrical activity and the recording equipment. Common types include:

- Wet electrodes (Ag/AgCl with conductive gel): High-quality signals, but require skin prep
- Dry electrodes: Easier setup, less stable impedance
- Active electrodes: Include pre-amplification at the site to reduce noise
- Invasive electrodes (e.g., ECoG): Used for surgical planning or research.

The 10–20 system is the international standard for placing electrodes at defined points based on head size, enabling reproducible and comparable recordings [30]. Neural signals recorded at the scalp are typically between 5–100 μV and must be amplified to a readable range. Differential amplifiers enhance signal quality by removing common-mode noise. Modern amplifiers often include:

- High input impedance
- Low-noise components
- Integrated bandpass filtering (e.g., 0.1–100 Hz).

EEG systems convert continuous analog voltages to digital values using ADCs. Important parameters include:

- Resolution (e.g., 16-bit or 24-bit)
- Sampling rate (commonly 250–1000 Hz).

According to the Nyquist theorem, the sampling rate must be at least twice the highest frequency of interest to prevent aliasing. Oversampling and antialiasing filters are commonly applied. Proper grounding minimizes power line interference (50/60 Hz) and helps stabilize baseline signals. Shielded cables, Faraday cages, and driven-right-leg circuits are used to reduce environmental noise and leakage currents further [27].

5.2.3 EEG Recording Techniques

EEG acquisition techniques vary based on the research or clinical objective. The choice of technique influences not only the quality of the data but also the type of inferences that can be drawn from it. Each recording mode imposes different requirements on electrode configuration, amplifier sensitivity, and artifact control.

Resting-state EEG involves recording spontaneous brain activity while the subject is at rest, typically with eyes closed or fixated. It is often used in clinical assessments of brain function, including evaluating consciousness in comatose patients, detecting neurodegeneration, and assessing cognitive reserve in aging populations. Studies have shown alterations in alpha and theta band power in patients with Alzheimer's disease and schizophrenia. Resting-state EEG is also pivotal in identifying biomarkers for mental health conditions and is widely used in neurofeedback training programs.

Event-related EEG recordings capture the brain's electrophysiological responses to specific sensory, motor, or cognitive events. These are typically extracted as event-related potentials (ERPs) by averaging time-locked EEG epochs across many trials. ERP components such as P300, N170, and N400 are well-established markers for attention, face processing, and semantic comprehension, respectively [31]. ERP-based methods are widely used in BCI applications, including spellers, lie detection, and cognitive workload assessments. The P300-based oddball paradigm remains a gold standard in non-invasive BCI design [32].

EEG is fundamental to sleep research and clinical sleep medicine. Polysomnography, the comprehensive study of sleep, integrates EEG with electrooculography (EOG), electromyography (EMG), and respiratory signals to identify sleep stages and diagnose disorders like obstructive sleep apnea and narcolepsy [33]. Sleep staging relies on detecting characteristic waveforms: sleep spindles, K-complexes, and slow-wave activity. Wearable and home-based sleep EEG devices are now increasingly available, enabling longitudinal studies of sleep health [34].

Ambulatory EEG systems allow for long-term recording (24–72 h) outside the laboratory or clinical environment. These systems are instrumental in diagnosing epilepsy when inpatient monitoring is inconclusive. They also enable EEG monitoring during daily activities, which is critical for ecological validity in psychiatric and cognitive studies [35]. Mobile EEG platforms integrate compact amplifiers, wireless transmission, and rechargeable batteries, making them suitable for research in real-world environments. These systems are especially valuable for neuroergonomics, sports neuroscience, and real-time neuroadaptive technologies.

High-density EEG (hdEEG) employs arrays of 64, 128, or 256 electrodes to improve spatial sampling of cortical activity. This enhanced resolution supports better source localization and topographic mapping, which are crucial for pre-surgical planning in epilepsy, functional brain mapping, and multimodal integration with MRI or MEG [27]. hdEEG can resolve the spatial distribution of brain rhythms and identify functionally distinct regions involved in cognitive tasks. However, it requires more sophisticated artifact rejection algorithms and signal processing techniques due to the increased data volume and inter-channel variability.

Recent advances in EEG acquisition include hybrid systems that combine EEG with other physiological modalities such as:

- fNIRS (functional near-infrared spectroscopy): Measures hemodynamic changes
- EMG: For simultaneous motor and neural activity
- Eye-tracking: Used in cognitive load and visual attention studies.

Such multimodal systems enhance the interpretability of EEG data and improve the robustness of machine learning models for diagnosis or BCI control [36].

5.2.4 Noise and Artifacts in Acquisition

Despite advancements in EEG acquisition technology, the quality of recorded signals remains highly vulnerable to various sources of noise and artifact contamination. These extraneous signals can distort EEG waveforms, obscure physiological patterns, and impair both clinical diagnostics and machine learning-based analyses. As such, identifying, minimizing, and managing artifacts is a critical component of EEG signal acquisition [37].

EEG signals are typically within the range of 5–100 μV, making them highly susceptible to interference. Noise sources can be categorized into the following domains:

- Physiological artifacts:
  - EMG activity from scalp and facial muscles, especially during jaw clenching, blinking, or facial movements
  - EOG artifacts from eye blinks and lateral eye movements
  - ECG contamination, particularly in recordings near the neck or chest
  - Respiratory and sweating-related potentials, especially during sleep or prolonged recording sessions
- Environmental and external sources:
  - Power line interference at 50 or 60 Hz
  - Electromagnetic interference from medical equipment, mobile phones, or wireless transmitters
  - Motion artifacts due to loose or moving electrodes and poor cable shielding
- Technical or instrumentation-related noise:
  - Impedance mismatch between the scalp and electrodes
  - Cable movements (microphonics)
  - Non-optimal grounding and shielding.

Each of these noise sources affects the signal differently, often in overlapping frequency ranges, which complicates preprocessing and analysis. For instance, EMG artifacts often contaminate the beta and gamma bands, while EOG primarily distorts the low-frequency delta and theta ranges.

Electrode–skin impedance plays a pivotal role in ensuring high-quality EEG recordings. Ideally, impedance should be below 5 kΩ for standard wet electrodes [18]. Elevated impedance can lead to thermal noise and reduce the signal-to-noise ratio (SNR), especially in multichannel systems where local impedance disparities can introduce differential signal artifacts. Proper skin preparation—using alcohol swabs, mild abrasion gels, or conductive pastes—helps reduce impedance and stabilizes the electrode–skin interface. Continuous monitoring of impedance during acquisition is critical in clinical environments, where signal integrity must be maintained for accurate diagnostic interpretation. Dry electrodes, while convenient and faster to deploy, often exhibit higher and more variable impedance levels. These challenges have driven recent innovations in conductive materials and active dry electrode designs, particularly for wearable EEG.

Effective electromagnetic shielding is essential for reducing ambient interference. EEG systems often incorporate:

- Driven right leg (DRL) circuits for common-mode noise rejection
- Shielded cables to prevent capacitive coupling with other devices
- Use Faraday cages in laboratory settings to minimize electromagnetic contamination.

Mobile EEG devices typically use wireless transmission modules that are susceptible to radiofrequency interference. Proper design of PCB layouts, isolation of analog and digital circuitry, and adherence to IEC 60,601–1 EMC standards are crucial in such systems [38]. Automated and manual artifact detection techniques are frequently used in clinical EEG interpretation. Clinicians are trained to identify characteristic patterns of:

- Eye blinks (slow, frontally dominant waves)
- Muscle tension (high-frequency bursts)
- Movement artifacts (large amplitude, irregular waveforms).

Software platforms now include real-time artifact recognition modules based on Independent Component Analysis (ICA), wavelet transforms, and deep learning [20]. These tools facilitate annotation, rejection, or correction of contaminated data segments without discarding entire epochs.

Recent studies have proposed novel artifact-resistant EEG acquisition strategies, including:

- Adaptive filtering algorithms for real-time EMG and EOG suppression
- Double-layered electrode arrays for improved spatial differentiation of accurate EEG vs. noise
- Reference electrode standardization techniques that minimize inter-channel contamination
- Motion-cancellation headgear for ambulatory EEG.

In regulatory terms, conformity assessments of EEG devices now frequently include evaluations of artifact susceptibility and recovery time.

5.2.5 *Signal Quality and Safety Standards*

The reliability and clinical relevance of EEG data depend heavily on ensuring high signal quality and adhering to rigorous safety and regulatory standards. EEG systems must operate under conditions that guarantee safe interaction with patients while maintaining the fidelity of signal acquisition. Signal quality, in this context, encompasses multiple dimensions including SNR, bit resolution, sampling rate, stability, and compliance with performance specifications under dynamic and often unpredictable clinical conditions.

SNR represents the proportion of the desired EEG signal amplitude to the background noise level. High SNR ensures that diagnostically relevant features such as spikes, slow waves, and oscillations are preserved and distinguishable. Ideal SNR values for clinical EEG systems typically range from 20–40 dB [39]. Modern amplifiers and preprocessing chains, including differential inputs and active shielding, have significantly enhanced achievable SNR levels.

Adaptive gain control and real-time feedback mechanisms have also been implemented to balance signal fidelity with environmental noise conditions dynamically. Particularly in mobile and wearable EEG setups, maintaining high SNR during movement or outside clinical settings is a significant engineering challenge.

The analog-to-digital conversion stage is vital for quantifying EEG signals. Higher resolution ADCs—typically 16-bit to 24-bit—are used to resolve subtle voltage changes in microvolt-range EEG signals. A 16-bit ADC offers 65,536 discrete levels, providing a voltage resolution of ~ 0.15 μV for a ± 5V input range, suitable for most neurophysiological recordings.

Sampling rate is also critical and must satisfy the Nyquist criterion. Most clinical systems sample at 256–1024 Hz, though research-grade devices may go up to 10 kHz for high-resolution studies. Oversampling and digital downsampling further improve effective resolution [40]. Quality indicators used in clinical-grade EEG acquisition include:

- Effective number of bits (ENOB)
- Common-mode rejection ratio (CMRR)
- Total harmonic distortion (THD)
- Drift rate and baseline stability.

High CMRR (>100 dB) is necessary to suppress ambient noise and maintain differential signal fidelity. ENOB should closely approximate the ADC resolution to indicate minimal internal system noise. Manufacturers are increasingly including automated self-check features to monitor these metrics in real-time, especially in wearable EEG platforms [22].

EEG equipment must comply with global medical device regulations and safety standards. Key regulatory frameworks include:

- IEC 60,601-2-26: Particular requirements for safety and performance of EEG equipment
- ISO 13485: Quality management systems for medical devices
- FDA 21 CFR Part 820: Quality system regulations for devices marketed in the United States
- European Medical Device Regulation [41].

In addition to electrical safety (e.g., leakage currents, isolation), standards also address electromagnetic compatibility (EMC), software validation, and biological safety of electrode materials. Conformity assessment ensures that EEG systems perform within required tolerances across their intended use scenarios. These include:

- Repetitive performance benchmarking under varied operating conditions
- Artifact resilience evaluations
- Longitudinal stability testing for wearable and home-use EEG.

Emerging EEG systems are increasingly adopting design principles that integrate safety features directly into acquisition hardware. Examples include:

- Electrode disconnect alarms
- High-impedance detection with auto-calibration
- Low-power isolation circuits to prevent shock hazards.

AI is also being used to assess device drift and anticipate hardware degradation. Predictive models can estimate electrode wear-out or ADC malfunction based on deviation from normative signal patterns. These models are now being integrated into regulatory approval pathways and remote diagnostics [42].

5.2.6 Advances in EEG Hardware

Technological innovation has significantly reshaped the design and performance capabilities of EEG acquisition systems. Modern EEG hardware is not only more compact, energy-efficient, and wireless but also features smart integration with AI-driven software and wearable platforms. These advances have improved accessibility, enabled mobile monitoring, and expanded EEGs' reach beyond traditional clinical environments into home settings, sports, rehabilitation, and consumer health.

One of the most transformative developments in EEG hardware has been the emergence of dry electrodes. Unlike traditional Ag/AgCl wet electrodes, dry electrodes require no conductive gel and minimal preparation. While initial versions suffered from higher impedance and poor signal quality, recent advances have made dry electrodes suitable for both research and clinical applications. Hybrid electrodes combine aspects of both dry and wet designs, often using hydrogel layers or spring-loaded contacts to improve skin conformity. Materials science advancements—such as conductive polymers, nanocomposites, and graphene-based films—have further enhanced contact reliability and biocompatibility [15]. Dry and hybrid systems are especially valuable in wearable and ambulatory EEG, where long-term usability, comfort, and maintenance-free operation are critical.

Wearable EEG systems are now designed for lightweight, unobtrusive recording. These devices incorporate flexible electronics that conform to the scalp, enabling motion-tolerant data acquisition without rigid headgear. Some commercially available systems weigh less than 100 g and integrate all acquisition circuitry in a behind-the-ear or headband module.

Flexible EEG technologies are being developed using stretchable circuit boards, textile electrodes, and printed sensor arrays. These configurations are beneficial for neonatal monitoring, elderly care, and long-duration sleep studies [43].

Wireless EEG systems have proliferated with the development of Bluetooth Low Energy (BLE), Wi-Fi, and custom RF protocols. These systems transmit digitized EEG signals in real-time to mobile devices, tablets, or cloud-based analytics platforms. They eliminate the constraints of tethered recordings and enhance mobility for patients and research subjects. Power efficiency is a key feature. Advances in ultra-low-power ADCs, system-on-chip (SoC) integration, and duty-cycled wireless protocols have allowed for multi-day continuous EEG monitoring with small battery footprints.

Next-generation EEG systems include smart sensors capable of local signal processing. These sensors perform initial tasks such as artifact rejection, band-pass filtering, and spike detection, reducing the data burden for downstream processing. Microcontrollers embedded within EEG modules can also trigger alerts when signal quality drops below a threshold or when artifact patterns such as eye blinks or muscle spikes are detected. AI-assisted smart sensors further enhance system responsiveness, especially in BCI applications [22].

High-density EEG systems have grown in electrode count (up to 512 channels) while reducing channel cross-talk, noise, and setup time. Innovations include modular caps, electrode arrays embedded in flexible polymers, and automated electrode positioning using 3D head models. Such systems improve spatial resolution for source localization, functional connectivity analysis, and clinical decision-making, particularly in pre-surgical epilepsy mapping and research on cortical networks [44].

AI is also being used to guide hardware design decisions. Predictive algorithms evaluate historical signal quality data to optimize electrode placement and hardware configurations for specific users or conditions. Adaptive systems can switch sampling rates or modify gain settings in real time, optimizing battery life and data integrity. AI-based models significantly improved hardware resilience to motion artifacts and signal drift over time. These findings are already informing design changes in post-market device iterations and regulatory frameworks.

EEG hardware is now increasingly embedded in multimodal headsets that capture heart rate (ECG/PPG), skin conductance, respiration, and eye movement. This integrative approach enhances contextual awareness in neuroadaptive systems and improves classification accuracy in applications like emotion detection and stress monitoring. Some commercial headsets integrate EEG with accelerometers and gyroscopes to compensate for motion and assess user activity. These systems are used in neuroergonomics, driver alertness monitoring, and gaming.

With rapid technological advancement, regulatory bodies are adapting standards to accommodate these new EEG platforms. Benchmarks for evaluating electrode performance, impedance tolerance, motion resilience, and electromagnetic emissions are evolving to reflect innovations in dry electrodes, wearable form factors, and AI-based controls. Conformity assessment studies have helped shape protocols for testing wearable EEG systems under real-world operating conditions. These include testing across humidity, temperature, and motion profiles, as well as fatigue and reusability assessments.

5.2.7 Emerging Trends and Future Directions

As EEG technology continues to evolve, several emerging trends are shaping the future of brain monitoring and neurotechnology. These trends reflect ongoing efforts to enhance accessibility, precision, intelligence, and integration of EEG systems into broader healthcare, research, and consumer applications. Future EEG platforms will be characterized by more intelligent analytics, seamless interoperability, and greater user-centric design.

Artificial intelligence is playing a transformative role in EEG interpretation. Machine learning models are now used to classify sleep stages, detect seizures, monitor anesthesia depth, and track mental workload in real time. Deep learning frameworks can extract complex patterns and biomarkers from raw EEG data with minimal preprocessing. Real-time AI integration is becoming standard in BCIs, where systems must recognize user intent from EEG signals with low latency. Adaptive models update continuously based on feedback, improving accuracy across sessions and users.

To reduce bandwidth and processing requirements, EEG systems are increasingly incorporating edge computing. Signal preprocessing, feature extraction, and inference models are embedded directly into wearable devices. This not only minimizes latency but also protects patient privacy by reducing cloud dependencies [45]. Edge-based systems support remote diagnostics, continuous cognitive monitoring, and decentralized neurotherapy, paving the way for more autonomous neurodevices.

Emerging form factors aim to make EEG acquisition invisible, integrating sensors into headbands, earphones, or even eyeglasses. These devices promise discreet, continuous monitoring for mental health tracking, stress detection, and neurofeedback. Applications range from workplace fatigue monitoring to at-home epilepsy management. Integration with smartphones and AR/VR platforms is also increasing, enabling immersive neurotechnology experiences.

EEG biomarkers are being actively developed for a range of clinical applications:

- Neurodegenerative diseases (e.g., Alzheimer's): Theta and alpha slowing patterns
- Major depressive disorder: Frontal alpha asymmetry
- Epilepsy: Interictal spikes and high-frequency oscillations
- ADHD: Theta/beta ratio.

Personalized neurotherapy approaches are emerging, where EEG-based phenotypes inform the selection of stimulation targets, medication regimens, or behavioral interventions. AI models trained on large-scale EEG databases are improving diagnostic stratification. Traditional BCIs involve unidirectional decoding of brain activity. In contrast, closed-loop BCIs provide feedback or stimulation based on real-time EEG interpretation. These systems are under investigation for:

- Stroke rehabilitation (adaptive movement training)
- Depression treatment (EEG-guided transcranial stimulation)
- Cognitive enhancement (neurofeedback).

Advances in control theory, signal modeling, and user adaptation have made these systems more viable. Real-time EEG decoding is critical to maintain user engagement and therapeutic effect. Future EEG platforms will not operate in isolation. Integration with fNIRS, ECG, motion capture, and eye tracking creates a holistic view of brain-body interactions. These multimodal biosensing systems support:

- More accurate cognitive state detection
- Enhanced emotion recognition
- Robust control interfaces in noisy environments.

Smart synchronization and data fusion algorithms allow for real-time contextual awareness, critical in applications like autonomous vehicles, surgery, and defense.

The rapid development of AI-driven EEG systems has prompted updates in regulatory policy. Regulatory agencies are now creating AI-specific evaluation frameworks that assess algorithm transparency, robustness, and bias. EEG system vendors must align with these standards to ensure clinical and commercial viability. Post-market surveillance (PMS) mechanisms are increasingly leveraging real-world data and predictive analytics to monitor system performance, as shown by recent AI-enhanced PMS models in EEG and patient monitors [46].

As demand for EEG systems grows, emphasis is shifting toward sustainable materials, recyclable components, and low-cost, high-volume production. This trend is critical for expanding EEG accessibility in low-resource settings and for population-scale brain health monitoring initiatives. Researchers are exploring biodegradable substrates, solar-charged EEG caps, and simplified designs that reduce e-waste. Open-source hardware models are also fostering collaboration and reducing costs.

5.3 EEG Signal Preprocessing

Electroencephalogram signal preprocessing is one of the most crucial steps in neurotechnology and biomedical signal analysis. Before EEG data can be reliably interpreted, it must be cleaned of noise and artifacts that originate from various internal and external sources. These contaminants distort the underlying neural information and can lead to inaccurate diagnoses or misleading experimental outcomes if not properly managed. The preprocessing stage ensures that only neural activity of interest is retained and enhanced, making subsequent steps like feature extraction and classification more effective [47]. EEG preprocessing is especially important in modern applications such as real-time BCIs, automated seizure detection, and cognitive workload monitoring, where system performance is directly dependent on the quality of input signals [15]. Furthermore, with the proliferation of wearable EEG and mobile neurotechnology, preprocessing has had to adapt to variable conditions, including motion, environmental noise, and varying electrode quality. This chapter provides a comprehensive examination of the theory, techniques, and tools involved in EEG preprocessing, contextualized with both clinical and research applications.

5.3.1 Objectives of EEG Preprocessing

The objectives of EEG preprocessing are multi-dimensional, reflecting the complexities of both neural activity and contamination sources. The foremost goal is to maximize the SNR. Since neural signals are generally weak—on the order of 5–100 μV—minor contaminations can easily overwhelm the brain's electrical patterns. Signal enhancement techniques focus on preserving amplitude and temporal dynamics critical to both time-domain and frequency-domain analyses.

Artifact suppression aims to reduce or eliminate non-cerebral contributions to the EEG signal. These include eye movements, muscle activity, and power line interference. Suppression must be selective and intelligent—complete elimination may lead to the loss of underlying brain signals, especially when artifact and EEG overlap in spectral characteristics [48]. EEG is used in large-scale studies across diverse populations and experimental conditions. Preprocessing should enable cross-subject and cross-session comparability. This requires protocols that are standardized yet adaptable to individual differences in scalp conductivity, head shape, and physiological behavior.

High-quality preprocessing ensures the reliability of downstream tasks such as:

- ERP analysis
- Power spectral density estimation
- Source localization
- Functional connectivity analysis
- Machine learning-based classification.

Poor preprocessing can lead to overfitting in AI models, false positive alarms in clinical monitors, or ineffective neurofeedback protocols.

5.3.2 Common Sources of Contamination

Understanding where noise originates is essential for designing effective preprocessing pipelines. EEG artifacts can be classified into physiological, environmental, and technical categories. Below is a more detailed look at these sources.

Ocular Artifacts (EOG)

Eye movements and blinks are among the most significant artifacts in EEG data. The corneo-retinal dipole primarily generates these artifacts and propagates through the frontal electrodes, such as Fp1 and Fp2. Vertical eye movements affect the entire frontal cortex, while horizontal movements affect lateral areas. These signals often overlap with delta (0.5–4 Hz) and theta (4–8 Hz) bands, potentially masking slow neural oscillations crucial for sleep staging and attention studies [39]. Eye blink frequency can also vary by task and participant state—rising with fatigue and cognitive load—making their detection and removal essential in cognitive neuroscience.

Muscle Artifacts (EMG)

Electromyographic contamination arises from the electrical activity of facial and neck muscles. These artifacts are particularly pronounced during speaking, chewing, or even subtle tension. Unlike ocular artifacts, EMG often spreads across many electrodes and dominates high-frequency ranges (>20 Hz), significantly affecting gamma-band studies and microstate analyses. In mobile EEG, motion artifacts and EMG often co-occur, complicating artifact removal. Smart preprocessing must be able to distinguish voluntary EEG activity (e.g., motor cortex engagement) from superimposed EMG.

Power Line Interference

Power line interference is a narrowband sinusoidal artifact typically centered at 50 Hz (Europe) or 60 Hz (North America). It originates from surrounding electrical devices, cables, and even fluorescent lighting. Despite good grounding and shielding practices, this artifact is often present in low-amplitude EEG systems. Notch filters or adaptive filters are commonly used to remove it, but excessive use may also degrade adjacent neural frequencies (e.g., beta band). Recent approaches include real-time adaptive filtering and phase-aware regression to maintain phase information for timing-sensitive applications like ERP studies [49].

Electrode and Cable Artifacts

Bad contact between the electrode and the skin can result in high-impedance noise, drift, or abrupt signal drops. Cable movement—particularly in mobile systems—can induce microphonic artifacts, where vibration causes electrical oscillation. These artifacts are typically low-frequency (<2 Hz) and can mimic cortical slow waves or epileptic discharges. Secure electrode placement, low-impedance contact (<5 kΩ), and use of active electrodes or shielded cables are effective prevention strategies. Continuous impedance monitoring systems have been integrated into modern EEG hardware to flag poor channels in real time [50].

Cardiac and Respiratory Artifacts

Electrocardiographic signals can contaminate EEG, especially in montages near the neck or chest. These artifacts are typically periodic and can be confused with sleep spindles or epileptic spikes. Respiratory activity, particularly in polysomnography, affects baseline stability due to mechanical shifts in scalp tension and thoracic movement. Artifact removal requires multi-sensor setups (e.g., concurrent ECG, respiration sensors) and regression-based correction techniques. ICA and PCA are often employed to isolate and subtract these components without distorting cortical signals.

5.3.3 Signal Filtering Techniques

Signal filtering is one of the foundational processes in EEG preprocessing, used to remove unwanted frequency components while preserving the integrity of the brain signals. Since EEG recordings are susceptible to a variety of noise sources distributed across the frequency spectrum, selecting appropriate filters is crucial for ensuring accurate analysis. Filtering can be implemented in both the analog and digital domains, but modern systems typically apply digital filters post-acquisition, allowing greater control over filter characteristics, such as cut-off frequencies, transition bands, and phase properties [13].

EEG preprocessing relies primarily on the following filter types:

- High-pass filters remove slow signal drifts and DC offsets, typically setting a cut-off between 0.1 Hz and 1 Hz. These are essential in removing slow trends caused by sweat, respiration, and electrode polarization.
- Low-pass filters are used to suppress high-frequency noise, such as EMG contamination. Cut-off frequencies range from 30–100 Hz, depending on the application. For example, ERP studies may apply a low-pass filter at 30 Hz, whereas gamma-band studies may retain frequencies up to 100 Hz.
- Band-pass filters combine high-pass and low-pass filters to isolate a specific frequency band. This is especially useful for analyzing standard EEG frequency ranges: delta (0.5–4 Hz), theta (4–8 Hz), alpha (8–13 Hz), beta (13–30 Hz), and gamma (>30 Hz).
- Notch filters are designed to eliminate narrowband interference, typically from power lines (50/60 Hz). However, their use should be limited when phase information is critical.

Filter design must carefully consider:

- Cut-off frequency: Too low may distort slow waves; too high may attenuate signal features
- Filter order: Higher-order filters yield steeper roll-offs but introduce more latency
- Phase response: Zero-phase filters (e.g., using forward and reverse filtering) avoid phase distortion but are unsuitable for real-time applications
- Causal versus non-causal filters: Real-time systems require causal filters; offline systems can use non-causal filters for better fidelity.

Butterworth, Chebyshev, and FIR filters are commonly used in EEG preprocessing, each offering a trade-off between sharpness, phase linearity, and computational efficiency [18]. The choice of filters often depends on the experimental or clinical goal:

- ERP studies: Require narrow band-pass filtering to preserve time-locked features; high-pass filters typically set at 0.1 Hz to retain slow potentials
- Sleep studies: Use filters to enhance delta and spindle activities; low-pass filters at 30 Hz are common

- BCI applications: Often use band-pass filters (e.g., 8–30 Hz) optimized for motor imagery or attention detection
- Seizure detection: Requires wideband filtering with minimal phase distortion; adaptive filters may be used for artifact suppression.

A growing number of EEG systems include programmable digital signal processors (DSPs) that allow dynamic filter configurations based on input signal quality and task-specific requirements [51]. Modern EEG preprocessing also employs advanced filtering strategies such as:

- Adaptive filtering: Adjusts filter parameters in real time based on incoming signal characteristics. Useful for mobile EEG where noise patterns vary rapidly.
- Wavelet filtering: Uses time–frequency decomposition to isolate and remove artifacts without traditional frequency-domain filtering. Particularly effective for short-lived transient noise.
- Empirical mode decomposition (EMD) and ensemble EMD: Nonlinear, data-driven methods that decompose signals into intrinsic mode functions (IMFs), helping isolate non-stationary noise components [44].
- Artifact subspace reconstruction (ASR): Detects and removes high-amplitude artifacts by reconstructing clean EEG subspaces. This method is well-suited for real-time applications and has been integrated into platforms like EEGLAB.

While filtering is indispensable, poor application can lead to significant signal distortion. Best practices include:

- Always visually inspect filtered vs. raw signals to ensure integrity
- Avoid high-pass filters > 1 Hz for ERP and delta-band analysis
- Combine filtering with ICA or PCA for more effective noise removal
- Document filter parameters for reproducibility and regulatory compliance.

Recent guidance from researchers and regulatory bodies emphasizes the importance of transparency in preprocessing steps, especially in clinical studies and AI model training [52].

5.3.4 Artifact Removal and Correction Techniques

Artifact removal and correction techniques are fundamental components of EEG signal preprocessing. While filtering reduces broad frequency-based noise, artifact removal methods target specific, often non-stationary sources of contamination such as eye blinks, muscle activity, and motion-induced distortions. These artifacts, if not addressed, can obscure clinically or cognitively relevant brain signals and skew analytical outcomes [22].

The most traditional approach to artifact management is manual inspection. Trained analysts visually review EEG recordings and mark segments containing artifacts for exclusion. This method, though time-consuming, allows fine control and is still used in clinical diagnostics, particularly for detecting subtle abnormalities or epileptiform discharges. However, manual rejection introduces subjectivity, lacks scalability for high-throughput applications, and results in data loss, especially if large segments are contaminated. Therefore, it is often used in conjunction with automated approaches.

ICA is one of the most widely adopted techniques for EEG artifact removal. It decomposes multichannel EEG signals into statistically independent components (ICs), some of which represent artifacts. Once identified—typically by their spatial, spectral, or temporal features—these components can be removed or reconstructed, preserving the remaining brain activity [53]. ICA is particularly effective for ocular, cardiac, and muscle artifacts. Automated ICA classification tools such as ICLabel and MARA use machine learning to identify components likely representing artifacts [54].

PCA is another blind source separation technique that identifies orthogonal components based on variance. Although less effective than ICA for complex sources, PCA can be useful for dimensionality reduction and for removing correlated noise across electrodes, such as power line interference.

Linear regression is often used to remove artifacts with known reference signals, such as EOG or EMG channels. The contaminating signal is modeled and subtracted from the EEG, minimizing residual noise. This method is straightforward and effective, but assumes linearity and requires accurate reference signals. Advanced regression variants include adaptive filtering and constrained least squares methods, which can dynamically adjust coefficients to match changing artifact profiles [20].

ASR is a data-driven technique that identifies periods of high-amplitude artifact and reconstructs the signal using a subspace learned from cleaner epochs. This technique preserves usable data that would otherwise be discarded and is highly suited for real-time and mobile EEG [28]. ASR is often integrated into pipelines like the PREP pipeline or Cleanline toolboxes in EEGLAB and is increasingly used in large-scale EEG studies and BCIs due to its automation and speed.

Wavelet decomposition provides a powerful method for isolating transient artifacts. It breaks the EEG into time–frequency coefficients, allowing selective removal of noise without affecting critical brain signals. This is especially effective for short-duration artifacts such as muscle twitches or sharp movement bursts [24]. Wavelet denoising is particularly favored in pediatric EEG, where patient movement is frequent and traditional filtering may fail. Combining multiple techniques often yields the best results. For example:

- ICA followed by regression
- Wavelet filtering after ASR
- PCA pre-processing prior to ICA for dimensionality reduction.

Ensemble pipelines are standard in research-grade preprocessing tools, allowing flexible customization for specific datasets or clinical populations. Recent years have seen an explosion of deep learning models applied to EEG artifact removal. Convolutional neural networks (CNNs), autoencoders, and recurrent neural networks (RNNs) have been trained to identify and suppress artifacts without explicit component separation. These models can generalize across subjects and conditions when trained on large, diverse datasets. While numerous artifact correction techniques exist, their performance depends on proper implementation. Best practices include:

- Combine methods where possible (e.g., ICA + ASR)
- Use validated datasets for model training
- Retain diagnostic signals by avoiding overcorrection
- Apply quality control metrics post-correction (e.g., retained variance, artifact-free rate).

Limitations still exist. ICA and PCA assume signal stationarity and may fail in rapidly changing environments. Regression may not capture non-linear interactions. AI methods require extensive, annotated training data and rigorous validation.

5.3.5 Channel Selection and Signal Referencing

Channel selection and referencing are essential steps in EEG preprocessing that influence data quality, interpretability, and computational efficiency. While most modern EEG systems can record from 32 to 256 channels, not all recorded electrodes contribute equally to a given study or clinical application. Strategic selection and referencing can reduce noise, enhance spatial resolution, and improve performance in automated classification systems. Channel selection involves identifying a subset of electrodes that carry the most relevant information for the target analysis. This process is driven by:

- The anatomical localization of brain regions of interest (e.g., frontal cortex for attention, motor cortex for BCIs)
- The task-specific signal characteristics (e.g., alpha oscillations in occipital regions)
- Computational constraints in mobile or real-time systems.

Selecting too many channels can introduce redundant information, increase processing load, and elevate noise due to poor electrode contact. On the other hand, selecting too few may miss critical spatial dynamics. Intelligent channel selection balances these considerations to optimize signal utility. Manual selection is based on domain expertise and standard montages like the 10–20 or 10–10 systems. For example, Fz, Cz, and Pz are often used in ERP studies due to their alignment with midline cortical structures. However, manual methods lack scalability for high-density recordings and large datasets. Automated methods use:

- Mutual information analysis to assess channel contribution to target classification
- Recursive feature elimination with SVMs or neural networks
- Genetic algorithms for identifying optimal channel subsets [9].

These approaches reduce dimensionality while preserving discriminative power in machine learning applications, particularly in EEG-based emotion recognition and cognitive workload detection. EEG measures voltage differences between electrodes, making the reference electrode a critical factor in shaping the observed signal. Typical referencing schemes include:

- Linked-ears (A1–A2): Traditional method, but susceptible to lateral noise
- Average reference: Subtracts the average signal across all electrodes, useful in high-density systems
- Common average reference (CAR): Excludes noisy or artifact-laden channels from the average
- Laplacian reference: Uses spatial derivatives to enhance local activity, suitable for source localization
- REST (Reference Electrode Standardization Technique): Computationally re-references signals to an ideal neutral point.

Selection depends on system configuration, number of electrodes, and type of analysis. For instance, average reference performs well in dense-array EEG, while Laplacian referencing enhances cortical mapping in motor studies. Improper referencing can introduce artificial signal distortions or attenuate actual neural activity. This is particularly problematic when comparing datasets across sites or systems. Researchers must document referencing methods meticulously and consider re-referencing in post-processing to standardize analysis pipelines [55].

Modern systems are exploring adaptive referencing strategies that dynamically adjust the reference based on signal quality. Algorithms assess noise levels and impedance across channels and select the most stable configuration in real time. AI-enhanced referencing has been proposed, using machine learning to identify optimal re-referencing strategies that maximize SNR and task sensitivity [10]. Both channel selection and referencing impact:

- Event-related potential morphology
- Source localization accuracy
- Classifier performance in BCI
- Consistency across subjects and sessions.

As EEG analysis becomes more automated and scalable, these preprocessing decisions must be encoded into reproducible pipelines with transparency and flexibility.

5.3.6 Signal Quality Assessment and Preprocessing Validation

Once EEG preprocessing steps are applied, it is essential to assess the signal quality and validate the effectiveness of the procedures. Poor signal quality can undermine the reliability of findings, whether in clinical diagnostics or cognitive neuroscience experiments. Systematic signal quality assessment enables early detection of hardware malfunctions, participant-related issues (e.g., excessive movement), or ineffective preprocessing pipelines. Several quantitative metrics are used to evaluate EEG signal quality:

- SNR: Measures the ratio between signal power and noise power. Higher SNR indicates cleaner data.
- Channel flatlining or dropout: Channels with abnormally low variance suggest electrode detachment or malfunction.
- Variance and standard deviation: Excessively high variance may indicate residual artifacts or faulty electrodes.
- Kurtosis and skewness: Used to detect non-Gaussian distributions indicative of transient noise bursts or artifacts.
- Impedance monitoring: High electrode–skin impedance (>10 kΩ) is linked to poor signal acquisition.

Real-time quality assessment tools often use these metrics to automatically flag bad channels or epochs before analysis [56]. Despite the availability of automated metrics, expert visual inspection remains indispensable, especially in clinical contexts. Human reviewers can spot subtle abnormalities, such as:

- Misplaced electrodes
- Residual EOG or EMG artifacts
- Rhythmic noise (e.g., from ventilators or tremors).

Visualization tools like time-domain plots, spectrograms, and topographical maps help assess preprocessing effectiveness and verify artifact suppression. Modern EEG platforms integrate signal quality validation into preprocessing pipelines. Examples include:

- EEGLAB's Clean_rawdata plugin: Performs automated channel rejection, ASR, and noise metrics
- PREP pipeline: Offers standardized referencing, artifact correction, and quality reporting
- BioSPPy and MNE-Python: Provide modular frameworks for signal inspection and validation.

These toolkits enhance reproducibility and reduce user bias. They are increasingly used in large-scale EEG studies and clinical trials. In regulatory contexts, EEG preprocessing workflows must meet performance benchmarks defined by authorities such as the FDA, IEC, or local health agencies. Metrics include:

- Accuracy of spike detection or seizure classification
- Robustness to motion artifacts
- Reproducibility across datasets.

These frameworks are especially crucial for wearable and AI-integrated EEG devices. Artificial intelligence is increasingly used to evaluate EEG quality by identifying noise patterns, drift, and anomalies across channels. Deep learning models trained on large EEG datasets can predict quality scores and suggest corrective actions, such as:

- Re-referencing
- Dynamic filter reconfiguration
- Channel replacement or interpolation.

Adequate validation requires a mix of automated tools and manual oversight. Best practices include:

- Document all preprocessing parameters and quality thresholds
- Combine quantitative and qualitative evaluations
- Use benchmark datasets for calibration and testing
- Perform inter-rater reliability checks for manual inspections
- Incorporate validation into real-time acquisition loops.

By rigorously validating EEG preprocessing, researchers and clinicians ensure that subsequent analyses rest on a foundation of trustworthy data. This strengthens the scientific and clinical utility of EEG across a growing range of applications.

5.3.7 Emerging Trends in EEG Preprocessing

The evolution of EEG technology and computational tools has catalyzed the emergence of innovative preprocessing methodologies. These approaches are redefining the boundaries of what EEG can capture, especially in mobile, wearable, and high-density settings. This section explores cutting-edge trends that are reshaping how EEG preprocessing is conducted across research, clinical, and consumer applications.

One of the most significant trends is the shift toward real-time, edge-based preprocessing. Wearable EEG devices are now incorporating embedded processors capable of executing filtering, artifact correction, and quality monitoring directly on the device. This development is crucial for:

- BCI systems requiring low-latency signal interpretation
- Telehealth and remote neurodiagnostics
- Consumer neurotechnology in gaming and wellness.

Studies have shown that such systems can maintain robust performance with sub-second latency, even under motion-rich conditions [57]. Adaptive preprocessing pipelines can dynamically adjust their parameters based on input signal conditions. These include:

- Varying filter bandwidth based on task demands
- Modifying artifact thresholds based on movement levels
- Re-weighting channels based on impedance fluctuations.

These context-aware pipelines use continuous feedback from auxiliary sensors (e.g., accelerometers, EOG, EMG) and AI models to adapt preprocessing in real time [22]. Transfer learning is gaining traction in EEG preprocessing, allowing models trained on one dataset or user to generalize to new subjects or tasks with minimal retraining. This reduces the need for extensive calibration sessions and accelerates deployment in clinical settings. Pretrained artifact classifiers, for instance, can detect blinks or EMG across datasets with high accuracy [58]. Toolkits like ICLabel leverage large databases to classify independent components automatically, using pretrained neural networks. End-to-end deep learning models are being developed to learn denoising mappings from raw EEG to clean signals. These include:

- Denoising autoencoders
- Recurrent neural networks (RNNs)
- Transformer-based architectures.

These models outperform traditional filters in preserving fine-grained temporal structure and suppressing nonstationary artifacts. However, they require large labeled datasets and careful validation to avoid overfitting or signal distortion. Integrating EEG with other physiological modalities (ECG, eye-tracking, fNIRS) allows for more precise artifact detection and signal segmentation. For example:

- EOG can improve blink removal
- EMG can inform muscle artifact suppression
- fNIRS can distinguish cortical from extracerebral activity.

These multimodal systems support hybrid pipelines where EEG preprocessing benefits from richer contextual awareness [59]. As AI becomes more embedded in EEG pipelines, explainability is essential. Efforts are underway to create preprocessing algorithms that:

- Provide justifications for data rejection
- Visualize learned filters or model attention
- Allow human override in ambiguous cases.

Such transparency is critical in clinical decision-making and regulatory approval, where black-box models are often unacceptable. To promote reproducibility, the EEG community is embracing:

- Open-source pipelines (e.g., MNE-Python, EEGLAB)
- Standardized data formats (e.g., BIDS-EEG)
- Public benchmark datasets (e.g., TUH, OpenBCI).

These efforts ensure that preprocessing algorithms are validated, comparable, and transparent. Standardization also facilitates cross-lab collaborations and meta-analyses. Preprocessing systems are being designed to comply with emerging standards for AI and medical devices. This includes:

- Documentation of preprocessing parameters for auditability
- Post-market surveillance of preprocessing performance
- AI monitoring frameworks that track drift and bias over time.

5.4 EEG Feature Extraction and Analysis

Electroencephalography is a cornerstone technique in neuroscience and neuroengineering. As a non-invasive method for capturing brain electrical activity via scalp electrodes, EEG provides rich temporal information at millisecond-level resolution. It is widely utilized in diverse fields such as clinical neurology, cognitive science, sleep research, human–computer interaction, and increasingly, in consumer-grade neurotechnology like wearables and BCIs.

Despite its ubiquity, the raw EEG signal is notoriously tricky to interpret. It is characterized by low signal-to-noise ratio, variability across sessions and subjects, and susceptibility to artifacts originating from muscle activity, eye movements, and environmental interference. Consequently, raw EEG waveforms must transform higher-level, structured representations—features—that encapsulate meaningful information relevant to specific applications [9, 60]. Feature extraction refers to the process of computing such descriptive attributes from the EEG signal. These features are designed to distill raw electrical activity into representations that capture essential physiological and cognitive characteristics. The ultimate goal is to facilitate interpretation, enable machine learning algorithms, and support decision-making in both clinical and research contexts. Feature extraction is central to applications such as:

- Clinical diagnostics and monitoring, including seizure detection in epilepsy, diagnosis of Alzheimer's disease, and sleep stage scoring.
- Cognitive and affective computing, where EEG is used to track attention, mental workload, drowsiness, or emotional arousal.
- BCIs, which translate neural signals into control commands for external systems such as robotic arms or software interfaces.
- Neurotherapeutic and rehabilitation monitoring, assessing changes in brain activity due to treatments like neurofeedback or transcranial stimulation.

This chapter systematically explores the different classes of EEG features—ranging from time-domain and frequency-based descriptors to nonlinear and connectivity measures—highlighting their computational principles, physiological significance, and practical applications [61].

5.4.1 Types of Features in EEG Analysis

EEG signals are inherently multidimensional and complex, prompting the development of a broad taxonomy of feature types. Features are typically categorized based on the domain in which they are computed (time, frequency, time–frequency), the mathematical approach (linear vs. nonlinear), or their physiological interpretation (e.g., oscillatory power vs. connectivity).

Time-Domain Features

Time-domain features are directly derived from the amplitude of EEG waveforms over time, without transforming the signal into other domains. These features are simple to compute, fast, and intuitive. They are instrumental in event-related potential (ERP) studies, where precise timing relative to a stimulus is crucial. Key examples include:

- Mean and median amplitude: Measures of central tendency.
- Variance and standard deviation: Indicators of signal dispersion.
- Peak-to-peak amplitude: Difference between local maxima and minima.
- Zero-crossing rate: A proxy for signal frequency content.
- Hjorth parameters: Statistical measures capturing activity, mobility, and complexity.
- Skewness and kurtosis: Higher-order moments that describe the waveform shape.

These features are critical in applications like ERP analysis, seizure detection, and sleep staging, where temporal precision and computational efficiency are prioritized.

Frequency-Domain Features

Frequency-domain features describe the spectral content of EEG signals. They are computed using techniques such as the FFT, Welch's method, or autoregressive models. Spectral features are particularly informative in characterizing neural oscillations within well-established frequency bands: delta, theta, alpha, beta, and gamma. Typical frequency-based features include:

- PSD: Overall energy distribution across frequencies.
- Band-specific power: Power within predefined bands (e.g., alpha power during rest).
- Relative band power: Normalized power as a fraction of total signal energy.
- Band ratios: Ratios such as theta/beta, used in ADHD assessment.
- Spectral entropy: A measure of randomness in frequency distribution.

These features are frequently employed in sleep scoring, drowsiness monitoring, cognitive load estimation, and emotion recognition systems [62, 63].

Time–Frequency Features

Because EEG is a non-stationary signal, its spectral properties often change over time. Time–frequency analysis captures this dynamic behavior by examining how frequency components evolve. Essential time–frequency techniques include:

- STFT: Computes FFT over sliding windows.
- Wavelet transform: Decomposes signals into time-localized frequency components.
- Hilbert-Huang transform (HHT) and Stockwell transform (ST): Advanced methods suited for transient dynamics.

Time–frequency representations are essential in detecting transient phenomena like epileptic spikes, sleep spindles, or cognitive responses to stimuli.

Nonlinear Features

Linear analyses may fail to fully describe the complex, sometimes chaotic, behavior of EEG. Nonlinear features capture signal irregularity, complexity, and unpredictability—important in understanding brain function under pathological or high-load conditions. Standard nonlinear features include:

- Entropy measures (approximate, sample, spectral, permutation).
- Fractal dimension: Describes self-similarity and complexity.
- Lyapunov exponent: Indicates chaotic behavior.
- Hurst exponent: Reflects signal memory and persistence.

Such features are critical in applications like seizure prediction, cognitive decline detection, and biometric identification.

Connectivity Features

Beyond isolated electrode activity, EEG can reveal patterns of interaction between brain regions. Functional connectivity features characterize synchronization or information flow across spatially distributed signals. Standard connectivity metrics include:

- Coherence: Frequency-specific correlation.
- Phase-locking value (PLV): Consistency of phase differences.
- Granger causality and transfer entropy: Measures of directional influence.

Connectivity features are pivotal in understanding brain networks in disorders like epilepsy, schizophrenia, and autism.

5.4.2 *Time-Domain Feature Extraction*

Time-domain analysis is one of the most fundamental and accessible approaches for extracting features from EEG signals. It involves directly analyzing the waveform's amplitude characteristics over time, without requiring transformation into other domains such as frequency or wavelet space. Time-domain features are particularly valued for their simplicity, low computational cost, and suitability for real-time applications such as brain-computer interfaces (BCIs), cognitive workload monitoring, and clinical diagnostics [10, 64]. Despite their apparent simplicity, time-domain features can provide rich insights into brain activity, especially in cases where signal timing or morphology is critical. Event-related potentials (ERPs), for example, are best captured using time-domain descriptors due to their precise temporal alignment with external stimuli.

The mean amplitude of an EEG segment reflects the average voltage over a given time window. The median amplitude, on the other hand, is the middle value when all amplitudes are ordered. It is less affected by transient spikes or artifacts and can offer a more robust measure in noisy recordings. Both metrics are often computed over specific epochs tied to cognitive events, such as in ERP studies where the average voltage from 300–600 ms post-stimulus might represent attention-related activation (e.g., the P300 component). Abnormal mean amplitudes may also signal neurological impairment or sedation. These metrics measure how widely the EEG signal varies within a time window. The standard deviation (SD) is the square root of the variance, offering a more interpretable measure in the original units. High variability in EEG amplitude can indicate active cognitive processing, epileptiform discharges, or muscle artifacts. Conversely, reduced variance might be observed in sedative states or neurodegenerative conditions.

Peak detection involves identifying local maxima and minima within a segment. This is essential for characterizing ERPs like N100 or P300, whose latencies and amplitudes are of clinical and research interest [65, 66]. The peak-to-peak amplitude is the difference between the highest and lowest points in a segment. In seizure detection, for instance, high-amplitude spikes or sharp waves are diagnostic indicators. Similarly, P2P amplitude is often used in motor imagery BCIs to gauge motor cortex activation.

The zero-crossing rate quantifies how often the EEG signal crosses the zero voltage line. ZCR indirectly reflects the dominant frequency of the signal. High ZCR values suggest faster oscillations, such as those seen during wakefulness or muscle activity. In contrast, low ZCR rates occur during slow-wave sleep or under general anesthesia [18]. ZCR is useful in automatic sleep staging algorithms and real-time monitoring of vigilance in operational environments. Introduced by Bo Hjorth in 1970, these parameters provide compact yet informative descriptors of EEG dynamics:

- Activity: Equivalent to signal variance, representing power.
- Mobility: Estimates the mean frequency of the signal.
- Complexity: Indicates how the waveform deviates from a pure sine wave, reflecting signal irregularity.

Hjorth parameters are especially prominent in BCI systems, where they are used for classifying motor imagery tasks with minimal processing overhead [22]. Time-domain waveforms can also be characterized by:

- Skewness: Indicates asymmetry of the signal distribution.
- Kurtosis: Describes the "peakedness" or tail heaviness.

These metrics are sensitive to rare, high-amplitude deviations such as epileptiform spikes. Elevated kurtosis may flag pathological activity, while skewness can indicate artifact-related distortions or lateralized brain activity. Computing the first and second derivatives of the EEG signal provides insight into:

- Signal slope or velocity: Rate of change in voltage.
- Acceleration: Useful for detecting rapid transitions or edge-like features in ERPs.

These derivatives are essential in identifying sharp epileptic discharges, ERP component onsets, and rapid fluctuations during cognitive transitions. Temporal asymmetry features compare the signal's properties when time is reversed. They can uncover underlying non-linear dynamics and hemispheric asymmetries associated with affective or psychiatric states [67, 68]. One common application is frontal alpha asymmetry, which has been linked to mood disorders like depression [69]. In such cases, asymmetry is computed by comparing mean or band-specific power across hemispheres.

The reliability and interpretability of time-domain features depend heavily on the selection of epochs—segments of EEG used for analysis. Key considerations include:

- Window length: Short windows offer better temporal precision but may increase noise sensitivity.
- Overlap: Overlapping windows smooth temporal transitions and improve statistical power.
- Alignment with events: For ERP studies, precise time-locking to stimulus or response events is critical.

Typical window sizes range from 200–600 ms for ERPs and 1–2 s for mental state classification tasks. Time-domain features are particularly well-suited to real-time systems due to their low latency and computational efficiency. They are often used as inputs for lightweight classifiers in:

- BCIs for motor control
- Neurofeedback systems for attention training
- Driver drowsiness detection
- Wearable EEG analytics.

Because these features can be computed on embedded devices with limited processing capacity, they form the backbone of many mobile EEG platforms.

5.4.3 Frequency-Domain Feature Extraction

While time-domain analysis provides insight into the amplitude characteristics of EEG signals, it does not explicitly capture the rhythmic or oscillatory activity that underpins much of brain function. Frequency-domain analysis fills this gap by decomposing EEG signals into their constituent frequencies and quantifying power distribution across canonical brain rhythms. This spectral perspective is essential for understanding brain states such as sleep, arousal, memory encoding, and pathological conditions like epilepsy or Alzheimer's disease [17]. Frequency-domain features are extracted using mathematical transforms, primarily the FFT and Welch's method, which convert time-series data into frequency representations. The resulting features can be absolute (total power in a band), relative (normalized power), or derived metrics (e.g., entropy, slope). These features are foundational in a wide range of EEG applications, from clinical diagnostics to real-time BCI systems [59, 70].

Power spectral density is the cornerstone of spectral EEG analysis. It quantifies how the power of a signal is distributed across different frequencies. PSD is typically estimated using:

- FFT: A direct, efficient method for obtaining the spectrum.
- Welch's method: An averaged periodogram approach that improves spectral stability by dividing the signal into overlapping segments and applying window functions.

PSD provides a window into neural oscillations that fall into well-established frequency bands:

- Delta (0.5–4 Hz): Associated with deep sleep and unconscious processing.
- Theta (4–8 Hz): Linked to drowsiness, memory consolidation, and meditative states.
- Alpha (8–13 Hz): Prominent during relaxed wakefulness; reduced with eyes open or cognitive effort.
- Beta (13–30 Hz): Reflects alertness, mental effort, and motor activity.
- Gamma (>30 Hz): Related to attention, sensory binding, and high-level cognition.

Changes in PSD are used to diagnose a variety of conditions, such as increased delta in traumatic brain injury or reduced alpha in Alzheimer's disease [24]. Frequency band features can be further refined into:

- Absolute band power: The total power within a specific band.
- Relative band power: The proportion of power in a given band relative to the total EEG power.

Relative band power is less sensitive to signal amplitude variations caused by electrode impedance or scalp thickness, making it ideal for inter-subject comparisons and mobile EEG devices. Applications:

- Sleep staging: Delta dominance in N3 sleep, spindle activity in stage 2.
- Stress detection: Elevated beta power in high-arousal states.

- ADHD assessment: Increased theta/beta ratio in children with attentional deficits.
- Driver alertness: Rising theta or falling alpha power as indicators of drowsiness.

Spectral entropy measures the unpredictability or complexity of the frequency distribution. High spectral entropy reflects a broad, flat distribution of power (e.g., during alert wakefulness or artifact contamination), while low entropy indicates concentrated, rhythmic activity (e.g., alpha dominance during rest).

Applications:

- Anesthesia monitoring: Entropy drops under sedation.
- Seizure detection: Sudden entropy changes around ictal events.
- Cognitive workload estimation: Mental effort may reduce spectral entropy due to narrowband activation.

Band ratio features compare power in one frequency band to another, capturing relationships between neural rhythms:

- Theta/alpha, theta/beta: Associated with attentional control and executive function.
- Alpha/beta: May indicate relaxation vs. cognitive engagement.
- Beta/alpha: Often used in stress monitoring.

These ratios provide compact, physiologically interpretable features and are robust to absolute amplitude scaling [48, 54]. Alpha asymmetry—the difference in alpha power between left and right hemispheres—is particularly relevant in affective neuroscience and clinical psychology. It has been linked to:

- Depression: Greater left frontal alpha (less activation) correlates with negative affect.
- Approach/avoidance behaviors: Frontal alpha asymmetry predicts motivational states.

In addition to total power, the location of frequency content is diagnostically meaningful:

- Peak frequency: The frequency at which maximum power occurs within a band.
- Center frequency: The weighted average frequency across the band.

Peak alpha frequency (PAF), for example, tends to decline with age and in neurodegenerative diseases such as Parkinson's and Alzheimer's [15]. A slower PAF can serve as a biomarker for cognitive decline. The spectral slope quantifies how EEG power decays with increasing frequency, often modeled as a 1/f distribution. The exponent α reflects the steepness of decay:

- A steep slope indicates regular, low-frequency dominance (e.g., sleep).
- A flat slope implies high-frequency activity or increased noise (e.g., hyperexcitability, schizophrenia).

Recent work links spectral slope to cortical excitation/inhibition balance—an emerging biomarker in psychiatric and developmental disorders [71]. Frequency-based features are pervasive in EEG analysis due to their robustness and physiological grounding. Applications include:

- Epilepsy: Burst detection in gamma or beta ranges.
- Sleep scoring: Delta, theta, and spindle analysis for automated hypnograms.
- Mental workload: Alpha suppression and beta activation under task conditions.
- Neurodegeneration: Theta power increases and alpha slowing in dementia.
- BCI systems: Motor imagery classification using mu (8–12 Hz) and beta rhythms.

Spectral features are also used as input to machine learning models for emotion recognition, workload estimation, and biometric authentication. Despite their strengths, frequency-domain features come with caveats:

- Assumption of stationarity: FFT assumes that signal characteristics remain constant over the window. This is often violated in EEG.
- Window size trade-off: Shorter windows improve temporal resolution but reduce frequency resolution.
- Spectral leakage: Power from neighboring frequencies may blur results, especially in short or improperly windowed segments.
- Artifact contamination: Muscle activity and eye blinks introduce high-frequency and low-frequency artifacts, respectively.

These limitations can be mitigated by:

- Applying overlapping windowing and tapering functions.
- Combining with time–frequency methods
- Using artifact rejection or independent component analysis (ICA) preprocessing.

5.4.4 Time–Frequency Feature Extraction

Traditional time-domain and frequency-domain analyses offer distinct advantages but also suffer inherent limitations—chiefly, a trade-off between temporal and spectral resolution. EEG signals are intrinsically non-stationary, meaning their frequency content evolves. Capturing this dynamic behavior is essential in many contexts, such as detecting transient neural events (e.g., epileptic spikes), analyzing stimulus-locked brain responses, or characterizing fluctuating cognitive states.

Time–frequency analysis addresses these challenges by providing joint temporal and spectral resolution. Rather than examining the signal solely in time or frequency, time–frequency methods represent EEG activity as a surface or matrix in which each point corresponds to a particular frequency at a specific time. These representations allow for fine-grained tracking of how neural oscillations change in response to internal or external events [72, 73].

Short-Time Fourier Transform (STFT)

The Short-Time Fourier Transform divides the EEG signal into overlapping time windows and applies the Fourier transform to each window. This results in a time-indexed series of frequency spectra.

Strengths:

- Simple and computationally efficient.
- Well-suited for signals with slowly varying frequency content.

Limitations:

- Fixed window size leads to a trade-off between time and frequency resolution (Heisenberg uncertainty principle).
- May not capture rapid transient events accurately.

Applications:

- Analysis of auditory evoked potentials.
- Motor imagery classification in BCIs.
- Seizure onset detection.

Wavelet Transform (WT)

The Wavelet Transform overcomes the fixed-resolution problem of STFT by using variable-sized windows:

- High frequencies: Analyzed with short, narrow windows (good temporal resolution).
- Low frequencies: Analyzed with wide, longer windows (better frequency resolution).

There are two primary forms:

- Continuous Wavelet Transform (CWT): Produces a high-resolution, redundant representation.
- Discrete Wavelet Transform (DWT): Offers compact, multi-level signal decomposition using orthogonal wavelets.

Wavelet-based features include:

- Wavelet energy: Power across decomposition levels.
- Wavelet entropy: Irregularity in the coefficient distribution.
- Statistical features: Mean, variance, skewness of wavelet coefficients.

Applications:

- Seizure detection (e.g., spike-wave complexes in temporal lobe epilepsy).
- Emotion recognition via frontal alpha/beta shifts.
- Sleep spindle identification.
- Compression and denoising of EEG data for wearable systems.

Hilbert-Huang Transform (HHT)

The Hilbert-Huang Transform (HHT) is designed specifically for non-linear and non-stationary signals. It consists of two steps:

1. Empirical Mode Decomposition (EMD): Decomposes the signal into intrinsic mode functions (IMFs), each representing a narrow-band component.
2. Hilbert Spectral Analysis: Applies the Hilbert transform to each IMF to derive instantaneous frequency and amplitude.

This yields a time–frequency-energy distribution with high adaptivity and resolution.

Strengths:

- Ideal for analyzing transient, irregular phenomena.
- Offers superior temporal resolution compared to FFT or STFT.

Limitations:

- Computationally intensive.
- Sensitive to noise; requires preprocessing.

Applications:

- Real-time workload tracking.
- Precise modeling of seizure dynamics.
- EEG signal decomposition in brain-state transition studies.

Stockwell Transform (ST)

The Stockwell Transform (S-transform) combines elements of STFT and WT, offering time–frequency decomposition with frequency-dependent resolution and preserved phase information [74]. It adapts the window length to frequency while retaining the ability to reconstruct the signal from the time–frequency representation.

Advantages:

- Good balance of time–frequency localization.
- Provides phase and magnitude information.
- Suitable for connectivity and coherence studies.

Applications:

- Real-time classification in BCIs.
- Transient oscillatory event tracking (e.g., bursts, suppressions).
- Auditory processing experiments.

Time–Frequency Representations (TFRs)

The output of STFT, WT, or ST can be visualized as Time–Frequency Representations (TFRs)—typically 2D matrices or images where axes correspond to time and frequency, and intensity represents power.

TFRs can be converted into feature vectors by:

- Extracting summary statistics (mean power, entropy) from regions of interest.
- Using band-specific energy curves over time.
- Employing dimensionality reduction techniques like PCA or t-SNE.

Additionally, TFRs can be fed directly into deep learning models:

- CNNs can classify mental states from TFR images.
- RNNs can process sequential slices of the TFR.

Use cases:

- Emotion recognition (e.g., arousal/valence classification).
- Mental workload estimation.
- Seizure type classification using TFR "signatures."

Event-Related Spectral Perturbation (ERSP)

ERSP quantifies stimulus-locked changes in spectral power relative to a pre-stimulus baseline:

It captures induced (non-phase-locked) and evoked (phase-locked) oscillatory activity. ERSP maps reveal task-related modulations in brain rhythms, such as:

- Theta bursts during memory encoding.
- Alpha suppression in visual attention.
- Beta rebounds in motor response.

Applications:

- ERP experiments with time-varying spectral features.
- Neurofeedback training in alpha or theta bands.
- Cognitive task profiling in working memory studies.

Implementing time–frequency analysis requires careful attention to:

- Computational cost: WT and HHT are more resource-intensive than FFT-based methods.
- Choice of wavelet: Morlet and Daubechies wavelets are popular for EEG; choice affects sensitivity and localization.
- Baseline normalization: Critical for ERSP and event-related designs to remove inter-trial variability.
- Artifact handling: Time–frequency representations are sensitive to muscle and blink artifacts, which often contaminate high-frequency components.

Selection of the optimal method depends on:

- The target signal features (e.g., transient vs. sustained activity).
- Application constraints (e.g., real-time BCI vs. offline diagnostics).
- Computational resources (embedded systems vs. cloud-based analysis).

5.4.5 Nonlinear Feature Extraction

Despite the utility of time-domain and frequency-domain analyses, they often fall short in capturing the complex, dynamic, and chaotic nature of brain activity. The human brain is a non-linear, non-stationary system with intricate interdependencies, self-organizing properties, and irregular behavior that defies purely linear models. To capture these higher-order dynamics, researchers have increasingly turned to nonlinear feature extraction techniques [75, 76].

Nonlinear features describe signal irregularity, complexity, self-similarity, and sensitivity to initial conditions—properties often altered in pathological states, cognitive transitions, or developmental processes. These features have gained traction in applications such as seizure prediction, dementia diagnosis, biometric authentication, and the study of consciousness.

This section outlines the major categories of nonlinear features, their computational underpinnings, and applications in EEG-based analysis.

Entropy quantifies the amount of disorder or unpredictability in a signal. In EEG analysis, entropy-based features reveal how "structured" or "random" the brain activity is—lower entropy suggests more regular, synchronized firing (as in seizures or deep sleep), while higher entropy indicates complex, desynchronized states (e.g., wakefulness, cognitive engagement).

ApEn estimates the likelihood that similar patterns of data remain similar over time. It provides a measure of signal regularity.

- Low ApEn: High regularity (e.g., epileptic seizures, sleep spindles).
- High ApEn: Irregular signals (e.g., awake, alert states).

An improved version of ApEn, SampEn, avoids bias from self-matches in pattern comparisons. It is more consistent across different data lengths and noise levels. Permutation Entropy evaluates complexity based on the relative ordering of values in a time series rather than their amplitude. It is robust to noise and ideal for short EEG segments. Derived from the frequency domain, spectral entropy measures how uniformly the power is distributed across frequency bands.

Applications:

- Sleep studies: Stage differentiation using entropy trends.
- Seizure prediction: Drops in entropy often precede ictal activity.
- Cognitive load monitoring: Increased mental workload tends to reduce entropy due to task-related synchronization.

Fractal dimension quantifies the geometric complexity or self-similarity of a waveform. EEG signals, particularly those generated during cognitive tasks, exhibit fractal-like properties, making FD a natural candidate for feature extraction [77].

Common algorithms:

- Higuchi's method: Computes FD by evaluating the waveform's curve length at multiple scales.
- Katz's method: Based on the signal's geometric length and amplitude.

- Petrosian FD: Uses binary conversions of the signal to estimate complexity efficiently.

Applications:

- Dementia detection: Reduced FD in Alzheimer's disease reflects loss of neural complexity.
- Workload estimation: FD decreases during fatigue or cognitive overload.
- Developmental neuroscience: Changes in FD correlate with brain maturation and age.

The Lyapunov exponent measures a system's sensitivity to initial conditions—a hallmark of chaos [78]. A positive Lyapunov exponent suggests that small perturbations in the signal's state will diverge rapidly, indicating dynamic instability.

Applications:

- Seizure prediction: Pre-ictal brain states often exhibit increasing Lyapunov exponents.
- Anesthesia depth monitoring: Reduced dynamical complexity is reflected in lower exponents.
- Consciousness studies: Different levels of awareness may correlate with changes in chaotic dynamics.

The Hurst exponent (H) assesses the long-term memory of a time series:

- $H > 0.5$: Persistent behavior; trends continue.
- $H < 0.5$: Anti-persistent behavior; trends reverse.
- $H = 0.5$: Random walk (white noise).

The Hurst exponent is particularly useful in resting-state EEG, where persistent oscillatory patterns may reflect internal cognitive or affective states.

Applications:

- Mood and affect detection.
- Aging research: Tracking degradation of memory in EEG signals.
- Attention disorders: Differentiating ADHD from controls.

Correlation dimension estimates the number of degrees of freedom (i.e., independent variables) in a system, derived from the reconstruction of the EEG's phase space. A high D2 implies a more complex, high-dimensional system, while low D2 suggests simplified or synchronized dynamics [79].

Applications:

- Epileptic seizures: Sudden drops in D2 near seizure onset.
- BCI systems: Differentiating mental states during motor imagery.
- Sleep cycle transitions: Capturing dynamic shifts in brain states.

NAR models extend linear autoregressive models by incorporating nonlinear terms to account for more complex dependencies in time-series data. These models predict future EEG values based on past values and nonlinear interactions.

Although computationally intensive, NAR models are effective in:

- Forecasting EEG signal trends.
- Closed-loop neurofeedback systems.
- Modeling pathological EEG patterns.

EEG signals have been shown to exhibit chaotic characteristics, especially during transitions between cognitive or physiological states. Additional metrics include:

- Kolmogorov entropy: Measures the rate of information production in a system.
- Correlation entropy: Similar to Kolmogorov entropy but based on correlation integrals.
- Largest Lyapunov exponent: Estimates the divergence of nearby trajectories in the phase space.

These features provide deep insights into the dynamical structure of brain activity but are typically reserved for offline analysis due to high computational cost and sensitivity to noise [80].

Applications of nonlinear features are broad and impactful:

- Neurodegenerative disorders: Reduced complexity and entropy in dementia.
- Seizure prediction: Early detection through dynamic instability.
- Cognitive experiments: Measuring mental effort or transitions.
- EEG biometrics: Using fractal and chaotic signatures for authentication.

Challenges include:

- Noise sensitivity: Nonlinear measures often respond disproportionately to artifacts.
- Parameter tuning: Many algorithms (e.g., SampEn, FD) require careful setting of parameters like window size and threshold.
- Interpretability: Unlike amplitude or power, nonlinear features lack direct physiological correspondence, complicating clinical use.
- Computational demand: Many methods are unsuitable for embedded or mobile systems without optimization.

Despite these limitations, nonlinear feature extraction provides unique and irreplaceable insights into brain function. When used in conjunction with linear features, they can significantly enhance the accuracy and robustness of EEG-based classification systems [81].

5.5 Clinical Applications of EEG

Electroencephalography has become a cornerstone of clinical neuroscience. Originally developed as a tool for exploring neural activity, EEG now plays a central role in diagnosing neurological disorders, managing treatments, and enabling therapeutic innovations such as brain-computer interfaces. With the growing accessibility of EEG technology and increasing precision in signal acquisition and interpretation, clinical EEG is evolving rapidly. This chapter explores the diverse clinical applications of EEG, emphasizing its role in neurology, rehabilitation, psychiatry, intensive care, and cutting-edge interfaces like brain-computer systems.

5.5.1 EEG in Neurology

EEG has proven indispensable in neurology, where its ability to detect real-time electrophysiological changes in the brain enables the diagnosis and management of multiple neurological conditions. In epilepsy, EEG is used to detect abnormal discharges, including spikes, sharp waves, and seizure activity. Long-term video-EEG monitoring allows for correlation of clinical events with EEG patterns, which is especially valuable in pre-surgical evaluation for epilepsy. For example, in temporal lobe epilepsy, characteristic interictal spikes in the anterior temporal region may help localize the epileptogenic zone. A notable case involved a 12-year-old boy with refractory seizures who underwent video-EEG and was later cured through surgical resection of the seizure focus, identified through EEG findings [9].

In sleep medicine, EEG is a fundamental component of polysomnography, used to analyze sleep stages and detect abnormalities such as sleep apnea, narcolepsy, and REM sleep behavior disorder. Specific patterns such as sleep spindles, K-complexes, and REM-associated rapid low-voltage activity help identify transitions between sleep stages. In narcolepsy, for instance, EEG helps confirm shortened REM latency, a key diagnostic feature. Portable EEG sleep monitors now allow for at-home studies, increasing patient compliance and diagnostic yield [82].

EEG also plays a critical role in diagnosing encephalopathies. In hepatic encephalopathy, for example, EEG typically shows generalized slowing and triphasic waves. In hypoxic-ischemic encephalopathy, post-cardiac arrest patients often display burst suppression patterns or electrocerebral inactivity, findings that assist in prognostication and treatment decisions. A real-world example includes an ICU patient whose EEG showed periodic discharges consistent with nonconvulsive status epilepticus, prompting anticonvulsant treatment and neurological recovery [18].

In neurodegenerative diseases, EEG provides supportive evidence for cognitive decline. Alzheimer's disease often presents with posterior alpha rhythm attenuation and increased theta or delta activity. In dementia with Lewy bodies, EEG may show fluctuating background rhythms, which can help distinguish it from Alzheimer's. Machine learning models are being trained to detect early EEG alterations in mild cognitive impairment (MCI), potentially allowing intervention before dementia fully develops. This integration of computational tools with clinical EEG is paving the way for precision neurology [15, 83].

5.5.2 EEG in Psychiatry

Psychiatric disorders traditionally lack robust physiological biomarkers, but EEG is beginning to fill that void. In major depressive disorder (MDD), studies have shown that left frontal alpha asymmetry is a potential trait marker. This finding has been used in neurofeedback therapies where patients attempt to modulate their frontal alpha activity. In a study at UCLA, patients receiving neurofeedback aimed at correcting alpha asymmetry showed significant mood improvements compared to controls.

In schizophrenia, EEG anomalies include reduced P300 amplitude and delayed latency in ERPs, especially during oddball paradigms. These markers correlate with cognitive deficits, including working memory impairment and attentional dysfunction. For instance, a patient with chronic schizophrenia might show diminished P300 responses even in the absence of acute psychosis, indicating enduring cognitive processing deficits. Functional connectivity analyses using EEG have revealed hypoconnectivity in frontal and temporal areas, insights that are now being explored for targeted neurostimulation therapies [71, 84].

In children with ADHD, EEG frequently shows elevated theta activity and reduced beta activity, especially in frontal and central regions. The theta/beta ratio is used both for diagnostic support and for guiding neurofeedback. Clinical trials have demonstrated that children undergoing EEG neurofeedback to reduce theta activity and enhance beta rhythms experience improved attention spans and reduced impulsivity. EEG is also being explored in autism spectrum disorders, where atypical patterns of gamma activity and altered connectivity may reflect sensory integration challenges. Interventions targeting these biomarkers through neurofeedback and transcranial stimulation are under active investigation [20, 85].

5.5.3 EEG in Rehabilitation

In neurorehabilitation, EEG offers unique insight into brain plasticity and reorganization. After a stroke, EEG can identify asymmetries in sensorimotor rhythms that reflect impaired cortical function. Motor imagery tasks recorded via EEG are used in BCI systems to facilitate motor relearning. A prominent study conducted in Germany involved chronic stroke patients using an EEG-BCI to control a robotic hand. Patients who visualized hand movement saw corresponding EEG changes, which triggered robotic assistance, reinforcing cortical activation. Over time, this led to improved voluntary movement and functional outcomes [86].

Neurofeedback based on EEG is also being utilized to enhance cognitive and emotional recovery. For patients with traumatic brain injury (TBI), training protocols aimed at increasing sensorimotor rhythm coherence have improved working memory and attentional control. In a rehabilitation center in Spain, EEG neurofeedback was used to treat patients with post-concussive syndrome. Those receiving targeted neurofeedback sessions demonstrated marked improvements in reaction time and mood regulation compared to standard therapy controls [24].

EEG-guided rehabilitation is not limited to motor recovery. In aphasia, patients have been trained to produce specific EEG rhythms linked to language production, improving naming and fluency. These emerging applications show that EEG is a dynamic tool, offering both real-time monitoring and therapeutic feedback in rehabilitation settings.

5.5.4 EEG in Intensive Care

In critical care environments, EEG serves as a real-time window into brain function. Continuous EEG (cEEG) monitoring is used in ICU settings to detect subclinical seizures, monitor sedation, and assess prognosis. For instance, in traumatic brain injury patients, cEEG can identify cortical spreading depolarizations, events associated with secondary brain injury. Intervention based on these findings has led to improved outcomes in neurocritical care units [87].

During surgical procedures, EEG is used to monitor anesthesia depth. The bispectral index (BIS) score, derived from EEG data, helps anesthesiologists adjust dosages to prevent intraoperative awareness or excessive sedation. In a study of over 500 patients undergoing general anesthesia, those monitored with BIS had lower incidences of postoperative cognitive dysfunction compared to those receiving standard care.

In cases of suspected brain death, EEG provides valuable confirmatory evidence. An isoelectric EEG over a specified period supports clinical findings such as the absence of brainstem reflexes. In jurisdictions where ancillary testing is required, EEG remains a widely accepted tool. A notable example involves a patient with suspected brain death following a drug overdose; EEG helped distinguish between actual brain death and drug-induced coma, preventing premature withdrawal of life support.

5.5.5 EEG in Brain-Computer Interfaces

The integration of EEG with BCI technology has opened new possibilities for assistive communication and control. EEG-based BCIs interpret specific neural signatures to command external devices. SMR-based BCIs rely on the modulation of mu and beta rhythms during motor imagery. These systems are used to control cursors, robotic arms, or prosthetic limbs. In a trial involving tetraplegic patients, those using SMR-BCIs achieved over 80% accuracy in two-dimensional cursor control after just four weeks of training [22].

P300-based BCIs are among the most reliable systems, particularly for spelling applications. These systems detect the P300 wave triggered when a user recognizes a target stimulus. For instance, ALS patients unable to speak or move can use a P300-based speller to select letters on a screen. One subject in a long-term home-use study achieved communication rates of up to five words per minute using only EEG and visual attention. This level of autonomy has profound implications for quality of life and social reintegration [88, 89].

SSVEP-based BCIs utilize the brain's natural response to flickering lights. By focusing on specific visual targets flickering at different frequencies, users can generate distinguishable EEG signals. These systems have been used for real-time control of smart home appliances. A recent project in Japan allowed users to operate lighting and television through SSVEP-BCI commands, demonstrating the system's versatility and real-world feasibility [22].

5.5.6 Technical Challenges and Future Directions

Despite its potential, EEG faces technical and user-specific challenges. Artifact contamination remains a significant barrier, with eye blinks, muscle movements, and environmental noise obscuring neural signals. Advanced filtering techniques and machine learning classifiers are being developed to isolate brain signals with greater accuracy. High-density EEG systems, now incorporating up to 256 channels, offer improved spatial resolution but demand greater computational power and data handling capacity.

User variability also presents challenges. Approximately one-quarter of individuals are considered BCI non-responders. This phenomenon, known as BCI illiteracy, is influenced by differences in cortical anatomy, attention span, and psychological state. Adaptive BCIs that personalize signal processing algorithms to each user are under development to address this issue [17].

The future of clinical EEG lies in hybrid systems. Combining EEG with EMG, eye-tracking, or near-infrared spectroscopy (NIRS) enhances both control reliability and interface usability. For example, in hybrid BCIs, users can initiate commands via EEG and refine them with eye movement, improving speed and accuracy [90].

Artificial intelligence is revolutionizing EEG analysis. Deep learning models can detect subtle patterns in raw EEG that elude traditional methods. In diagnostic neurology, convolutional neural networks have been trained to differentiate epileptic seizures from other paroxysmal events with over 90% accuracy. Integrating EEG with imaging techniques such as fMRI further bridges the gap between electrophysiology and structural neurobiology.

References

1. Luan L, Robinson JT, Aazhang B, Chi T, Yang K, Li X, Xie C (2020) Recent advances in electrical neural interface engineering: minimal invasiveness, longevity, and scalability. Neuron 108(2):302–321
2. Kong NW, Gibb WR, Tate MC (2016) Neuroplasticity: insights from patients harboring gliomas. Neural Plast 2016:1–13
3. Bonner AM (2015) The use of neurodiagnostic technologies in the 21st century neuroscientific revolution. Neurodiagn J. 55(2):103–109
4. Wintermark M, Colen R, Whitlow CT, Zaharchuk G (2018) The vast potential and bright future of neuroimaging. Br J Radiol 91(1086):20170640
5. Yang J, Sui SF, Liu Z (2022) Brain structure and structural basis of neurodegenerative diseases. Biophys Rep. 8(2):67–79
6. Barha CK, Nagamatsu LS, Liu-Ambrose T (2016) Basics of neuroanatomy and neurophysiology. Handb Clin Neurol 138:3–15
7. Khadria A (2022) Tools to measure membrane potential of neurons. Biomed J 45(6):1028–1038
8. Campbell A, Wu C (2018) Chronically implanted intracranial electrodes: tissue reaction and electrical changes. Micromachines. 9(11):561
9. Badnjević A, Pokvić LG, Smajlhodžić-Deljo M, Spahić L, Bego T, Meseldžić N, Prnjavorac L, Prnjavorac B, Bedak O (2024) Application of artificial intelligence for the classification of the clinical outcome and therapy in patients with viral infections: the case of COVID-19. Technol Health Care 32(3):1859–1870
10. Hafizovic L, Causevic A, Deumić A, Bećirović LS, Gurbeta L, Badnjević A (2021) The use of artificial intelligence in diagnostic medical imaging: systematic literature review. In: Proc 2021 IEEE 21st Int Conf bioinformatics bioeng (BIBE). 1–6
11. Medjedovic E, Stanojevic M, Jonuzovic-Prosic S, Ribic E, Begic Z, Cerovac A, Badnjevic A (2024) Artificial intelligence as a new answer to old challenges in maternal-fetal medicine and obstetrics. Technol Health Care 32(3):1273–1287
12. Cao M, Huang H, He Y (2017) Developmental connectomics from infancy through early childhood. Trends Neurosci 40(8):494–506
13. Magjarević R, Badnjević A (2023) Inspection and testing of electrocardiographs (ECG) devices. In: Badnjević A, Cifrek M, Magjarević R, Džemić Z (eds) Inspection of Medical Devices: For Regulatory Purposes. Springer Nature Switzerland, Cham, pp 139–171
14. Nayak L, Dasgupta A, Das R, Ghosh K, De RK (2018) Computational neuroscience and neuroinformatics: recent progress and resources. J Biosci 43(2):223–243
15. Pokvić LG, Deumić A, Softić A, Badnjević A (2024) Application of artificial intelligence for management of maintenance and prediction of performances. In: Badnjević A, Cifrek M, Magjarević R, Džemić Z (eds) Inspection of Medical Devices. Springer, Cham, pp 101–117
16. Xia M, He Y (2017) Functional connectomics from a "big data" perspective. Neuroimage 160:152–167
17. Merdović N, Spahić L, Hundur M, Pokvić LG, Badnjević A (2024) Advancement of post-market surveillance of medical devices leveraging artificial intelligence: infusion pumps case study. Technol Health Care 33(2):915–921
18. Bećirović F, Spahić L, Merdović N, Gurbeta Pokvić L, Badnjević A (2025) Advancement of post-market surveillance of medical devices leveraging artificial intelligence: patient monitors case study. Technol Health Care 33(2):974–980
19. Go Y (2025) Neural circuit research using molecular barcode technology. Neurosci Res 190:44–51
20. Hundur M, Spahić L, Bećirović F, Gurbeta Pokvić L, Badnjević A (2025) Advancement of post-market surveillance of medical devices leveraging artificial intelligence: ECG devices case study. Technol Health Care. 33(2):??-??. [Epub ahead of print]
21. Berndt A, Cai D, Cohen A, Juarez B, Iglesias JT, Xiong H, Slesinger PA (2023) Current status and future strategies for advancing functional circuit mapping in vivo. J Neurosci 43(47):8824–8833

22. Spahić L, Pokvić LG, Badnjević A (2024) Inspection and testing of infusion pumps. In: Badnjević A, Cifrek M, Magjarević R, Džemić Z (eds) Inspection of Medical Devices. Springer, Cham, pp 309–327
23. Giansanti D (2023) The neural odyssey: unveiling the potential of neuroscience-ICT convergence. Biology (Basel). 12(6):775
24. Martinović M, Kosović M, Spahić L, Softić A, Pokvić LG, Badnjević A (2025) Machine learning for improved medical device management: a focus on dialysis machines. Technol Health Care. 33
25. Mamat M, Wang Z, Jin L, He K, Li L, Chen Y (2024) Beyond nodes and edges: a bibliometric analysis on graph theory and neuroimaging modalities. Front Neurosci 18:123456
26. Guidolin D, Tortorella C, Marcoli M, Maura G, Agnati LF (2022) Intercellular communication in the central nervous system as deduced by chemical neuroanatomy and quantitative analysis of images: impact on neuropharmacology. Int J Mol Sci 23(11):5983
27. Badnjevic A, Deumic A, Imamovic E, Pokvic LG (2023) A novel method for conformity assessment testing of defibrillators for post-market surveillance purposes. Technol Health Care 31(1):317–325
28. Nicolae IE, Sultana AE, Aursulesei R, Fulop S (2021) Treating electrical and biopotential artifacts in an EEG pilot study experiment. Conf Proc IEEE Eng Med Biol Soc 2021:3065–3068
29. Mihajlovic V, Grundlehner B, Vullers R, Penders J (2015) Wearable, wireless EEG solutions in daily life applications: what are we missing? IEEE J Biomed Health Inform 19(1):6–21
30. Pinho F, Cerqueira J, Correia J, Sousa N, Dias N (2017) MyBrain: a novel EEG embedded system for epilepsy monitoring. J Med Eng Technol 41(7):555–564
31. Brundavani P, Vardhan DV (2023) A novel approach for minimising anti-aliasing effects in EEG data acquisition. Open Life Sci. 18(1):20230042
32. Qian G, Suo Y, Cai Q, Lian Y, Zhao Y (2024) A 382 nVrms 100 GΩ@50 Hz active electrode for dry-electrode EEG recording. IEEE Trans Biomed Circuits Syst.??(??):??-??. [Epub ahead of print]
33. Xie Y, Oniga S (2024) A comprehensive review of hardware acceleration techniques and convolutional neural networks for EEG signals. Sensors 24(10):3325
34. Wang G, Liu H, Wang W, Kang H (2024) Methods for processing physiological artifacts in single/few-channel EEG signals. Chin J Med Instrum. 48(3):273–278
35. Ehteshamzad S (2024) Assessing the potential of EEG in early detection of Alzheimer's disease: a systematic comprehensive review (2000–2023). J Alzheimers Dis Rep. 8(1):175–190
36. Tohidi M, Madsen JK, Moradi F (2019) Low-power high-input-impedance EEG signal acquisition SoC with fully integrated IA and signal-specific ADC for wearable applications. IEEE Trans Biomed Circuits Syst 13(5):1192–1202
37. Cassani R, Estarellas M, San-Martin R, Fraga FJ, Falk TH (2018) Systematic review on resting-state EEG for Alzheimer's disease diagnosis and progression assessment. Dis Markers 2018:5174815
38. Li R, Zhang Y, Fan G, Li Z, Li J, Fan S, Lou C, Liu X (2023) Design and implementation of high sampling rate and multichannel wireless recorder for EEG monitoring and SSVEP response detection. Front Neurosci 17:1170803
39. Badnjević A, Pokvić LG, Deumić A, Bećirović LS (2022) Post-market surveillance of medical devices: a review. Technol Health Care 30(6):1315–1329
40. He C, Chen YY, Phang CR, Stevenson C, Chen IP, Jung TP, Ko LW (2023) Diversity and suitability of the state-of-the-art wearable and wireless EEG systems: review. IEEE J Biomed Health Inform 27(5):2107–2120
41. European Parliament. Medical Device Regulation (MDR) 2017/745. Off J Eur Union. 2017
42. Kim M, Yoo S, Kim C (2022) Miniaturization for wearable EEG systems: recording hardware and data processing. Biomed Eng Lett 12:259–274
43. Qin Y, Zhang Y, Zhang Y, Liu S, Guo X (2023) Application and development of EEG acquisition and feedback technology: a review. Biosensors. 13(10)
44. Islam MK, Rastegarnia A, Yang Z (2016) Methods for artifact detection and removal from scalp EEG: a review. Neurophysiol Clin 46(4–5):287–305

45. Uktveris T, Jusas V (2018) Development of a modular board for EEG signal acquisition. Sensors 18(7):2148
46. Xu J, Mitra S, Van Hoof C, Yazicioglu RF, Makinwa KAA (2017) Active electrodes for wearable EEG acquisition: review and electronics design methodology. IEEE Rev Biomed Eng 10:187–196
47. Waser M, Garn H, Benke T, Dal-Bianco P, Ransmayr G, Schmidt R, Sorensen HBD (2017) A flexible method for the automated offline detection of artifacts in multi-channel electroencephalogram recordings. In: Conf Proc IEEE Eng Med Biol Soc
48. Chaddad A, Wu Y, Kateb R, Bouridane A (2023) Electroencephalography signal processing: a comprehensive review and analysis of methods and techniques. Sensors. 23
49. Robbins KA, Touryan J, Mullen T, Kothe C, Bigdely-Shamlo N (2020) How sensitive are EEG results to preprocessing methods: a benchmarking study. IEEE Trans Neural Syst Rehabil Eng 28(7):1573–1583
50. Widmann A, Schröger E, Maess B (2015) Digital filter design for electrophysiological data—a practical approach. J Neurosci Methods 250:34–46
51. Desjardins JA, van Noordt S, Huberty S, Segalowitz SJ, Elsabbagh M (2021) EEG-IP-L preprocessing pipeline for signal quality assessment incorporating annotation and blind source separation. J Neurosci Methods 358:109179
52. Barthélemy Q, Mayaud L, Renard Y, Kim D, Kang SW, Gunkelman J, Congedo M (2017) Online denoising of eye-blinks in electroencephalography. Neurophysiol Clin 47(2):93–102
53. Alhaddad MJ, Kamel MI, Makary MM, Hargas H, Kadah YM (2014) Spectral subtraction denoising preprocessing block to improve P300-based brain-computer interfacing. Biomed Eng Online 13(1):33
54. Saeidi M, Karwowski W, Farahani FV, Fiok K, Taiar R, Hancock PA, Al-Juaid A (2021) Neural decoding of EEG signals with machine learning: a systematic review. Brain Sci 11(12):1688
55. Ouyang G, Li Y (2025) Protocol for semi-automatic EEG preprocessing incorporating independent component analysis and principal component analysis. STAR Protoc
56. Nejedly P, Kremen V, Sladky V, Cimbalnik J, Klimes P, Plesinger F, Worrell GW (2020) Multi-center intracranial EEG dataset for classification of graphoelements and artifactual signals. Sci Data. 7
57. Bisht A, Singh P, Kaur C, Agarwal S, Ajmani M (2022) Progress and challenges in physiological artifacts' detection in electroencephalographic readings. Curr Med Imaging
58. Gao X, Gui K, Wu X, Metcalfe B, Zhang D (2025) Effects of different preprocessing pipelines on motor imagery-based brain-computer interfaces. IEEE J Biomed Health Inform
59. Sun C, Mou C (2023) Survey on the research direction of EEG-based signal processing. Front Neurosci
60. Hamzah HA, Abdalla KK (2024) EEG-based emotion recognition systems: comprehensive study. Heliyon
61. Singh AK, Krishnan S (2022) Trends in EEG signal feature extraction applications. Front Artif Intell
62. Mujkić A, Baralić E, Ombašić A, Bećirović LS, Pokvić LG, Badnjević A (2022) Machine intelligence in biomedical data modeling, processing, and analysis. In: Proc 2022 11th mediterranean conf embedded computing (MECO). 1–10
63. Riaz F, Hassan A, Rehman S, Niazi IK, Dremstrup K (2016) EMD-based temporal and spectral features for the classification of EEG signals using supervised learning. IEEE Trans Neural Syst Rehabil Eng 24(1):28–35
64. Amin HU, Mumtaz W, Subhani AR, Saad MNM, Malik AS (2017) Classification of EEG signals based on pattern recognition approach. Front Comput Neurosci 11:103
65. Thangarajoo RG, Reaz MBI, Srivastava G, Haque F, Ali SHM, Bakar AAA, Bhuiyan MAS (2021) Machine learning-based epileptic seizure detection using wavelet and EMD-based decomposition techniques: a review. Sensors. 21
66. Antony MJ, Sankaralingam BP, Mahendran RK, Gardezi AA, Shafiq M, Choi JG, Hamam H (2022) Classification of EEG using adaptive SVM classifier with CSP and online recursive ICA. Sensors. 22

67. He C, Liu J, Zhu Y, Du W (2021) Data augmentation for deep neural networks model in EEG classification task: a review. Front Hum Neurosci. 15
68. Torabi A, Daliri MR (2021) Applying nonlinear measures to the brain rhythms: an effective method for epilepsy diagnosis. BMC Med Inform Decis Mak. 21
69. Spahić L, Sredović U, Kurpejović Z, Mrdanović E, Pokvić G, Badnjević A (2025) Machine learning for improved medical device management: a focus on infant incubators. Technol Health Care. 33
70. Browarczyk J, Kurowski A, Kostek B. Analyzing the effectiveness of the brain-computer interface for task discerning based on machine learning. Sensors. 2020; 20.
71. Peruničić Ž, Lalatović I, Spahić L, Ašić A, Pokvić LG, Badnjević A (2024) Enhancing mechanical ventilator reliability through machine learning-based predictive maintenance. Technol Health Care. 32
72. Lian J, Shi Y, Zhang Y, Jia W, Fan X, Zheng Y (2020) Revealing false positive features in epileptic EEG identification. Int J Neural Syst 30(7):2050036
73. Tan X, Wang D, Chen J, Xu M (2023) Transformer-based network with optimization for cross-subject motor imagery identification. Bioengineering (Basel). 1 0
74. Pu Y, Tian P, Huang Z (2024) Characteristics of the top 100 cited electroencephalography articles on aging: a bibliometric analysis. Am J Transl Res. 16
75. Peng G, Nourani M, Harvey J, Dave H (2020) Feature selection using F-statistic values for EEG signal analysis. Conf Proc IEEE Eng Med Biol Soc. 2020
76. Renfrew M, Cheng R, Daly JJ, Cavusoglu M (2008) Comparison of filtering and classification techniques of electroencephalography for brain-computer interface. Conf Proc IEEE Eng Med Biol Soc. 2008
77. Ozmen NG, Gumusel L, Yang Y (2018) A biologically inspired approach to frequency domain feature extraction for EEG classification. Comput Math Methods Med. 2018
78. Wu T, Kong X, Zhong Y, Chen L (2022) Automatic detection of abnormal EEG signals using multiscale features with ensemble learning. Front Hum Neurosci. 16
79. Zheng J, Liang M, Sinha S, Ge L, Yu W, Ekstrom A, Hsieh F (2022) Time-frequency analysis of scalp EEG with Hilbert-Huang transform and deep learning. IEEE J Biomed Health Inform 26(4):1549–1557
80. Guan S, Dong T, Cong LK (2025) Method for EEG signal recognition based on multi-domain feature fusion and optimization of multi-kernel extreme learning machine. Sci Rep. 15
81. Myrden A, Chau T (2016) Feature clustering for robust frequency-domain classification of EEG activity. J Neurosci Methods
82. Alkhachroum A, Appavu B, Egawa S, Foreman B, Gaspard N, Gilmore EJ, Hirsch LJ, Kurtz P, Lambrecq V, Kromm J, Vespa P, Zafar SF, Rohaut B, Claassen J (2022) Electroencephalogram in the intensive care unit: a focused look at acute brain injury. Intensive Care Med 48(9):1211–1225
83. Frauscher B, Rossetti AO, Beniczky S (2024) Recent advances in clinical electroencephalography. Curr Opin Neurol 37(2):223–231
84. Tsiamalou A, Dardiotis E, Paterakis K, Fotakopoulos G, Liampas I, Sgantzos M, Siokas V, Brotis AG (2022) EEG in neurorehabilitation: a bibliometric analysis and content review. Neurol Int 14(3):517–528
85. He W, Wang D, Meng Q, He F, Xu M, Ming D (2024) Applications and prospects of electroencephalography technology in neurorehabilitation assessment and treatment. J Biomed Eng. 41
86. Liu XY, Wang WL, Liu M, Chen MY, Pereira T, Doda DY, et al (2025) Recent applications of EEG-based brain-computer interface in the medical field. Mil Med Res. 12(1)
87. Vărbu K, Muhammad N, Muhammad Y (2022) Past, present, and future of EEG-based BCI applications. Sensors. 22
88. Akras Z, Jing J, Westover MB, Zafar SF (2025) Using artificial intelligence to optimize anti-seizure treatment and EEG-guided decisions in severe brain injury. Neurotherapeutics

89. Liao XY, Jiang YE, Xu RJ, Qian TT, Liu SL, Che Y (2025) A bibliometric analysis of electroencephalogram research in stroke: current trends and future directions. Front Neurol. 16
90. Tremblay S, Rogasch NC, Premoli I, Blumberger DM, Casarotto S, Chen R, Daskalakis ZJ (2019) Clinical utility and prospective of TMS-EEG. Clin Neurophysiol 130(4):802–819

Chapter 6
Processing Ultrasound Signals and Imaging

This section examines the physical principles, acquisition methods, processing techniques, and clinical uses of ultrasound. It begins with the fundamentals of acoustic wave propagation, reflection, and impedance, alongside the design and function of ultrasound transducers (Sect. 6.1). Signal acquisition (Sect. 6.2) is described through the pulse–echo principle and Doppler ultrasound, demonstrating how physiological information is translated into imaging data. Image processing techniques (Sect. 6.3) address speckle and random noise reduction, spatial and wavelet filtering, segmentation, and edge detection for enhanced visualization. Clinical applications (Sect. 6.4) are explored across cardiovascular imaging, fetal monitoring, and organ assessment, emphasizing ultrasound's role as a noninvasive, real-time diagnostic tool. This section highlights ultrasound as a cornerstone of biomedical imaging, combining physics, engineering, and clinical utility.

6.1 Principles of Ultrasound Imaging

Ultrasound imaging, also known as sonography, is a medical imaging technique that utilizes high-frequency sound waves to produce images of the internal body structures. This non-invasive procedure is widely used in various medical fields, including obstetrics, cardiology, and musculoskeletal imaging. Modern-day medicine and diagnostics would be unimaginable without ultrasound imaging. The invention of ultrasound imaging is an example of interdisciplinary cooperation between engineer Tom Brown and obstetrician Ian Donald. They were the first ones to prototype and test the system primarily intended for use in scanning ships for flaws [1, 2].

The first instance of using an ultrasound for diagnostic imaging was in 1957 when a woman diagnosed with inoperable gastric cancer was examined [3]. Ultrasound imaging showed that she did not have cancer, but an ovarian cyst. Soon after,

© The Author(s), under exclusive license to Springer Nature Switzerland AG 2026
A. Badnjević and L. Spahić, *Biomedical Signals and Systems*, Series in BioEngineering,
https://doi.org/10.1007/978-3-032-10982-8_6

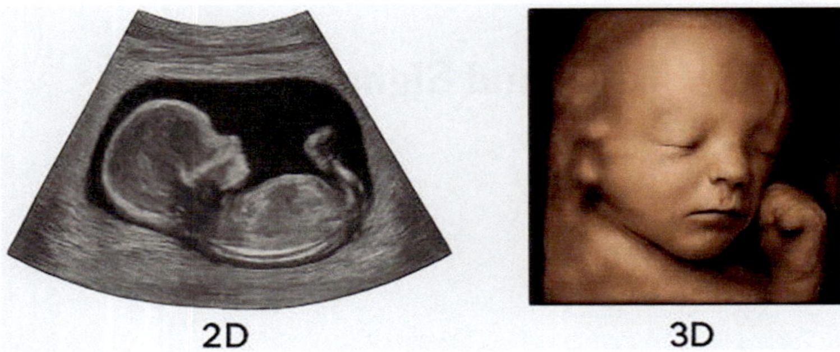

Fig. 6.1 2D and 3D ultrasound image comparison

Ian Donald started using ultrasound for the examination of pregnant women, thus enabling prenatal diagnosis of fetal abnormalities.

Ultrasound technology has continued developing ever since with the advent of 3D ultrasound technology that was first used for taking an image of a fetus in 1986. The debut of 4D ultrasound followed with its first utilization in obstetrics and gynecology in 1990s [1, 4]. The primary difference between conventional 2D ultrasound and sophisticated 3D and 4D ultrasound lies in the range of angles that can be recorded, accompanied by a vast amount of data that can be collected (Fig. 6.1).

As it can be inferred from Fig. 3.3, the advancements in the field of ultrasound happened at a very fast pace because the need for safe cutting-edge medical imaging was constantly increasing. The first 2D ultrasounds provided insight into the function of internal structures and were promptly advanced into real-time 2D ultrasounds. However, even the 2D ultrasound did not allow for simultaneous imaging of complex movements such as facial movements. The introduction of 3D and 4D technologies was welcomed as the most advanced and cutting-edge technologies in sonography. This form of imaging has been in development for more than 20 years, but it is only now becoming applicable in clinical settings. This type of imaging first saw clinical usage thanks to advancements in computing power and the development of sonography-specific programming [5, 6]. Although each manufacturer's method of processing acquired 3D data differs, the basic concepts remain the same. As long as the sonographer is familiar with the equipment's functions, three-dimensional technology can be both advantageous and simple. The "fourth dimension" of 4D ultrasound is its capability to record movements, which makes it particularly interesting for use in obstetrics to evaluate the health of the fetus. The 4D ultrasound combines the benefits of spatial imaging of the fetal face with the addition of time, providing a tool for monitoring the fetal face. As a result of this breakthrough technology, it is now possible to determine and measure the appearance and duration of each facial movement and expression [7–9] (Fig. 6.2).

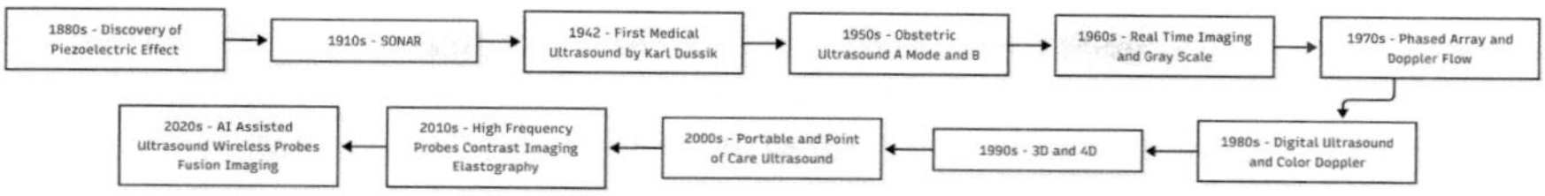

Fig. 6.2 Timeline of ultrasound technology development for usage in clinical practice

6.1.1 Basics of Ultrasound Wave Propagation and Reflection

Ultrasound imaging begins with the generation of ultrasound waves. A transducer, a handheld device, is used to send high-frequency sound waves into the body. These waves are typically in the range of 1–20 MHz. The transducer contains piezoelectric crystals that vibrate when an electric current is applied, producing sound waves. The sound waves then travel through the body, encountering various tissues, organs, and bones. The speed and direction of the waves are influenced by the density and composition of the structures they encounter. Gel is often applied to the skin to enhance the transmission of ultrasound waves by reducing air pockets that can deflect the waves. As the sound waves travel through the body, they are reflected to the transducer by different internal structures at varying intensities. This phenomenon is known as an echo. The amount and pattern of the echoes depend on the properties and composition of the tissues. The transducer also acts as a receiver, capturing the reflected sound waves or echoes. The piezoelectric crystals in the transducer convert the echoes into electrical signals. The time delay between the transmission of sound waves and the reception of echoes is measured [10–12].

Once the ultrasound waves are generated and transmitted, they must be compiled into a comprehensible output. Ultrasound image formation is a complex process that involves the interaction of sound waves with tissues, the reception of echoed signals, and the translation of these signals into visual images. This process is influenced by various factors, including the properties of tissues, the characteristics of ultrasound waves, and the technology used in ultrasound machines. As the ultrasound waves propagate through the body, they encounter different tissues, each with distinct acoustic properties. The density and elasticity of the tissues influence the speed and attenuation of waves. The body absorbs some waves, while others are reflected as echoes, depending on the acoustic impedance of the tissues. The transducer also functions as a receiver, capturing the echoes reflected by internal body structures. The time it takes for the echoes to return and their intensity are critical data points [13, 14]. The depth and nature of the reflecting structures influence these. The returned echoes induce a vibration in the piezoelectric crystals of the transducer, converting them into electrical signals. The frequency and amplitude of these signals are proportional to the characteristics of the echoes. Modern ultrasound machines are equipped with sophisticated signal processing systems. They amplify, filter, and process the electrical signals to extract meaningful data for image formation. Algorithms and computational models play a crucial role in enhancing the signal-to-noise ratio and improving image quality. The processed signals are then used to reconstruct images of the internal body structures [15]. This involves mapping the amplitude and time delay of the received echoes to corresponding pixel values in the image. The intensity of the echo determines the brightness of each pixel, while the time delay and the direction of the sound beam determine the position. The reconstructed images are displayed in real-time on a monitor. Advanced technologies allow for various modes of imaging, including 2D, 3D, and Doppler imaging, each offering unique insights into the anatomy and function of body organs [16, 17].

Ultrasound waves are sound waves with frequencies above the range of human hearing, typically defined as above 20 kHz. These high-frequency waves are mechanical longitudinal waves, meaning that the particles of the medium through which they travel vibrate in the direction of wave propagation. When ultrasound waves travel through a medium, they create pressure variations that propagate at a speed determined by the properties of the medium, specifically its density and elasticity. The basic equation for the propagation speed c of a sound wave in a medium is given by Eq. 6.1.

$$c = \sqrt{\frac{E}{\rho}} \tag{6.1}$$

where:

- c is the speed of sound,
- E is the **elastic modulus** (which measures the stiffness of the medium),
- P is the **density** of the medium.

In medical ultrasound, the waves typically propagate through soft tissues, which are primarily composed of water, although their physical properties (density and elasticity) vary depending on the type of tissue.

When a transducer emits ultrasound waves, they travel through tissues. Still, the behavior of the waves depends on whether the medium is homogeneous (uniform properties throughout) or heterogeneous (having varying properties in different regions). In a homogeneous medium, where the density and elastic properties are constant throughout, ultrasound waves propagate in a uniform direction, without scattering or distortion. In this idealized situation, the speed of sound remains constant, and the waves retain their shape and amplitude as they propagate. However, in biological tissues, which are generally heterogeneous, sound waves encounter changes in tissue composition and physical properties, causing them to reflect, refract, and scatter.

Tissues like muscle, fat, bone, and blood vessels exhibit varying acoustic impedances, which can cause significant changes in how ultrasound waves propagate; thus, they are considered heterogeneous media. The presence of air interfaces, such as those found in the lungs, or bone surfaces, such as those around joints, results in complex scattering and attenuation of ultrasound waves. The heterogeneous nature of human tissue presents both challenges and opportunities for distinguishing different tissue types in medical ultrasound imaging [18, 19].

The reflection of ultrasound waves occurs when the wave moves from one medium into another with a different acoustic impedance. The amount of reflection depends on the difference in acoustic impedance between the two media, and it is governed by the reflection coefficient R, which is defined as Eq. 6.2.

$$R = \left(\frac{Z_2 - Z_1}{Z_2 + Z_1}\right)^2 \tag{6.2}$$

where:

- Z_1 and Z_2 are the acoustic impedances of the two media.

If the impedance mismatch is significant (such as between soft tissue and bone), a high percentage of the sound is reflected, resulting in a high reflection coefficient and strong echoes, which are useful for imaging purposes. Conversely, more minor impedance differences, such as between fat and muscle, result in weaker reflections and less contrast in imaging [20].

Transmission and Refraction

When an ultrasound wave passes from one medium into another, it is not only reflected but also transmitted into the second medium. The degree of transmission and the direction of wave propagation after entering the new medium is governed by Snell's Law of refraction, which describes how waves change direction when moving between materials with different speeds of sound (Eq. 6.3):

$$\frac{\sin\theta_1}{\sin\theta_2} = \frac{c_1}{c_2} \tag{6.3}$$

where:

- θ_1 and θ_2 are the angles of incidence and refraction, respectively.
- c_1 and c_2 are the speeds of sound in the first and second media.

The change in direction due to refraction can be significant in cases where the contrast between tissues is high, such as when ultrasound waves pass from muscle to fat or soft tissue to bone. As ultrasound waves travel through tissue, their intensity gradually decreases due to various phenomena, including absorption, scattering, and reflection. This reduction in wave intensity is known as attenuation. The amount of attenuation A is given by Eq. 6.4

$$A = \alpha d \tag{6.4}$$

where:

- α is the **attenuation coefficient** of the tissue (which depends on the tissue type and frequency),
- d is the distance the wave travels through the tissue.

Higher frequency ultrasound waves experience more rapid attenuation, which limits their depth of penetration into tissues. Lower frequencies, while providing deeper penetration, have lower resolution. Thus, the choice of frequency is a trade-off between resolution and penetration depth [17]. Ultrasound waves interact differently with various tissues, and these interactions are used to generate contrast in ultrasound images. For example:

- Muscle has a moderate acoustic impedance and leads to some reflection and scattering.
- Fat has a lower impedance than muscle, leading to less reflection but still noticeable contrast.
- Bone has a very high impedance, causing most of the ultrasound wave to be reflected, which results in poor penetration and strong reflection.
- Air-filled spaces, like the lungs, result in almost complete reflection, severely limiting ultrasound penetration.

These differences are critical in creating diagnostic images and in assessing the tissue composition of organs. Doppler ultrasound utilizes the Doppler effect to assess the movement of tissues or blood. When an ultrasound wave reflects off a moving object (e.g., red blood cells in a vessel), the frequency of the reflected wave changes based on the velocity of the moving object. The Doppler shift Δf is given by Eq. 6.5.

$$\Delta f = \frac{2f_0 v}{c} \tag{6.5}$$

where:

- f0 is the frequency of the transmitted ultrasound wave,
- v is the velocity of the moving object (e.g., blood flow),
- c is the speed of sound in the tissue.

This shift enables Doppler ultrasound to measure the velocity and direction of blood flow, which is crucial in assessing conditions such as vascular disease, heart valve dysfunction, and arterial blockages.

6.1.2 Acoustic Impedance and Transducers

Acoustic impedance (Z) is a fundamental property of any material through which sound waves propagate. It determines how sound waves interact with a given medium, especially when transitioning between different materials with varying properties. The acoustic impedance is defined as the product of the medium's density (ρ) and the speed of sound (c) in that medium (Eq. 6.6):

$$Z = \rho c \tag{6.6}$$

where:

- Z is the **acoustic impedance**,
- ρ is the **density** of the medium (in kg/m^3),
- c is the **speed of sound** in the medium (in m/s).

For medical ultrasound applications, acoustic impedance plays a crucial role in determining the extent to which ultrasound waves are reflected at interfaces between different tissues or structures. A significant difference in impedance between two tissues results in a greater reflection of the sound wave, which is crucial for forming high-contrast images. Conversely, more minor impedance differences result in less reflection and more transmission [21].

The principle of impedance mismatch governs the amount of reflection that occurs at the boundary between two media. For example, when ultrasound waves travel from soft tissue (with relatively low impedance) to bone (with high impedance), a significant portion of the wave is reflected, creating a strong echo. This principle is harnessed in ultrasound imaging to generate contrasts between various anatomical structures [22, 23]. At the interface between two different media (e.g., muscle and fat), the ultrasound wave undergoes both reflection and transmission. The reflection coefficient (R) quantifies the proportion of the wave that is reflected from the interface and is given by the following Eq. 6.7.

$$R = \left(\frac{Z_2 - Z_1}{Z_2 + Z_1}\right)^2 \tag{6.7}$$

where:

- Z_1 and Z_2 are the acoustic impedances of the first and second media, respectively.

The reflection coefficient ranges from 0 (no reflection) to 1 (complete reflection). A significant impedance difference leads to a high reflection coefficient, while a slight impedance difference results in a low reflection and greater transmission.

As an example, consider the interface between bone and soft tissue. Bone has a high acoustic impedance, and soft tissue has a significantly lower impedance. When an ultrasound wave strikes this boundary, a large portion of the sound wave is reflected, making bone structures highly visible on ultrasound images [24].

In contrast, the reflection of ultrasound waves between two soft tissues (e.g., liver and kidney) with similar acoustic impedances is much less pronounced. This results in a more uniform transmission of waves through the tissue, which makes it more challenging to visualize soft tissue structures in isolation.

6.1.3 Transducers

The transducer is the key device in ultrasound systems that both generates and detects ultrasound waves. It converts electrical energy into mechanical energy (sound waves) during transmission and converts the returning mechanical energy (reflected sound waves) back into electrical energy for processing. A typical ultrasound transducer consists of multiple key components, each playing a vital role in wave propagation and reflection [17, 25]. At the heart of most ultrasound transducers are piezoelectric crystals, which exhibit the ability to generate an electrical charge in response to mechanical stress. When an alternating voltage is applied to the piezoelectric material, it vibrates at the same frequency as the applied voltage, generating ultrasound waves. Conversely, when ultrasound waves strike the piezoelectric crystal, it produces an electrical signal proportional to the frequency and intensity of the wave [26, 27].

The choice of piezoelectric material has a significant impact on the performance of the transducer. Commonly used materials include lead zirconate titanate (PZT) and quartz, both of which possess strong piezoelectric properties. The frequency of the ultrasound waves produced by the transducer is determined by the thickness of the piezoelectric crystal: thinner crystals produce higher-frequency waves, which are ideal for imaging superficial structures, while thicker crystals produce lower-frequency waves, which are better suited for deeper penetration.

The frequency of the ultrasound waves is a crucial factor that impacts the resolution and penetration of the ultrasound signal. High-frequency ultrasound (e.g., 10–15 MHz) is typically used for imaging superficial tissues such as the skin, muscles, and tendons, as it provides high resolution but limited penetration. On the other hand, low-frequency ultrasound (e.g., 2–5 MHz) is used for imaging deeper tissues or organs, as it penetrates more effectively, but at the expense of resolution (Fig. 6.3).

To achieve real-time imaging, array transducers are often employed. These consist of multiple piezoelectric elements arranged in a grid pattern, enabling the beam steering and focusing of ultrasound waves. By controlling the timing of the electrical signals sent to each element, the ultrasound system can steer and focus the beam, allowing for more precise imaging of structures at various depths. As ultrasound waves travel through different tissues, their behavior is primarily governed by the acoustic impedance of each tissue. The amount of reflection at tissue boundaries is crucial for generating contrast in ultrasound images, as previously discussed. The interaction between the ultrasound wave and the tissue can be further classified into several effects: refraction, scattering, and attenuation [17, 28]. When an ultrasound wave passes from one medium into another with a different speed of sound, it undergoes refraction, meaning it changes direction. The degree of refraction depends on the angle of incidence and the difference in the speed of sound between the two media. Refraction is particularly important in cases involving curved surfaces or angled interfaces between tissues.

Scattering occurs when ultrasound waves encounter small structures or heterogeneities within the tissue, such as blood cells, fibers, or minor tissue irregularities.

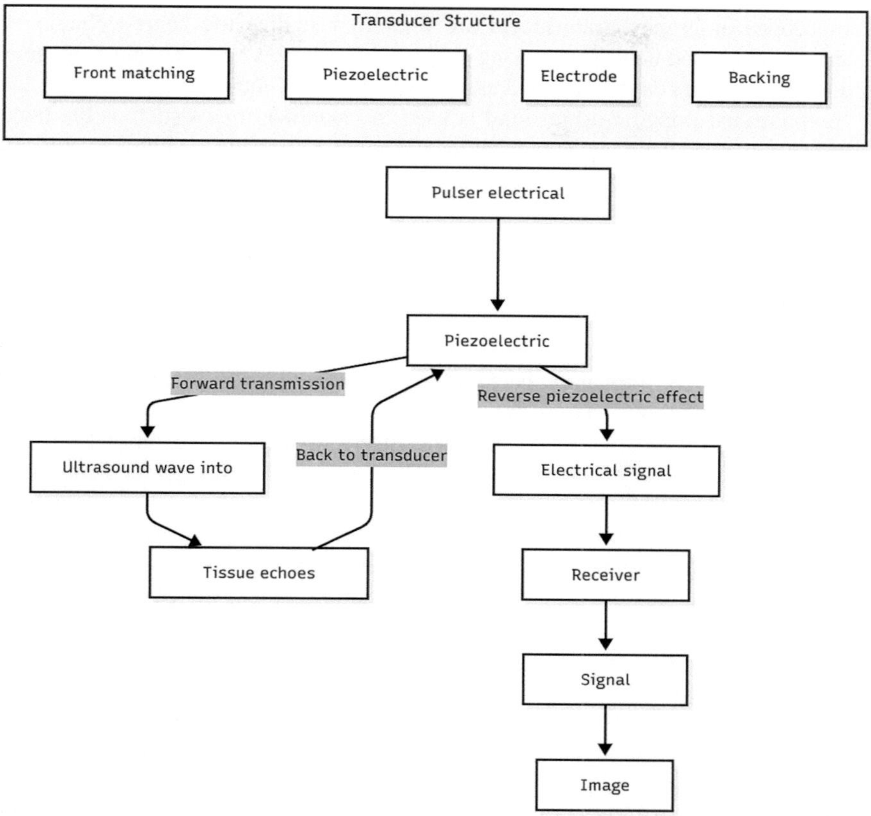

Fig. 6.3 Schematic of a piezoelectric transducer and its operation in an ultrasound system

This scattering can affect the clarity of the ultrasound image and contribute to the background noise. The Rayleigh scattering theory is often used to describe the relationship between the size of the scatterers and the wavelength of the ultrasound wave. As ultrasound waves propagate through tissue, they undergo attenuation, meaning their intensity decreases due to absorption and scattering. High-frequency waves tend to attenuate more rapidly, limiting their depth of penetration. This phenomenon must be considered when designing transducers and selecting imaging frequencies for various applications [29, 30]. Understanding the principles of acoustic impedance and transducer functionality is fundamental for various medical imaging applications:

- In obstetrics, ultrasound is used to visualize the fetus, placenta, and amniotic fluid. The significant difference in acoustic impedance between the soft tissues of the fetus and the surrounding fluid creates a strong contrast for imaging.

- In echocardiography, transducers are used to visualize the heart's chambers, valves, and blood flow. The varying acoustic impedances between blood, muscle, and valve tissues enable detailed assessment of heart function.
- In abdominal imaging, ultrasound is used to visualize organs such as the liver, kidneys, and gallbladder. The difference in acoustic impedance between the soft tissue organs and surrounding fluids (like bile or urine) helps in distinguishing anatomical structures.

6.2 Ultrasound Signal Acquisition

Signal acquisition in ultrasound is essential for translating the interaction of sound waves with tissues into diagnostic images. Ultrasound systems rely on high-frequency sound waves, generated by transducers, which are transmitted through the body and reflected by tissue interfaces. These reflected waves are then converted into electrical signals for processing and imaging [9, 31]. The key component of signal acquisition is the ultrasound transducer, which contains piezoelectric elements that convert electrical signals into sound waves and vice versa. As ultrasound waves encounter tissues, they undergo reflection, refraction, and scattering based on the acoustic impedance of the tissue. These reflections are captured by the transducer and used to reconstruct images, providing crucial information about the structure and condition of tissues [32].

Signal processing plays a vital role in ultrasound imaging, where the received signals are amplified, filtered, and processed to enhance image quality. Techniques such as beamforming and time-of-flight measurement enable precise focusing and depth calculation, thereby contributing to high-resolution imaging. Additionally, signal processing algorithms are employed to reduce noise, improve contrast, and reconstruct images that are easy to interpret clinically [15]. Ultimately, the goal of ultrasound signal acquisition is to obtain high-quality, accurate representations of internal structures, aiding in the diagnosis of a wide range of medical conditions. Advances in transducer design, signal processing techniques, and computational algorithms continue to improve the resolution, depth penetration, and real-time imaging capabilities of ultrasound systems.

6.2.1 Transducers

Ultrasound transducers are essential components in medical imaging, responsible for emitting sound waves and capturing their echoes. The type of transducer chosen for a given diagnostic procedure plays a significant role in the quality of the images obtained, as well as the penetration capabilities and resolution. Among the various transducer types, linear and convex transducers are the most commonly used due to their versatile applications and varying degrees of resolution and penetration [33].

1. Introduction to Ultrasound Transducer Types

Transducers are often designed with different geometries that influence their field of view, imaging depth, and resolution. The selection of a particular transducer depends on the clinical need, anatomical region being examined, and the trade-off between image resolution and penetration depth. The most widely used types of ultrasound transducers include linear, convex, curved array, and phased array transducers. Each of these transducers comes with a set of advantages and limitations that cater to specific applications. Here, we will focus on linear and convex transducers, which are the most frequently utilized in clinical ultrasound imaging due to their broad applications in both superficial and deep tissue examination.

Linear Transducers

Linear transducers are characterized by their straight-line arrangement of piezoelectric crystals, which generate high-frequency sound waves that travel in parallel. The linear array produces a rectangular field of view, enabling imaging along a straight line. The reflected sound waves are received by the transducer and processed to create high-resolution images of the tissue structures [34, 35]. The high-frequency nature of linear transducers, typically operating in the range of 5–15 MHz, enables exceptional resolution, making them ideal for imaging superficially located structures. However, the trade-off is that high-frequency waves suffer from limited penetration depth, meaning linear transducers are not suited for visualizing deeper structures [33, 36]. The linear transducer functions by emitting short pulses of sound waves in the range of 2–10 MHz, with the echoes being reflected from interfaces in the body (e.g., tissue boundaries, organ surfaces). The time-of-flight of these reflected sound waves is used to calculate the distance of the tissue boundaries, thus creating real-time, detailed images (Fig. 6.4).

Equation 6.8 The time-of-flight of the sound wave is given by the formula:

$$d = \frac{c \cdot t}{2} \tag{6.8}$$

where:

- d is the distance to the reflecting object,
- c is the speed of sound in the tissue (approximately 1540 m/s),
- t is the time taken for the echo to return to the transducer.

The high resolution and superficial imaging capabilities of linear transducers make them ideal for several clinical applications:

1. Musculoskeletal imaging: Linear transducers are widely used for examining superficial muscles, tendons, and ligaments. Their high frequency provides excellent resolution, enabling detailed visualization of tissue injuries, such as tears, inflammation, or strains. Linear transducers are also used in joint imaging and soft tissue evaluation, offering insight into the integrity of muscles and tendons.

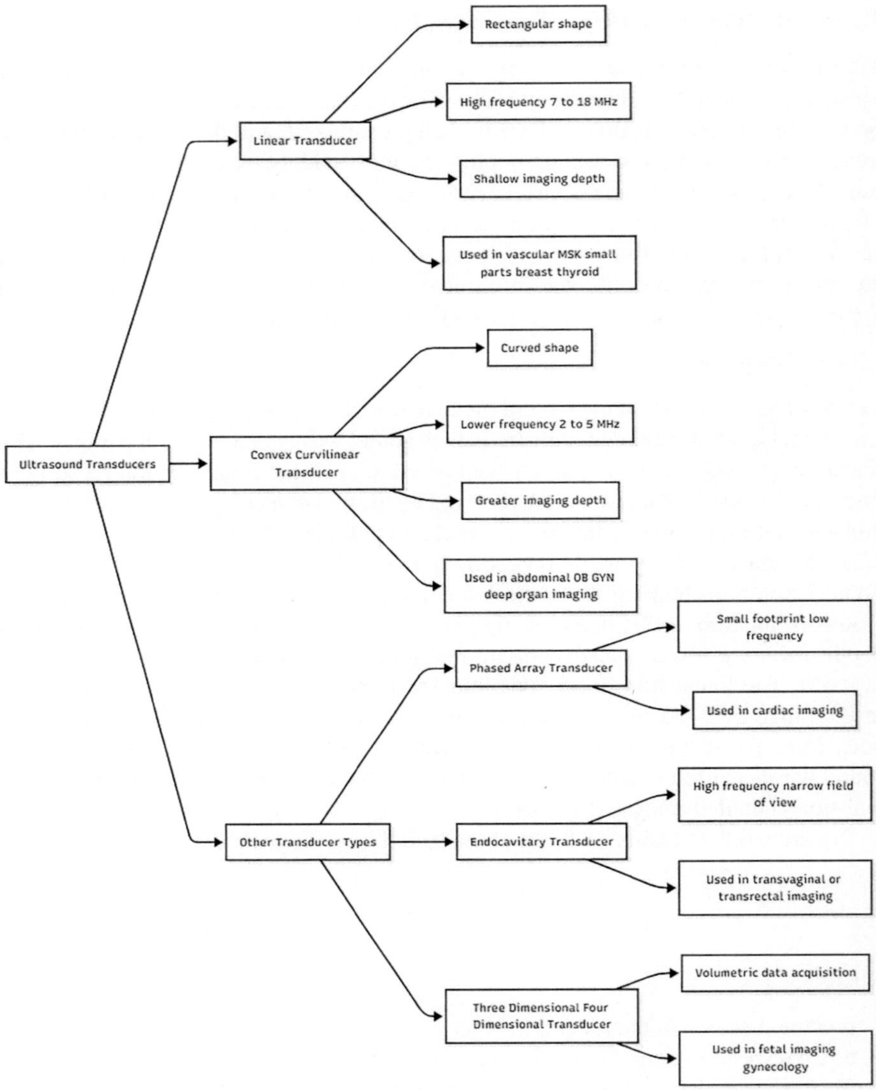

Fig. 6.4 Overview of ultrasound transducer types, highlighting the distinction between linear and convex transducers

2. Vascular imaging: Linear transducers are highly effective in visualizing superficial blood vessels, such as the carotid artery, venous structures, and peripheral veins. The detailed resolution offered by linear transducers enables the detection of vascular abnormalities, including plaques, thrombosis, and stenosis.

3. Thyroid and breast imaging: The ability of linear transducers to capture high-resolution images makes them essential for evaluating thyroid and breast tissue. Linear transducers aid in detecting and assessing cysts, nodules, and tumors in the thyroid gland and breast, where high precision is crucial for distinguishing between benign and malignant growths.
4. Dermatological imaging: Linear transducers are well-suited for dermatological imaging due to their ability to visualize skin and subcutaneous tissue. They are used in evaluating skin lesions, moles, and subcutaneous fat, helping to guide biopsy procedures and monitor wound healing.
5. Lymphatic imaging: Linear transducers are also used to visualize lymph nodes, especially in oncological imaging, where they assist in evaluating metastatic spread or inflammation of lymphatic structures.

While linear transducers are excellent for high-resolution imaging of superficial structures, their penetration depth is limited. This restriction makes them unsuitable for imaging deeper tissues such as abdominal organs or the heart, as the sound waves cannot effectively reach deeper layers of tissue. In such cases, other transducer types, such as convex or phased array transducers, are more appropriate [17, 37].

Convex Transducers

Convex transducers, also known as curved array transducers, feature a curved array of piezoelectric crystals. The curved shape of the transducer allows for the emission of sector-shaped ultrasound beams, resulting in a broader field of view compared to linear transducers. Convex transducers typically operate at lower frequencies, ranging from 2 to 5 MHz, which enables greater penetration into the body [17, 27]. The lower frequency of convex transducers facilitates the visualization of deeper structures by providing a trade-off between penetration depth and image resolution. The sector-shaped beam is ideal for imaging deep organs, such as those found in the abdomen or pelvis, and it offers an advantage in situations where expansive field views are required [30, 38].

The broader field of view and greater depth penetration of convex transducers make them ideal for imaging structures deep within the body. Key applications of convex transducers include:

1. Abdominal imaging: Convex transducers are commonly used in abdominal ultrasound to visualize organs, including the liver, gallbladder, spleen, kidneys, and pancreas. The ability of convex transducers to penetrate deep tissues allows for high-quality imaging of these organs and the detection of tumors, cysts, stones, or inflammation.
2. Obstetric and gynecological imaging: Convex transducers are essential tools in obstetric ultrasound, enabling the visualization of the fetus, placenta, and amniotic fluid. These transducers also provide images of the uterus and ovaries for the assessment of pregnancy and gynecological conditions, such as fibroids or ovarian cysts.
3. Cardiac imaging: In echocardiography, convex transducers are used to image deeper cardiac structures, including the left and right ventricles, atria, valves,

and large vessels such as the aorta. The sector-shaped beam provides a comprehensive view of the heart, enabling the detection of valvular defects, chamber enlargement, and cardiac anomalies.
4. Pediatric imaging: In pediatric patients, convex transducers are helpful in examining deep structures such as the kidneys, bladder, and liver. Due to the reduced tissue density in children, convex transducers can effectively visualize these organs for conditions like congenital anomalies or abdominal infections.
5. Pelvic imaging: Convex transducers are often used in pelvic ultrasound to visualize the bladder, uterus, and ovaries, especially in patients with pelvic pain or infertility issues. They help evaluate conditions such as endometriosis, fibroids, or ovarian cysts.

The main limitation of convex transducers is their lower resolution when compared to linear transducers. Due to the lower frequency used in convex transducers, the imaging resolution is lower, which may affect the ability to visualize fine details, especially in superficial tissues. Moreover, the sector-shaped beam can lead to some distortion at the edges of the image, which might compromise the clarity of certain areas [21, 39].

Comparison Between Linear and Convex Transducers

While both linear and convex transducers serve different purposes in clinical ultrasound, understanding their strengths and weaknesses allows clinicians to select the most suitable device for a given situation (Table 6.1).

In recent years, transducer technology has undergone considerable advancements. Key innovations in both linear and convex transducers include:

- Matrix arrays: These enable better focusing and image quality across different depths, thereby improving both resolution and penetration.
- 1D and 2D arrays: These arrays enable dynamic focusing, providing more explicit images at different depths in real-time.

Table 6.1 Comparison between linear and convex transducers

Feature	Linear transducer	Convex transducer
Frequency range	5–15 MHz	2–5 MHz
Resolution	High resolution for superficial tissues	Lower resolution for deeper tissues
Field of view	Rectangular, parallel	Sector-shaped, wide-angle
Penetration depth	Shallow depth	Greater penetration for deeper tissues
Common applications	Musculoskeletal, vascular, breast, thyroid	Abdominal, obstetric, gynecological, cardiac
Limitations	Limited depth penetration	Lower resolution for superficial structures

- Multifrequency transducers: Some transducers can now operate over a wide range of frequencies, enabling better adaptability in various clinical scenarios.

Additionally, the integration of adaptive beamforming algorithms has improved the precision and clarity of ultrasound images by compensating for tissue attenuation and improving the signal-to-noise ratio [6].

6.2.2 *Pulse-Echo Principle and Doppler Ultrasound*

The two core principles that underpin most ultrasound applications are the pulse-echo principle and Doppler ultrasound. Both principles form the foundation of various clinical applications, including diagnostic imaging, vascular assessment, and cardiac monitoring. In this section, we explore these two principles in depth, discussing their mechanisms, applications, and mathematical underpinnings.

The pulse-echo principle is the basic concept underlying the operation of most diagnostic ultrasound systems. It involves the emission of a pulse of high-frequency sound waves, which propagate through the body and interact with different tissues. These sound waves are reflected by the interfaces between different tissue types, with varying amounts of sound energy being reflected to the transducer, which then converts the echo signals into an image [24, 40]. This principle is based on the time-of-flight measurement of sound waves, meaning that the time it takes for the sound waves to travel to the tissue and back is directly proportional to the distance to the tissue. By analyzing the echo and its time delay, the ultrasound system can reconstruct an image of the tissue structures.

The pulse-echo method involves the following steps:

1. Transmission: The transducer emits short bursts or pulses of high-frequency sound waves.
2. Propagation: These sound waves travel through the body tissues and reflect off surfaces, such as the boundaries between different tissues or organs.
3. Reflection: Part of the sound wave is reflected to the transducer, while the rest of the energy continues deeper into the body.
4. Reception: The reflected sound waves are received by the transducer, which detects the returning echo signals.
5. Signal processing: The time it takes for the echoes to return is used to calculate the distance to the reflecting tissue interface, which is then used to create a real-time image.

The frequency of the sound waves largely influences the resolution of the pulse-echo technique. Higher frequencies result in better resolution and are more suitable for imaging superficial structures, while lower frequencies penetrate deeper into the body but sacrifice resolution. The relationship between frequency and resolution is a key factor in transducer selection for different clinical applications [17, 41]

The time delay of the reflected echoes can be used to calculate the depth of the tissue being imaged, with the speed of sound in tissue being a constant factor. The pulse duration and the frequency of the transducer determine the resolution. Higher-frequency transducers produce shorter wavelengths, resulting in finer details, but with reduced tissue penetration. Applications of pulse-echo ultrasound are:

1. Abdominal imaging: The pulse-echo principle is widely used in abdominal ultrasound for visualizing organs such as the liver, kidneys, gallbladder, and pancreas. The ability to measure the time delay of the echoes enables the creation of high-resolution images, which are crucial for diagnosing conditions such as tumors, stones, cysts, or inflammatory diseases.
2. Obstetrics and gynecology: The pulse-echo technique is crucial in obstetric ultrasound, where it is used to monitor the fetus, placenta, and amniotic fluid. It also assists in assessing the uterus and ovaries for fibroids, cysts, and other conditions. The time-of-flight measurements provide information on the size and position of these structures, allowing for accurate diagnosis and monitoring.
3. Cardiac imaging: Echocardiography relies heavily on the pulse-echo principle for visualizing the heart chambers, valves, and major blood vessels. By emitting pulses through the chest wall and receiving the echoes, clinicians can assess cardiac function, detect valvular defects, and measure the sizes of the chambers.
4. Musculoskeletal imaging: The pulse-echo method is effective in evaluating muscles, tendons, and ligaments, especially in cases of tears, inflammation, or degenerative conditions. The ability to detect subtle variations in tissue boundaries makes this method ideal for soft tissue imaging.
5. Vascular imaging: Pulse-echo ultrasound is used to examine blood vessels, allowing the visualization of arterial and venous structures. This is crucial in detecting stenosis, thrombosis, or aneurysms. In vascular ultrasound, the time-of-flight of the echoes provides information on the depth and diameter of blood vessels.

Doppler Ultrasound

The Doppler effect is a phenomenon that occurs when the frequency of a wave changes due to the relative motion between the source of the wave and the observer. In the context of ultrasound, the Doppler effect is used to measure the velocity of moving blood cells, allowing for the assessment of blood flow within blood vessels [42, 43]. When an ultrasound wave is reflected from a moving object, such as red blood cells, the frequency of the returning wave is altered depending on the direction of movement. If the blood is moving toward the transducer, the frequency increases (known as a positive shift), and if it is moving away, the frequency decreases (known as a negative shift). By analyzing this frequency shift, the Doppler ultrasound system can measure the speed and direction of blood flow [17, 44].

Equation 6.9 The Doppler shift frequency (f_d) is given by the following equation:

$$f_d = \frac{2\, v\, f_0 \cos(\theta)}{c} \tag{6.9}$$

where:

- f_d is the Doppler shift frequency,
- v is the velocity of the blood flow,
- f_0 is the transmitted frequency,
- θ is the angle between the ultrasound beam and the direction of blood flow,
- c is the speed of sound in blood (approximately 1500 m/s).

Types of Doppler ultrasound:

1. Continuous-wave Doppler ultrasound: Continuous-wave Doppler uses two transducer elements: one for transmitting sound waves and the other for receiving the echoes. This method enables the detection of high-velocity blood flow but lacks depth resolution, making it suitable for imaging high-flow vessels, such as the aorta.
2. Pulsed-wave Doppler ultrasound: Pulsed-wave Doppler ultrasound involves the transmission of short pulses of sound waves and the detection of the returning echoes. This method offers improved depth resolution, enabling the measurement of blood flow velocities at specific locations within blood vessels, such as the carotid arteries or renal arteries.
3. Color Doppler ultrasound: In color Doppler, the Doppler shift frequency is used to create a color-coded map that overlays on the conventional grayscale ultrasound image. This enables real-time visualization of blood flow and is particularly useful in detecting vascular abnormalities, such as stenosis or regurgitation. The colors typically represent the direction of velocity (red for flow toward the transducer, blue for flow away from the transducer).
4. Power Doppler ultrasound: Power Doppler focuses on the amplitude of the Doppler shift rather than the frequency shift. It is more sensitive to slow-flowing blood and provides better visualization of small vessels and areas with low blood flow, such as tumors or areas of inflammation.

Doppler ultrasound has found its application in a variety of diagnostic niches, most importantly:

1. Cardiovascular imaging: Doppler ultrasound is extensively used in echocardiography to assess heart valve function, detect stenosis, and evaluate blood flow velocities in the aorta, mitral valve, and pulmonary arteries. It also aids in measuring cardiac output and left ventricular function.
2. Vascular imaging: Doppler ultrasound plays a crucial role in detecting vascular disorders such as deep vein thrombosis (DVT), arterial stenosis, and aneurysms. It provides real-time assessment of blood flow and can be used to guide the treatment of vascular pathologies.
3. Obstetric and gynecological imaging: Doppler ultrasound is used in obstetric imaging to assess fetal well-being, particularly the blood flow in the umbilical artery and placenta. It is also used to evaluate the uterine arteries and ovarian flow, helping in diagnosing conditions like pre-eclampsia or gestational hypertension.

4. Renal imaging: Doppler ultrasound is valuable in evaluating the renal arteries for stenosis, arterial insufficiency, and hypertension. It is also used to assess renal transplant function and kidney perfusion in patients with kidney disease.

While Doppler ultrasound is a powerful tool for assessing blood flow and tissue motion, it has certain limitations:

1. Angle dependency: The accuracy of Doppler measurements depends on the angle between the ultrasound beam and the direction of blood flow. The optimal angle is 0° (directly along the flow direction), as the Doppler shift is maximized at this angle. At angles greater than 60°, the Doppler shift becomes less accurate.
2. Inability to visualize deep structures: Doppler ultrasound has limited penetration depth, mainly when used with higher-frequency probes. This limits its ability to assess deep structures, such as the heart or abdominal vessels, in obese patients.
3. Artifact sensitivity: Doppler ultrasound is sensitive to artifacts caused by motion, tissue interfaces, and poor signal-to-noise ratio, which may impact the clarity of the measurements.

The pulse-echo principle and Doppler ultrasound are fundamental to modern diagnostic ultrasound, providing complementary advantages in various clinical applications. Pulse-echo ultrasound provides high-resolution images by analyzing the time-of-flight of sound waves. In contrast, Doppler ultrasound provides critical insights into blood flow, enabling the assessment of flow velocities, vascular conditions, and cardiac function. Together, these techniques form the backbone of non-invasive diagnostic imaging, offering physicians powerful tools to assess the structure and function of organs and tissues across a wide range of medical fields [28, 45].

6.3 Ultrasound Image Processing Techniques

Unlike other imaging modalities, ultrasound images are produced by sound waves rather than electromagnetic radiation, leading to unique challenges in the acquisition and processing of these images. The inherent properties of ultrasound, such as low resolution, high noise levels, and limited tissue contrast, make image processing techniques essential for enhancing the quality and accuracy of ultrasound images, thus improving visual clarity, signal-to-noise ratio (SNR), and diagnostic value. The resolution of ultrasound images is often lower than that of CT scans or MRI, and artifacts such as speckle noise, shadowing, and reverberation can obscure key anatomical details [6]. Moreover, the depth of penetration and the frequency of the ultrasound waves significantly impact image quality, requiring tailored processing techniques depending on the region of interest. Ultrasound images also often suffer from low contrast between different tissues. For instance, the distinction between soft tissues such as muscles and fat or between solid organs and fluid-filled cavities can be difficult to interpret without enhanced contrast. Therefore, image enhancement and contrast adjustment play an essential role in improving the clarity of ultrasound images (Fig. 6.5).

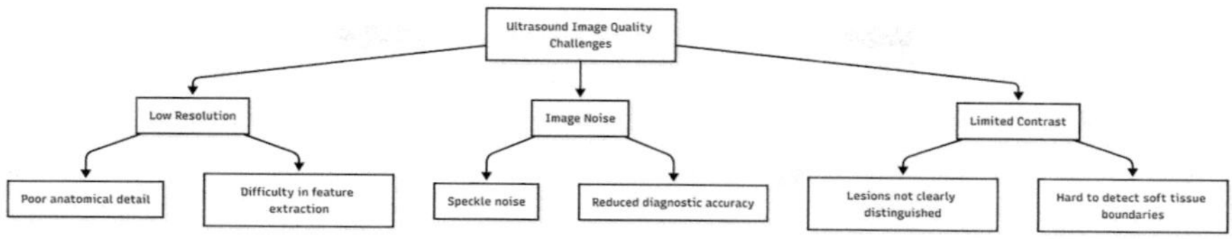

Fig. 6.5 Diagram showing challenges in ultrasound image quality, including low resolution, noise, and limited contrast

The goal of image processing in ultrasound is multifaceted as it involves noise reduction, image enhancement, image segmentation, and motion correction, among other tasks. These processes aim to highlight important features of the image, such as blood vessels, organ structures, and pathological areas, while minimizing or eliminating irrelevant details. Image processing not only improves the visualization of internal structures but also plays a critical role in quantifying tissue properties, such as elasticity and blood flow, which are crucial for diagnostic purposes. By applying advanced algorithms, mathematical models, and computational tools, ultrasound images can be transformed to provide more precise, clear, and informative representations of underlying anatomical structures [21].

In ultrasound imaging, the preprocessing stage plays a pivotal role in improving image quality, making it a crucial component in the ultrasound signal processing pipeline. Two of the most significant challenges in ultrasound image acquisition are speckle noise and random noise, which can severely hinder the accuracy and clarity of the images. The goal of preprocessing is to reduce these unwanted artifacts and enhance the features of interest, thus improving diagnostic accuracy [46, 47]. This subchapter delves into the mechanisms of speckle reduction and noise removal, discussing various techniques employed to mitigate these issues and the principles behind their effectiveness.

6.3.1 Speckle Noise in Ultrasound Imaging

Speckle noise refers to a granular or salt-and-pepper appearance that can obscure the true details of the tissue structures in ultrasound images. The interference of multiple sound waves causes it to be reflected from tissue interfaces, leading to constructive and destructive interference patterns. Speckle is particularly prevalent in high-frequency imaging, where the wavelength of sound waves is comparable to the size of tissue inhomogeneities, such as small blood vessels and muscle fibers. Speckle noise can be a significant challenge, as it creates the illusion of additional fine detail in the image, which makes it difficult for clinicians to distinguish between real tissue structures and the unwanted noise. In addition, speckle can obscure small objects or pathological regions, limiting the effectiveness of ultrasound as a diagnostic tool [9].

Several techniques are employed to reduce speckle noise while preserving important tissue features such as edges and boundaries. The key challenge in speckle reduction is to smooth the image without blurring or distorting these features.

6.3.2 Spatial Filtering Techniques

Spatial filters operate by modifying the pixel values based on the local neighborhood in the image. The idea is to smooth out the variations caused by speckle noise while retaining significant structures, such as edges or boundaries, in the image.

Median filtering is a nonlinear filter that replaces each pixel value with the median of its surrounding pixels. This technique is particularly effective for removing speckle noise, as it preserves edges while reducing the noise. Median filters are handy in reducing salt-and-pepper noise and are often applied in a 3×3 or 5×5 neighborhood window.

Equation 6.10 For a median filter, the new pixel value at position (x,y) is given by:

$$I_{\text{new}}(x, y) = \text{Median}(I(x+i, y+j))\text{for} - k \leq i, j \leq k \tag{6.10}$$

where I(x,y) is the original pixel value, and the median function selects the middle value of the intensity distribution in the neighborhood.

Gaussian filters are linear filters that smooth images by averaging pixel values based on a Gaussian kernel. The Gaussian function places more weight on pixels near the center of the kernel, thus reducing the influence of distant pixels and smoothing out high-frequency noise while preserving edges. This filter works well for reducing high-frequency noise without blurring significant features [24, 48].

Equation 6.11 The filtered image, I_{filtered}(x,y) can be computed using the convolution operation with a Gaussian kernel G(x,y):

$$I_{\text{filtered}}(x, y) = \sum_{i=-k}^{k} \sum_{j=-k}^{k} I(x+i, y+j) \cdot G(i, j) \tag{6.11}$$

where G(i,j) is the Gaussian kernel function, which is defined as Eq. 6.12:

$$G(i, j) = \frac{1}{2\pi\sigma^2}\exp\left(-\frac{i^2+j^2}{2\sigma^2}\right) \tag{6.12}$$

Here, σ controls the spread of the filter, and the filter smooths out small-scale variations in the image.

6.3.3 *Wavelet-Based Denoising*

Wavelet transforms offer a more advanced method for speckle reduction by decomposing the image into various frequency components. Wavelets can capture both high-frequency details (such as edges) and low-frequency smooth areas (such as tissue regions) in different scales. By using a multi-scale decomposition approach, wavelet denoising can reduce speckle noise while retaining critical information in the image.

Discrete Wavelet Transform (DWT) transforms an image into multiple frequency bands, separating the low-frequency components (smooth regions) and the high-frequency components (details, including noise). By thresholding the high-frequency coefficients, noise can be suppressed while preserving the important image features.

Equation 6.13 The wavelet transform W of an image I is expressed as:

$$W = \sum_{i,j} \alpha_{ij}\, \phi_{ij}(x, y) \tag{6.13}$$

where:

- α_{ij} are the wavelet coefficients
- ϕ_{ij} are the scaling functions or wavelet basis functions.

A soft thresholding function can then be applied to reduce high-frequency noise (Eq. 6.14):

$$\alpha_{\text{thresholded}} = \text{sign}(\alpha) \cdot \max(|\ \alpha\ | - \lambda, 0) \tag{6.14}$$

where λ is the threshold value, which controls the level of noise reduction.

6.3.4 *Non-Local Means (NLM) Filtering*

Non-Local Means (NLM) filtering is an advanced method that compares patches of pixels rather than individual pixels. This approach is beneficial for preserving structural details in the image while reducing noise. The NLM filter computes a weighted average of all pixels in the image based on the similarity of their neighboring patches, providing effective speckle reduction without significant blurring. The NLM filter works by calculating a similarity measure between patches in a local region and averaging the pixels accordingly [13, 49].

Equation 6.15 The NLM filter output at pixel (x,y) is:

$$I_{\text{NLM}}(x, y) = \frac{\sum_{(i,j)} w(x, y, i, j)\, I(i, j)}{\sum_{(i,j)} w(x, y, i, j)} \tag{6.15}$$

where:

- w(x,y,i,j) is the weight based on the similarity between patches centered at (x,y)
- (i,j) and I(i,j) is the pixel intensity.

6.3.5 *Random Noise in Ultrasound Imaging*

In addition to speckle noise, random noise—resulting from electronic interference, signal degradation, or environmental factors—can also impair ultrasound image quality. Random noise tends to add unwanted variations to the signal, which are often difficult to distinguish from tissue structures. Effective noise removal is essential to ensure the most transparent possible image for diagnostic purposes [31, 50].

Adaptive filters are handy for dealing with random noise, as they adjust their behavior based on the local characteristics of the image. Wiener filtering is an adaptive filter that optimizes the trade-off between smoothing and edge preservation by using local image statistics such as the variance and mean. The Wiener filter uses the local signal variance to adapt the filtering process. In regions with high signal variance (edges and important features), the filter applies less smoothing, whereas in regions with low variance (background noise), more smoothing is applied.

Equation 6.16 The Wiener filter can be defined as:

$$H(x, y) = \frac{\mathbb{E}[I_{\text{local}}(x, y)]}{\mathbb{E}[I_{\text{local}}^2(x, y)] + \lambda} \tag{6.16}$$

where:

- $E[I_{local}(x,y)]$ is the local mean of the image at pixel (x,y)
- λ is a constant that controls the amount of smoothing.

Bilateral filtering is another noise removal technique that uses a combination of spatial distance and intensity difference to filter out noise. This method preserves edges by using both geometric and intensity-related criteria to smooth the image. The bilateral filter applies a weighted average of neighboring pixels, where the weight depends not only on the distance between pixels but also on the intensity difference. This ensures that edges remain sharp while noise is reduced in uniform regions.

Equation 6.17 The bilateral filter output is given by:

$$I_{\text{bilateral}}(x, y) = \frac{\sum_{(i,j)} G_{\sigma_s}(x-i, y-j)\, G_{\sigma_r}(I(x, y) - I(i, j))\, I(i, j)}{\sum_{(i,j)} G_{\sigma_s}(x-i, y-j)\, G_{\sigma_r}(I(x, y) - I(i, j))} \tag{6.17}$$

where:

- $G_{\sigma s}$ is the spatial Gaussian kernel
- $G_{\sigma r}$ is the range Gaussian kernel based on intensity differences.

Preprocessing steps such as speckle reduction and noise removal are critical for improving the quality and interpretability of ultrasound images. By applying techniques such as median filtering, Gaussian smoothing, wavelet denoising, and adaptive filtering, ultrasound images can be enhanced, enabling more accurate and reliable diagnosis. The challenge in ultrasound image preprocessing lies in striking the right balance between noise reduction and detail preservation, ensuring that important tissue structures remain visible while unwanted artifacts are minimized [30].

6.3.6 Feature Extraction, Edge Detection, and Segmentation

Feature extraction plays a critical role in ultrasound imaging as it enables the identification and isolation of relevant anatomical features, which are essential for diagnosis, monitoring, and planning treatments. Ultrasound images provide valuable information about tissue characteristics, blood flow, and other physiological phenomena, but the data is often cluttered with noise and artifacts [51]. To interpret this information effectively, advanced computational methods are used to process and extract features from the image. Among these, edge detection and segmentation are two of the most essential techniques for isolating key anatomical features in ultrasound images. These methods are instrumental in delineating borders between tissues, organs, or pathologies. For instance, edge detection identifies boundaries such as those between a tumor and surrounding tissue, while segmentation divides an image into distinct regions that correspond to anatomical structures [52, 53].

Edge Detection in Ultrasound Imaging

Edge detection is one of the fundamental operations used in image processing to identify sharp intensity changes in an image. In ultrasound imaging, these changes in intensity typically represent boundaries between different tissues, such as muscle, fat, or organs. Accurate edge detection is crucial for visualizing structures such as blood vessels, tumors, and organs, which helps in diagnosing diseases, planning interventions, and monitoring progression.

Edge detection techniques are based on the observation that the transition between tissues with different acoustic properties causes a significant change in the intensity values in the ultrasound image. This intensity change can be detected by evaluating the gradient of the image, which measures how much the pixel intensity changes in space. In regions with high gradient values, a clear boundary between tissues is likely to exist. The most common approach to edge detection is by calculating the gradient magnitude at each pixel in the image. This is done by computing the changes in the image intensity in both the horizontal and vertical directions. Significant changes in intensity signify an edge [15, 27].

Several edge detection techniques are used in ultrasound imaging, depending on the complexity of the image and the quality of the data. Here, we discuss some of the most widely used methods.

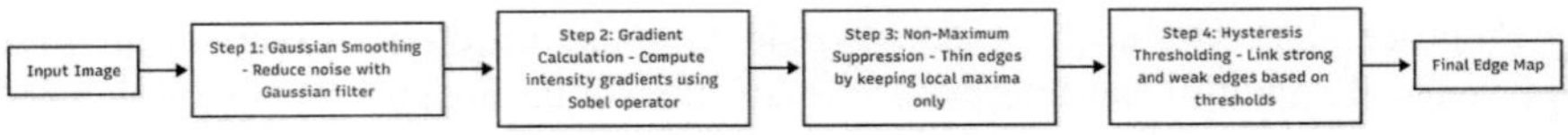

Fig. 6.6 A multi-step flow diagram illustrating the steps of the Canny edge detection process, including Gaussian smoothing, gradient calculation, non-maximum suppression, and hysteresis thresholding

The Sobel operator is a gradient-based edge detection method that calculates the gradient magnitude using convolution with the Sobel kernels for the x and y directions. These kernels are used to approximate the spatial derivatives in the image, which can then be combined to compute the gradient.

Equation 6.18 The gradient magnitude G at a pixel (x,y) is given by:

$$G(x, y) = \sqrt{(G_x(x, y))^2 + (G_y(x, y))^2} \tag{6.18}$$

where Gx and Gy are the gradients in the **horizontal** and **vertical** directions, respectively, calculated from the Sobel operator. These kernels are defined as:

$$G_x = \begin{bmatrix} -1\,0\,1 & & -1 & -2 & -1 \\ -2\,0\,2], & G_y = [& 0 & 0 & 0 \\ -1\,0\,1 & & 1 & 2 & 1 \end{bmatrix}$$

The Sobel method is fast, simple, and effective for detecting edges in ultrasound images, especially when the edges are relatively sharp.

The Canny edge detector is an advanced edge detection algorithm that provides superior edge detection by applying a multi-step process. It includes Gaussian smoothing, gradient calculation, non-maximum suppression, and hysteresis thresholding. This method is especially useful in ultrasound imaging, where noise reduction and precise edge detection are critical for identifying small or subtle boundaries [17] (Fig. 6.6).

The Canny edge detection algorithm involves the following steps:

Gaussian smoothing:

The image is first smoothed using a Gaussian filter to reduce noise. The filter is applied using convolution (Eq. 6.19):

$$I_{\text{smooth}} = I * G_\sigma \tag{6.19}$$

where Gσ is the Gaussian kernel with standard deviation σ

Gradient calculation:

The gradient is calculated in the horizontal and vertical directions using the image derivatives (Eq. 6.20):

$$G_x = \frac{\partial I}{\partial x}, G_y = \frac{\partial I}{\partial y} \tag{6.20}$$

Edge tracing by hysteresisensures that weak edges are only considered true edges if they are connected to firm edges. The edge magnitude is compared to two threshold values T1 and T2, where T2 is higher, and edges with a gradient magnitude higher than T2 are kept as strong edges. In contrast, weak edges with a gradient magnitude between T1 and T2 are included if connected to strong edges.

The Prewitt operator is similar to the Sobel operator but uses a different convolution kernel. It is commonly used when a simpler, less computationally intensive method is needed for edge detection. The Prewitt operator emphasizes edges and contours, but it may not detect fine details as well as more complex algorithms like Canny.

Equation 6.21: The Prewitt operator uses the following kernels for horizontal and vertical gradients:

$$G_x = \begin{bmatrix} -1\,0\,1 & -1 & -1 & -1 \\ -1\,0\,1], G_y = [& 0 & 0 & 0 \\ -1\,0\,1 & 1 & 2 & 1 \end{bmatrix} \tag{6.21}$$

Segmentation in Ultrasound Imaging

Segmentation involves dividing an ultrasound image into distinct regions that represent specific structures or tissues. The goal is to isolate areas of interest such as organs, blood vessels, tumors, and other anatomical features. Since ultrasound images often contain a variety of tissues with similar grayscale intensities and complex noise patterns, segmentation can be a challenging task. Segmentation in ultrasound images is typically driven by features such as intensity, texture, shape, and spatial coherence. The process may involve either region-based methods or edge-based methods, or even a combination of both. The key objective is to identify boundaries between tissues or distinguish regions of interest from the background.

Effective segmentation techniques also take into account the complexity and quality of ultrasound data. Ultrasound images often suffer from speckle noise, shadowing, and acoustic artifacts, which can hinder segmentation accuracy. Therefore, the chosen method must be robust to such distortions while providing accurate delineation of tissue boundaries [54, 55].

One of the simplest segmentation techniques is thresholding, where pixel intensity values are compared to a pre-defined threshold to classify pixels as either belonging to a structure of interest or the background. Thresholding works well when there is a clear contrast between tissues but can fail in the presence of noise or when there is insufficient contrast.

Equation 6.22 In binary thresholding, the segmented image S(x,y) is defined as:

$$S(x, y) = \begin{cases} 1, & I(x, y) \geq T \\ 0, & I(x, y) < T \end{cases} \tag{6.22}$$

where:

- T is the chosen threshold value
- I(x,y) represents the intensity at pixel (x,y).

Region growing is a method that starts with an initial seed point and iteratively adds neighboring pixels that meet certain similarity criteria. The growth continues until the region satisfies the stopping condition, such as the intensity value exceeding a defined threshold or the homogeneity of the region being violated (Fig. 6.7).

Equation 6.23 In region growing, the new region R_{new} is formed by adding neighboring pixels whose intensity values are within a certain tolerance of the seed pixel:

$$R_{\text{new}} = R_{\text{old}} \cup \{(x, y) \mid | I(x, y) - I(x_0, y_0) | < T\} \tag{6.23}$$

where:

- I(x,y) is the intensity value at pixel (x,y)
- $I(x_0,y_0)$ is the intensity at the seed point.

Active contour models, or snakes, are used to extract object boundaries by evolving an initial curve or contour to fit the desired object boundary. The contour is influenced by image features such as gradient magnitude (which represents edges) and smoothness constraints to ensure the boundary remains well-defined [56, 57].

Equation 6.24 The energy functional E of a snake is defined as:

$$E = \int_{\Gamma} (\alpha \mid \nabla I|^2 + \beta \mid \kappa|^2) ds \tag{6.24}$$

where:

- Γ represents the contour,
- ∇I is the gradient of the image,
- κ is the curvature of the contour.
- Parameters α and β control the trade-off between edge fidelity and smoothness.

With the rise of deep learning, CNNs have become the dominant method for segmentation in medical imaging, including ultrasound. Models like U-Net have been particularly successful for segmentation tasks in ultrasound due to their ability to capture both fine-grained details and large-scale features. CNNs learn to map the pixel intensities of the input to the corresponding labels or regions (Fig. 6.8).

Equation 6.25 In CNN-based segmentation, the segmentation mask Y is produced by the network as:

$$\text{Y} = \text{CNN(X)} \tag{6.25}$$

where:

- X is the input ultrasound image
- Y is the resulting segmented image.

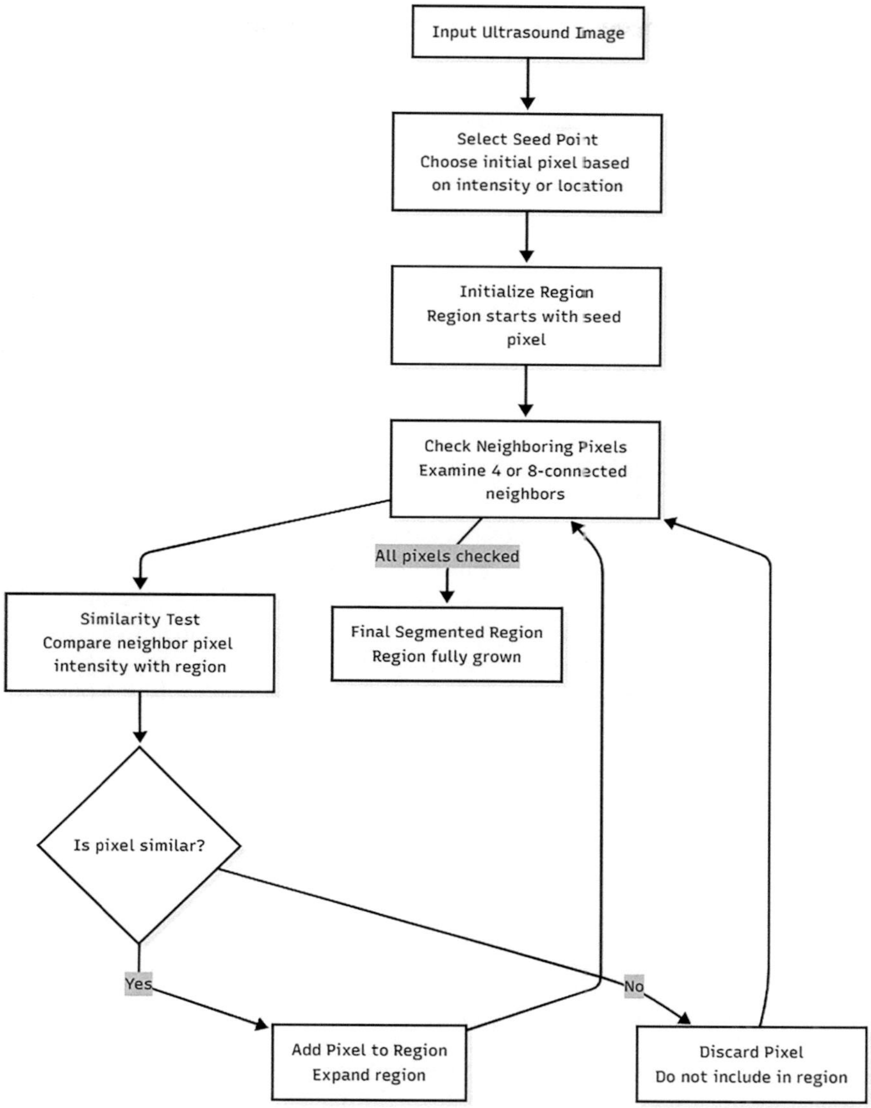

Fig. 6.7 A step-by-step illustration of the region-growing algorithm applied to an ultrasound image, showing the initial seed point and how the region expands to encompass the neighboring pixels

The techniques of edge detection and segmentation form the backbone of ultrasound image processing, enabling the extraction of key features necessary for diagnosis and treatment planning. Whether using simple gradient-based methods like the Sobel operator, advanced methods such as Canny edge detection, or more sophisticated techniques like region growing and deep learning, feature extraction helps

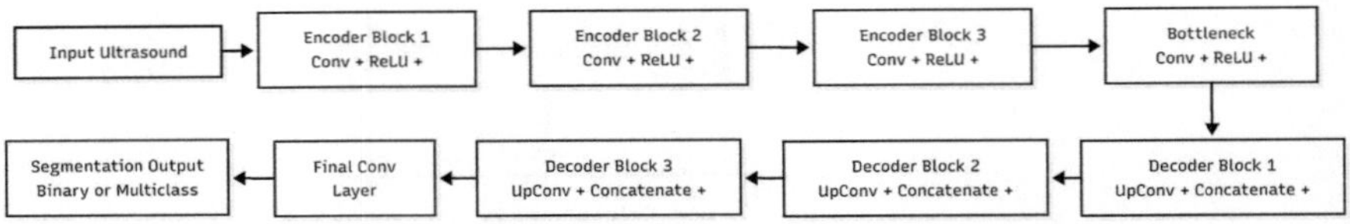

Fig. 6.8 A flowchart showing the U-Net architecture applied to ultrasound image segmentation

identify and isolate critical anatomical structures within ultrasound images [58, 59]. These methods allow clinicians to observe and evaluate anatomical features with high precision, leading to more accurate diagnoses, treatment strategies, and real-time monitoring of diseases and conditions.

6.4 Clinical Applications of Ultrasound

Ultrasound has revolutionized clinical practice by offering real-time, non-invasive visualization of internal structures with no ionizing radiation. Its accessibility, affordability, and diagnostic versatility have made it indispensable across nearly all medical disciplines. This chapter explores in depth three major clinical domains where ultrasound plays a vital role: cardiovascular imaging, fetal imaging, and organ monitoring. These applications underscore the breadth of ultrasound's utility, from acute diagnostics to long-term disease management.

6.4.1 Cardiovascular Imaging

Ultrasound is foundational in cardiovascular diagnostics. Echocardiography, in particular, is the cornerstone of non-invasive heart imaging. Its value lies in the ability to visualize cardiac anatomy, assess function, and guide interventions in real-time.

Transthoracic Echocardiography (TTE)

TTE is typically the first-line imaging modality for evaluating cardiac pathology. A transducer placed on the chest transmits high-frequency sound waves that create dynamic images of the heart. TTE allows for a comprehensive assessment of the heart's anatomy and function. This includes evaluating the size and structure of the heart chambers, measuring myocardial thickness, and analyzing the heart's contractility [6]. One of its most critical applications is measuring the ejection fraction, which provides an estimate of how well the heart pumps blood. Additionally, wall motion abnormalities can be identified, aiding in the diagnosis of ischemic heart disease or prior myocardial infarction. These capabilities make TTE essential in assessing patients with suspected or confirmed heart failure, and it plays a central role in guiding medical management and monitoring therapeutic response.

In the evaluation of valvular disease, TTE provides clear images of valve anatomy and motion. It helps diagnose valvular stenosis, where valve openings are narrowed, and regurgitation, where valves fail to close properly and blood leaks backward. Doppler ultrasound, often used in conjunction with TTE, quantifies the severity of these abnormalities by measuring flow velocities and pressure gradients across valves. This non-invasive evaluation is crucial for making timely decisions about valve repair or replacement. TTE is also highly effective in detecting pericardial effusions, which are accumulations of fluid around the heart. If the effusion leads to cardiac tamponade—a life-threatening condition—TTE findings guide emergency pericardiocentesis to remove the fluid. Furthermore, in pediatric cardiology, TTE is indispensable for identifying congenital heart defects such as septal defects, patent ductus arteriosus, and abnormal connections between cardiac chambers or vessels. Overall, TTE provides critical insights that shape diagnostic clarity and influence the direction of treatment in a wide range of cardiovascular conditions [15].

Transesophageal Echocardiography (TEE)

Transesophageal echocardiography (TEE) offers superior imaging resolution, especially for posterior cardiac structures and in patients with poor transthoracic windows. By introducing a probe into the esophagus—positioned directly behind the heart—TEE provides close-range, high-definition images that reveal subtle structural details often missed by TTE. It is beneficial in patients who are obese, intubated, or have chest wall deformities that interfere with transthoracic imaging.

TEE is particularly valuable in diagnosing infective endocarditis. It enables clear visualization of valvular vegetations, abscesses, and other complications like leaflet perforation or prosthetic valve dehiscence. In many cases, TEE is the only reliable modality for detecting these high-risk conditions, which require prompt treatment. It also provides critical insight in identifying atrial septal defects, patent foramen ovale, and left atrial appendage thrombi, especially in patients with atrial fibrillation undergoing evaluation for cardioversion or ablation procedures [20, 60].

In the intraoperative setting, TEE is routinely used during cardiac surgery to assess the success of repairs, monitor left ventricular function, and detect complications in real time. For patients undergoing valve replacement, TEE ensures accurate prosthesis placement and function. In trauma and critical care, it assists in evaluating hemodynamic instability when other imaging modalities are impractical. TEE's role in guiding both diagnosis and therapeutic decision-making has solidified its status as a cornerstone in modern cardiology and cardiac surgery.

Stress Echocardiography

Stress echocardiography is a specialized technique that evaluates myocardial function under physical or pharmacological stress. Its primary purpose is to uncover areas of the heart muscle that do not receive adequate blood supply during exertion, indicative of ischemic heart disease. During the test, ultrasound images are acquired at rest and again following stress, allowing clinicians to compare myocardial performance under both conditions.

Exercise stress testing is often used in conjunction with treadmill or bicycle protocols. However, when patients cannot exercise adequately, pharmacologic agents such as dobutamine are administered to increase heart rate and contractility. Stress-induced changes in wall motion, such as new or worsening hypokinesia, signify ischemia and guide further management decisions like coronary angiography or revascularization [1, 21].

Beyond ischemia detection, stress echocardiography can assess valvular function under stress conditions, especially in patients with borderline stenosis or regurgitation at rest. It also helps determine contractile reserve in heart failure patients, assisting in prognostication and therapy planning. The test's non-invasive nature, combined with its diagnostic and prognostic accuracy, makes it a vital tool in the comprehensive evaluation of suspected or known coronary artery disease.

Contrast Echocardiography

Contrast echocardiography involves the use of intravenous microbubble contrast agents that enhance the visualization of cardiac structures. These microbubbles are composed of inert gas encased in a lipid or protein shell. They are small enough to traverse capillaries, allowing them to highlight endocardial borders and vascular structures without exiting the bloodstream. This technique significantly improves the accuracy of left ventricular volume and ejection fraction measurements, particularly in patients with suboptimal acoustic windows. In cases where apical segments or endocardial delineation is poor, contrast agents provide more transparent, more consistent images. Additionally, contrast echocardiography can detect left ventricular thrombi, intracardiac masses, and structural abnormalities with greater precision.

In patients with suspected patent foramen ovale (PFO) or atrial septal defects (ASDs), contrast agents are injected during a Valsalva maneuver to detect right-to-left shunting. The appearance of microbubbles in the left atrium within three cardiac cycles confirms the diagnosis [3, 61]. Contrast-enhanced studies also play a role in myocardial perfusion imaging, though this application is less widespread. Overall, contrast echocardiography offers a valuable enhancement to standard techniques, especially when image quality is compromised or when more detailed anatomic resolution is required.

Vascular Ultrasound

Vascular ultrasound is integral to the diagnosis and management of diseases affecting both the arterial and venous systems. It provides real-time visualization of blood flow and vessel wall characteristics using B-mode imaging, color Doppler, and spectral waveform analysis. In the arterial system, carotid duplex ultrasound is used to assess the degree of stenosis in the carotid arteries, which supply blood to the brain. Plaque morphology, location, and hemodynamic impact are carefully evaluated to determine stroke risk and guide surgical or medical management. In peripheral arterial disease, ultrasound identifies sites of obstruction, evaluates flow dynamics, and helps stratify the severity of ischemia. It also aids in planning and monitoring revascularization procedures.

In the venous system, duplex ultrasound is the gold standard for diagnosing deep vein thrombosis (DVT), a potentially life-threatening condition. It assesses vein compressibility, intraluminal thrombus, and flow patterns. Vascular ultrasound is also used to evaluate for venous insufficiency, venous malformations, and to guide catheter placements. Its non-invasive nature, wide availability, and absence of radiation make it ideal for repeated use and long-term monitoring of vascular health.

6.4.2 Fetal Imaging

Ultrasound plays an indispensable role in modern obstetrics, providing a safe, real-time method for monitoring fetal development, detecting congenital anomalies, and guiding prenatal interventions. From the earliest stages of pregnancy through delivery, fetal ultrasound supports clinical decisions and reassures expecting families. Its non-invasive nature and the absence of ionizing radiation make it ideal for repeated use throughout gestation, allowing for ongoing evaluation of both the fetus and the intrauterine environment [45, 62].

First Trimester

In the first trimester, ultrasound is primarily used to confirm intrauterine pregnancy, determine gestational age, and evaluate fetal viability. A transvaginal approach offers superior resolution in early pregnancy and is particularly useful when assessing for ectopic pregnancy, molar pregnancy, or other complications. The presence of a gestational sac, yolk sac, and fetal pole with cardiac activity confirms viability. Crown-rump length (CRL) is measured to estimate gestational age with a high degree of accuracy, especially during the first 12 weeks.

First-trimester ultrasound also allows for early detection of multiple gestations, chorionicity, and amnionicity, which are essential for anticipating risks associated with twin pregnancies. Additionally, between 11 and 14 weeks, the nuchal translucency (NT) scan is performed. NT refers to the fluid-filled space at the back of the fetal neck and, when increased, may be associated with chromosomal abnormalities such as trisomy 21, trisomy 18, and Turner syndrome. When combined with maternal serum screening, first-trimester ultrasound contributes to early risk assessment for aneuploidy and structural defects.

Second Trimester

The second-trimester anatomy scan, typically performed between 18 and 22 weeks of gestation, is a cornerstone of prenatal care. This detailed assessment systematically evaluates the fetus from head to toe, examining major organ systems, the spine, extremities, and the placenta. It enables the diagnosis of congenital anomalies such as neural tube defects, congenital heart disease, abdominal wall defects, and limb abnormalities. Any findings can prompt further investigation or early intervention planning [31, 63].

During this scan, biometric measurements—including biparietal diameter, head circumference, abdominal circumference, and femur length—are taken to estimate fetal weight and growth trajectory. These parameters are essential for monitoring development and identifying growth restriction or macrosomia. The position of the placenta is also assessed to rule out placenta previa or abnormal placental attachment, which may impact the mode of delivery. Amniotic fluid volume is estimated using either the amniotic fluid index (AFI) or the single deepest pocket method, both of which help assess the fetal environment.

In addition, fetal sex is often determined during the second-trimester scan, if desired by the parents. While this information is not medically necessary, it can contribute to parental bonding and preparation. Most importantly, the second-trimester scan serves as a key decision point for the ongoing management of pregnancy, helping clinicians tailor follow-up imaging, genetic counseling, or referrals to maternal–fetal medicine specialists as needed [24, 64].

Third Trimester

Third-trimester ultrasound continues the surveillance of fetal growth, positioning, and well-being as delivery approaches. Fetal biometry remains important in this stage, especially in pregnancies complicated by conditions like gestational diabetes, hypertension, or intrauterine growth restriction (IUGR). Serial measurements are often necessary to track growth velocity and assess for signs of placental insufficiency or fetal macrosomia.

Fetal presentation—whether cephalic, breech, or transverse—is evaluated to inform delivery planning. In breech or transverse presentations, ultrasound is used to guide external cephalic version procedures. Assessment of the cervix and lower uterine segment can provide additional context for labor readiness or risk of preterm birth. The placenta is re-evaluated for location and signs of aging or calcification, and amniotic fluid levels are measured to ensure an appropriate intrauterine environment.

Biophysical profile (BPP) combines ultrasound parameters such as fetal movement, tone, breathing, and amniotic fluid volume with non-stress testing to assess overall fetal well-being. Doppler flow studies—especially of the umbilical artery, middle cerebral artery, and ductus venosus—offer insight into fetal oxygenation and placental resistance. Abnormal Doppler findings may necessitate expedited delivery or additional surveillance [27, 65]. Ultrasound also guides third-trimester procedures like amniocentesis, intrauterine transfusions, and selective reduction in multifetal pregnancies.

6.4.3 Organ Monitoring

Ultrasound's application extends far beyond the cardiovascular system and pregnancy care. Its ability to non-invasively monitor organ structure and function makes it a central tool in internal medicine, surgery, and critical care. From assessing acute conditions like gallstones or hydronephrosis to chronic disease surveillance in the liver and kidneys, ultrasound has become essential for timely diagnosis and management.

Abdominal Organs

Abdominal ultrasound provides valuable insights into the health of solid and hollow organs, including the liver, gallbladder, pancreas, kidneys, spleen, and bladder. It is often the first-line imaging modality for patients presenting with abdominal pain, jaundice, or abnormal liver function tests. Liver ultrasound can detect steatosis, cirrhosis, masses, and hepatic congestion. Doppler imaging of the hepatic vasculature can reveal portal hypertension and hepatic vein thrombosis. In addition, ultrasound-guided liver biopsy and elastography allow for non-invasive fibrosis staging.

Gallbladder ultrasound is the gold standard for detecting cholelithiasis and cholecystitis. Thickening of the gallbladder wall, pericholecystic fluid, and a positive sonographic Murphy's sign are hallmark features. Pancreatic imaging is more challenging due to bowel gas but can identify pancreatitis, pseudocysts, and tumors in experienced hands. Renal ultrasound is invaluable for diagnosing hydronephrosis, nephrolithiasis, cystic disease, and chronic kidney changes. It is also used to assess renal artery flow and to guide percutaneous nephrostomy [17, 66]. Evaluation of the spleen for size and focal lesions complements assessments for hematologic or metastatic disease. For the bladder, ultrasound provides a non-invasive measurement of post-void residual volume and identification of bladder wall irregularities or masses.

Lung and Pleura

Lung ultrasound is increasingly used in both emergency and critical care settings due to its rapid, bedside application and diagnostic accuracy. Though lungs are air-filled and traditionally thought unsuitable for ultrasound, pathologic changes such as fluid accumulation or loss of aeration allow excellent visualization. Pneumothorax is diagnosed by the absence of lung sliding and the presence of the "barcode sign" on M-mode. Pleural effusions appear as anechoic or hypoechoic collections, often with floating lung tissue or fibrin strands. Thoracentesis can be safely performed under ultrasound guidance, reducing complication rates.

In pulmonary edema, lung ultrasound identifies multiple B-lines—comet-tail artifacts caused by fluid-filled interlobular septa. These findings, along with pleural irregularities, help differentiate cardiogenic from non-cardiogenic causes. In pneumonia, consolidated lung tissue appears hypoechoic with visible air bronchograms, and ultrasound can assist in monitoring treatment response [67]. Lung ultrasound is also effective in diagnosing atelectasis, pleural thickening, and guiding chest tube placement. Its portability and high sensitivity make it an indispensable tool in respiratory care.

Musculoskeletal System

Musculoskeletal ultrasound offers high-resolution imaging of soft tissues, tendons, ligaments, and joints. It is commonly used in sports medicine and rheumatology to diagnose tendon tears, bursitis, and joint effusions. Dynamic imaging allows for real-time assessment of tendon gliding, impingement, or subluxation. Rotator cuff pathology, Achilles tendinopathy, and lateral epicondylitis (tennis elbow) are well visualized with ultrasound [68].

Ultrasound is beneficial for guiding joint injections, aspirations, and tendon sheath injections, increasing procedural accuracy and reducing patient discomfort. In rheumatologic diseases like rheumatoid arthritis, ultrasound detects early synovitis, effusions, and erosions that may not be visible on X-ray. It is also employed in evaluating muscle injuries, hematomas, foreign bodies, and soft tissue masses. Its portability and ability to provide immediate feedback during physical examination enhance clinical decision-making and patient care.

Thyroid and Neck

Ultrasound is the imaging modality of choice for evaluating the thyroid gland and surrounding neck structures. It provides detailed information about the size, echotexture, and vascularity of thyroid nodules. Suspicious features such as microcalcifications, irregular margins, and increased intranodular flow can suggest malignancy and guide the need for biopsy. Ultrasound is also used for surveillance of known benign nodules and postoperative monitoring for thyroid cancer recurrence [52].

In addition to the thyroid, ultrasound evaluates cervical lymph nodes, salivary glands, and parathyroid glands. It helps localize parathyroid adenomas in patients with primary hyperparathyroidism, guiding surgical excision. Ultrasound-guided fine-needle aspiration (FNA) increases diagnostic yield and safety during biopsy of suspicious neck masses. In head and neck oncology, ultrasound plays a key role in staging, surveillance, and treatment planning.

Genitourinary System

Renal and bladder ultrasound is widely used to assess urinary tract obstruction, hydronephrosis, nephrolithiasis, and bladder wall abnormalities. It is the first-line tool for evaluating hematuria and can detect renal masses, cortical thinning, and congenital anomalies. Doppler ultrasound of renal arteries aids in diagnosing renovascular hypertension and post-transplant complications.

Transabdominal and transvaginal ultrasound are essential in gynecology for evaluating the uterus, ovaries, and adnexa. They detect uterine fibroids, endometrial abnormalities, ovarian cysts, and ectopic pregnancies. Endometrial thickness measurements are crucial in evaluating abnormal uterine bleeding, especially in postmenopausal women [69]. In male patients, testicular ultrasound helps diagnose torsion, epididymitis, varicoceles, and testicular tumors. Transrectal ultrasound (TRUS) is used in urology to guide prostate biopsies and assess prostate size in cases of benign prostatic hyperplasia.

References

1. Wang S, Hossack JA, Klibanov AL (2020) From anatomy to functional and molecular biomarker imaging and therapy: ultrasound is safe, ultrafast, portable, and inexpensive. Invest Radiol 55(9):559–572
2. Wells PN, Liang HD, Young TP (2011) Ultrasonic imaging technologies in perspective. J Med Eng Technol 35(6–7):289–299
3. Rix A, Lederle W, Theek B, Lammers T, Moonen C, Schmitz G, Kiessling F (2018) Advanced ultrasound technologies for diagnosis and therapy. J Nucl Med 59(5):740–746
4. Peruničić Ž, Lalatović I, Spahić L, Ašić A, Pokvić LG, Badnjević A (2024) Enhancing mechanical ventilator reliability through machine learning based predictive maintenance. Technol Health Care
5. Cohen M (2005) Recherche technologique et applications futures [Technical R & D and future applications]. J Radiol 86(12 Pt 2):1934–1936
6. Pokvić LG, Deumić A, Softić A, Badnjević A (2024) Application of artificial intelligence for management of maintenance and prediction of performances. In: Badnjević A, Cifrek M, Magjarević R, Džemić Z (eds) Inspection of Medical Devices. Springer, Cham, pp 67–81
7. Younas U, Muhammad J, Ali Q, Sediqmal M, Kedzia K, Jan AZ (2024) On the study of solitary wave dynamics and interaction phenomena in the ultrasound imaging modelled by the fractional nonlinear system. Sci Rep 14(1):26080
8. Qin YX (2025) Ultrasound imaging and regulated mechanotransduction for characteristics, regeneration, and therapeutics of bone. Mechanobiol Med. 3(1):100116
9. Mujkić A, Baralić E, Ombašić A, Bećirović LS, Pokvić LG, Badnjević A (2022) Machine intelligence in biomedical data modeling, processing, and analysis. Conf Proc IEEE MECO. 11:1–10
10. Adler RS (2023) Musculoskeletal ultrasound: a technical and historical perspective. J Ultrasonogr. 23(95):e172–e187
11. Harvey CJ, Pilcher JM, Eckersley RJ, Blomley MJ, Cosgrove DO (2002) Advances in ultrasound. Clin Radiol 57(3):157–177
12. Badnjević A, Pokvić LG, Džemić Z, Bečić F (2020) Risks of emergency use authorizations for medical products during outbreak situations: a COVID-19 case study. Biomed Eng Online 19(1):75
13. Martinović M, Kosović M, Spahić L, Softić A, Pokvić LG, Badnjević A (2025) Machine learning for improved medical device management: a focus on dialysis machines. Technol Health Care
14. Stidham RW, Xu J, Johnson LA, Kim K, Moons DS, McKenna BJ, Rubin JM, Higgins PD (2011) Ultrasound elasticity imaging for detecting intestinal fibrosis and inflammation in rats and humans with Crohn's disease. Gastroenterology 141(3):819-826.e1
15. Badnjević A, Pokvić LG, Smajlhodžić-Deljo M, Bećirović LS, Imamović E, Deumić A (2024) Application of artificial intelligence for the classification of the clinical outcome and therapy in patients with viral infections: the case of COVID-19. Technol Health Care 32(3):1859–1870
16. Mantziaras G, Luvoni GC (2020) Advanced ultrasound techniques in small animal reproduction imaging. Reprod Domest Anim 55(Suppl 2):17–25
17. Badnjević A (2023) Evidence-based maintenance of medical devices: current shortage and pathway towards solution. Technol Health Care 31(1):293–305
18. Hoskins PR, Kenwright DA (2015) Recent developments in vascular ultrasound technology. Ultrasound 23(3):158–165
19. Feleppa EJ, Alam SK, Deng CX (2002) Emerging ultrasound technologies for early markers of disease. Dis Markers 18(5–6):249–268
20. Spahić L, Sredović U, Kurpejović Z, Mrdanović E, Pokvić G, Badnjević A (2025) Machine learning for improved medical device management: a focus on infant incubators. Technol Health Care

21. Spahić L, Jeremić L, Lalatović I, Muratović T, Džuho A, Gurbeta Pokvić L, Badnjević A (2024) Machine learning for improved medical device management: a focus on defibrillator performance. Technol Health Care
22. Lee W, Roh Y (2017) Ultrasonic transducers for medical diagnostic imaging. Biomed Eng Lett 7(2):91–97
23. Maulik D, Heitmann E, Maulik D (2010) New horizons in Doppler ultrasound technology: relevance for obstetrical applications. Clin Obstet Gynecol 53(4):831–843
24. Medjedovic E, Stanojevic M, Jonuzovic-Prosic S, Ribic E, Begic Z, Cerovac A, Badnjevic A (2024) Artificial intelligence as a new answer to old challenges in maternal-fetal medicine and obstetrics. Technol Health Care 32(3):1273–1287
25. Nagueh SF, Quiñones MA (2014) Important advances in technology: echocardiography. Methodist Debakey Cardiovasc J 10(3):146–151
26. Gennisson JL, Deffieux T, Fink M, Tanter M (2013) Ultrasound elastography: principles and techniques. Diagn Interv Imaging 94(5):487–495
27. Hafizović L, Čaušević A, Deumić A, Bećirović LS, Pokvić LG, Badnjević A (2021) The use of artificial intelligence in diagnostic medical imaging: systematic literature review. Conf Proc IEEE BIBE. 21:1–6
28. Powers J, Kremkau F (2011) Medical ultrasound systems. Interface Focus. 1(4):477–489
29. Seo J, Kim YS (2017) Ultrasound imaging and beyond: recent advances in medical ultrasound. Biomed Eng Lett 7(2):57–58
30. Badnjević A, Pokvić LG, Deumić A, Bećirović LS (2022) Post-market surveillance of medical devices: a review. Technol Health Care 30(6):1315–1329
31. Bećirović F, Spahić L, Merdović N, Gurbeta Pokvić L, Badnjević A (2025) Advancement of post-market surveillance of medical devices leveraging artificial intelligence: patient monitors case study. Technol Health Care 33(2):974–980
32. Hasegawa H (2021) Advances in ultrasonography: image formation and quality assessment. J Med Ultrason 48(4):377–389
33. Chen Z, Basarab A, Kouamé D (2016) Compressive deconvolution in medical ultrasound imaging. IEEE Trans Med Imaging 35(3):728–737
34. Chen Y, Liu J, Luo J, Luo X (2021) A data-driven approach for high frame rate synthetic transmit aperture ultrasound imaging. In: Lect Notes Comput Sci. 412–420
35. Long J, Trahey G, Bottenus N (2022) Spatial coherence in medical ultrasound: a review. Ultrasound Med Biol 48(6):975–996
36. Elango T, Kalalii S, Rajalakshmi P (2024) A novel approach to ultrasound beamforming using synthetic transmit aperture with low complexity and high SNR for medical imaging. arXiv preprint arXiv:2401.12345
37. Demi L (2018) Practical guide to ultrasound beam forming: beam pattern and image reconstruction analysis. Appl Sci 8(9):1544
38. Luijten B, Chennakeshava N, Eldar YC, Mischi M, van Sloun RJG (2023) Ultrasound signal processing: from models to deep learning. Ultrasound Med Biol 49(3):677–698
39. Göbl R (2022) Receive beamforming in medical ultrasound: a review of aperture data processing. La Trobe University
40. Vos HJ, van Neer PL, Mota MM, Verweij MD, van der Steen AF, Volker AW (2016) F-k domain imaging for synthetic aperture sequential beamforming. IEEE Trans Ultrason Ferroelectr Freq Control 63(1):60–71
41. Mirzaei M, Asif A, Rivaz H (2021) Virtual source synthetic aperture for accurate lateral displacement estimation in ultrasound elastography. IEEE Trans Ultrason Ferroelectr Freq Control 68(5):1687–1695
42. Synnevag JF, Austeng A, Holm S (2009) Benefits of minimum-variance beamforming in medical ultrasound imaging. IEEE Trans Ultrason Ferroelectr Freq Control 56(9):1868–1879
43. Tang MX, Mulvana H, Gauthier T, Lim AK, Cosgrove DO, Eckersley RJ, Stride E (2011) Quantitative contrast-enhanced ultrasound imaging: a review of sources of variability. Interface Focus. 1(4):520–539

44. Peralta L, Ramalli A, Reinwald M, Eckersley RJ, Hajnal JV (2020) Impact of aperture, depth, and acoustic clutter on the performance of coherent multi-transducer ultrasound imaging. Appl Sci 10(21):7655
45. Hundur M, Spahić L, Bećirović F, Gurbeta Pokvić L, Badnjević A (2025) Advancement of post-market surveillance of medical devices leveraging artificial intelligence: ECG devices case study. Technol Health Care
46. Kang H, Park C, Yang H (2024) Evaluation of denoising performance of ResNet for ultrasound images. Bioeng (Basel)
47. Lee H, Lee MH, Youn S, Lee K, Lew HM, Hwang JY (2022) Speckle reduction via deep content-aware image prior. IEEE Trans Ultrason Ferroelectr Freq Control
48. Yi J, Kang HK, Kwon JH, Kim KS, Park MH, Seong YK, et al (2021) Technology trends and applications of deep learning in ultrasonography. Ultrasonography
49. Monkam P, Lu W, Jin S, Shan W, Wu J, Zhou X, et al (2023) US-Net: a lightweight network for simultaneous speckle suppression and texture enhancement. Comput Biol Med
50. Zhang B, Li Z, Hao Y, Wang L, Li X, Yao Y (2025) A review of lightweight convolutional neural networks for ultrasound signal classification. Front Physiol
51. Xu L, Huang Y, Zhou H, Mao Q, Yin W (2025) A semi-supervised ultrasound image segmentation network integrating enhanced mask learning. IEEE J Biomed Health Inform
52. Li X, Wang Y, Zhao Y, Wei Y (2022) Fast speckle noise suppression algorithm in breast ultrasound image using 3D deep learning. Front Physiol
53. Xiao X, Zhang J, Shao Y, Liu J, Shi K, He C, Kong D (2025) Deep learning-based medical ultrasound image and video segmentation methods. Sensors (Basel)
54. Zhang L, Zhang J, Li Z, Song Y (2020) A multiple-channel and atrous convolution network for ultrasound image segmentation. Med Phys
55. Singh P, Mukundan R, De Ryke R (2020) Feature enhancement in medical ultrasound videos using contrast-limited adaptive histogram equalization. J Digit Imaging
56. Cai L, Li Q, Zhang J, Zhang Z, Yang R, Zhang L (2023) Ultrasound image segmentation based on Transformer and U-Net with joint loss. PeerJ Comput Sci
57. Zhang L, Zhang J (2022) Ultrasound image denoising using GANs with residual dense connectivity and weighted joint loss. PeerJ Comput Sci
58. Cammarasana S, Nicolardi P, Patanè G (2022) Real-time denoising of ultrasound images based on deep learning. Med Biol Eng Comput
59. Zhang C, Wang L, Wei G, Kong Z, Qiu M (2024) A dual-branch and dual attention transformer and CNN hybrid network for ultrasound image segmentation. Front Physiol
60. Villemain O, Baranger J, Friedberg MK, Papadacci C, Dizeux A, Messas E et al (2020) Ultrafast ultrasound imaging in pediatric and adult cardiology: techniques, applications, and perspectives. JACC Cardiovasc Imaging 13(7):1625–1639
61. Merdović N, Spahić L, Hundur M, Pokvić LG, Badnjević A (2024) Advancement of post-market surveillance of medical devices leveraging artificial intelligence: infusion pumps case study. Technol Health Care 33(2):915–921
62. Kühle H, Cho SKS, Barber N, Goolaub DS, Darby JRT, Morrison JL, et al (2023) Advanced imaging of fetal cardiac function. Front Cardiovasc Med
63. Girard M, Deschamps J, Razzaq S, Lavoie N, Denault A, Beaubien-Souligny W (2023) Emerging applications of extracardiac ultrasound in critically ill cardiac patients. Can J Cardiol
64. Rafailidis V, Sidhu PS (2018) Vascular ultrasound: the potential of integration of multiparametric ultrasound into routine clinical practice. Ultrasound
65. Kenny JES (2022) Assessing fluid intolerance with Doppler ultrasonography: a physiological framework. Med Sci (Basel)
66. Dave JK, McDonald ME, Mehrotra P, Kohut AR, Eisenbrey JR, Forsberg F (2018) Recent technological advancements in cardiac ultrasound imaging. Ultrasonics
67. Davis LM, Martinez-Correa S, Freeman CW, Adams C, Sultan LR, Le DQ, et al (2024) Ultrasound innovations in abdominal radiology: techniques and clinical applications in pediatric imaging. Abdom Radiol (NY)

68. Lux H, Edling M, Siemroth P, Schrader S (2018) Ultrasound: medical imaging and beyond. Materials 11(5):812
69. Shen YT, Chen L, Yue WW, Xu HX (2021) Artificial intelligence in ultrasound. Eur J Radiol 142:109862

Chapter 7
Blood Pressure Signal Acquisition and Analysis

This section focuses on blood pressure measurement, analysis, and clinical significance. It begins with the physiological basis of blood pressure and its relation to the cardiac cycle (Chap. 7.1), followed by an overview of measurement techniques, including direct and indirect methods. The technical aspects of sensor design, validation standards, and accuracy requirements are addressed, alongside the challenges of continuous and noninvasive monitoring. Future directions emphasize the integration of advanced technologies for real-time assessment and improved predictive value. The section underscores blood pressure as a critical biomedical signal with far-reaching implications for diagnosis, risk stratification, and healthcare monitoring.

7.1 Blood Pressure Signal Analysis

Blood pressure measurements have been part of the basic clinical examination since the earliest days of modern medicine. The origin of blood pressure is the pumping action of the heart, and its value depends on the relationship between cardiac output and peripheral resistance. Therefore, blood pressure is considered one of the most critical physiological variables with which to assess cardiovascular hemodynamics. Venous blood pressure is determined by vascular tone, blood volume, cardiac output, and the force of contraction of the chambers of the right side of the heart. Since venous blood pressure must be obtained invasively, the term blood pressure most commonly refers to arterial blood pressure, which is the pressure exerted on the arterial walls when blood flows through the arteries [1, 2]. The highest value of pressure, which occurs when the heart contracts and ejects blood to the arteries, is called the systolic pressure (SP). The diastolic pressure (DP) represents the lowest value occurring between the ejections of blood from the heart. Pulse pressure (PP) is the difference between SP and DP, i.e., PP = SP – DP. The period from the end of one heart contraction to the end of the next is called the cardiac cycle. Mean pressure (MP) is

© The Author(s), under exclusive license to Springer Nature Switzerland AG 2026

A. Badnjević and L. Spahić, *Biomedical Signals and Systems*, Series in BioEngineering,
https://doi.org/10.1007/978-3-032-10982-8_7

the average pressure during a cardiac cycle. Mathematically, MP can be decided by integrating the blood pressure over time. When only SP and DP are available, MP is often estimated by an empirical formula (Eq. 7.1):

$$MP = DP + \frac{PP}{3} \tag{7.1}$$

Note that this formula can be very inaccurate in some extreme situations. Although SP and DP are most often measured in the clinical setting, MP has particular importance in some situations because it is the driving force of peripheral perfusion. SP and DP can vary significantly throughout the arterial system, whereas MP is almost uniform in everyday situations.

The values of blood pressure vary significantly during 24 h according to an individual's activity. Three factors, namely, the diameter of the arteries, the cardiac output, and the state or quantity of blood, are mainly responsible for the blood pressure level. When the tone increases in the muscular arterial walls so that they narrow or become less compliant, the pressure becomes higher than usual [2, 3]. Unfortunately, increased blood pressure does not ensure proper tissue perfusion. In some instances, such as certain types of shock, blood pressure may seem appropriate when peripheral tissue perfusion has all but stopped. Nevertheless, observation or monitoring of blood pressures affords dynamic tracking of pathology and physiology affecting the cardiovascular system. This system, in turn, has profound effects on the other organs of the body.

7.1.1 Mean Arterial Pressure and the Cardiac Cycle

Mean arterial pressure (MAP) is the average pressure in the arteries during a single cardiac cycle. While systolic and diastolic pressures fluctuate with each heartbeat, MAP offers a more stable representation of tissue perfusion. It is especially crucial in critical care and anesthesia, where it is used as a target variable to ensure adequate oxygen delivery to vital organs.

The MAP is not simply the arithmetic mean of systolic and diastolic pressures, due to the nonlinear nature of the cardiac cycle. Diastole occupies a longer duration than systole, especially at resting heart rates. Hence, the commonly used empirical formula to estimate MAP is (Eq. 7.2):

$$MAP = DP + \frac{1}{3}(SP - DP) \tag{7.2}$$

This estimation, while convenient, loses accuracy under certain conditions such as tachycardia, bradycardia, or abnormal vascular compliance. For precise assessment, integration of the pressure waveform over time is used, particularly in invasive settings.

The cardiac cycle itself includes multiple phases: atrial systole, isovolumetric contraction, ventricular ejection, isovolumetric relaxation, and ventricular filling. Each phase contributes to the pressure dynamics within the heart and the arterial system. The cyclical nature of blood pressure variations can be captured through continuous monitoring, providing insights into cardiovascular performance and compensatory mechanisms in disease states [2, 4].

7.1.2 Measurement Techniques

The basis of any physiological measurement is the biological signal, which is first sensed and transduced or converted from one form of energy to another. The signal is then conditioned, processed, and amplified. Subsequently, it is displayed, recorded, or transmitted (in some ambulatory monitoring situations). Blood pressure sensors often detect mechanical signals, such as blood pressure waves, to convert them into electric signals for further processing or transmission. They work on a variety of principles, for example, resistance, inductance, and capacitance. For accurate and reliable measurements, a sensor should have good sensitivity, linearity, and stability.

Indirect Blood Pressure Measurement

Indirect measurement is often called noninvasive measurement because the body is not entered into the process. The upper arm, containing the brachial artery, is the most common site for indirect measurement because of its closeness to the heart and convenience of measurement. However, many other sites may have been used, such as the forearm or radial artery, finger, etc. Distal sites such as the wrist, although convenient to use, may give much higher systolic pressure than brachial or central sites as a result of the phenomena of impedance mismatch and reflective waves. An occlusive cuff is usually placed over the upper arm and is inflated to a pressure greater than the systolic blood pressure. The cuff is then gradually deflated, while a detector system simultaneously employed determines the point at which the blood flow is restored to the limb. The detector system does not need to be a sophisticated electronic device. It may be as simple as manual palpation of the radial pulse. The most commonly used indirect methods are auscultation and oscillometry (Fig. 7.1).

Auscultatory Method

The auscultatory method most commonly employs a mercury column, an occlusive cuff, and a stethoscope. The stethoscope is placed over the blood vessel for auscultation of the Korotkoff sounds, which define both SP and DP. The Korotkoff sounds are mainly generated by the pulse wave propagating through the brachial artery. The Korotkoff sounds consist of five distinct phases. The onset of Phase I Korotkoff sounds (first appearance of clear, repetitive, tapping sounds) signifies SP, and the onset of Phase V Korotkoff sounds (sounds disappear completely) often defines DP.

Observers may differ significantly in their interpretation of the Korotkoff sounds. A simple mechanical error can occur in the form of air leaks or obstruction in the cuff,

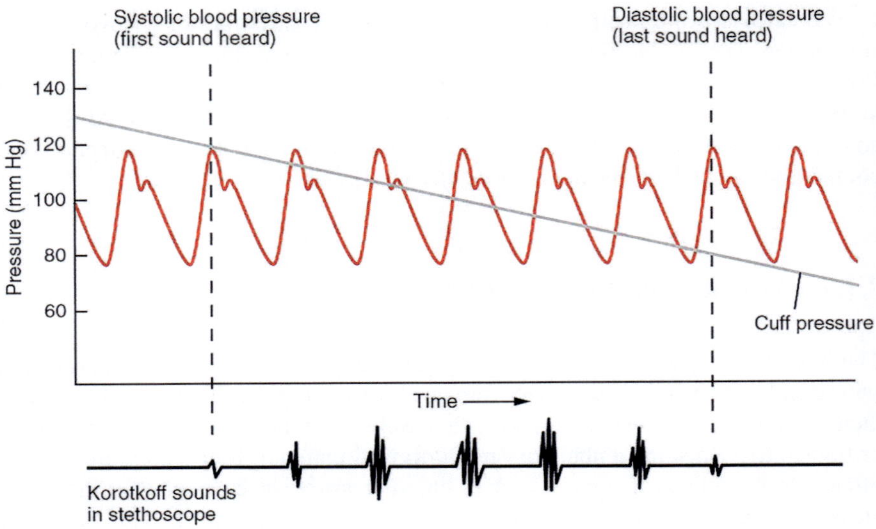

Fig. 7.1 Blood pressure measurement (https://commons.wikimedia.org/wiki/File:2111_Blood_Pressure_Graph.jpg)

coupling tubing, or Bourdon gage. Mercury can leak from a column gauge system. Despite the errors inherent in such simple systems, more mechanically complex systems have come into use. The impetus for the development of more elaborate detectors has come from the advantage of reproducibility from observer to observer and the convenience of automated operation [5, 6]. Examples of this improved instrumentation include sensors using plethysmographic principles, pulse-wave velocity sensors, and audible as well as ultrasonic microphones.

The readings by auscultation do not always correspond to those of intra-arterial pressure. The differences are more pronounced in certain special occasions such as obesity, pregnancy, arteriosclerosis, shock, etc. Experience with the auscultation method has also shown that determination of DP is often more difficult and less reliable than SP. However, the situation is different for the oscillometric method, where oscillations caused by the pressure pulse amplitude are interpreted for SP and DP according to empirical rules.

Oscillometric

Method In recent years, electronic pressure and pulse monitors based on oscillometry have become popular for their simplicity of use and reliability. The principle of blood pressure measurement using the oscillometric technique is dependent on the transmission of intra-arterial pulsation to the occluding cuff surrounding the limb. An approach using this technique could start with a cuff placed around the upper arm and rapidly inflated to about 30 mmHg above the systolic blood pressure, occluding blood flow in the brachial artery. A sensor measures the pressure in the cuff. The pressure is then gradually decreased, often in steps, such as 5 to 8 mmHg. The oscillometric signal is detected and processed at each step of pressure. The cuff pressure can also be deflated linearly in a similar fashion to the conventional auscultatory method.

Arterial pressure oscillations are superimposed on the cuff pressure when the blood vessel is no longer fully occluded. Separation of the superimposed oscillations from the cuff pressure is accomplished by filters that extract the corresponding signals. Signal sampling is carried out at a rate determined by the pulse or heart rate. The oscillation amplitudes are most often used with an empirical algorithm to estimate SP and DP [7]. Unlike the Korotkoff sounds, the pressure oscillations are detectable throughout the whole measurement, even at cuff pressures higher than SP or lower than DP. Since many oscillometric devices use empirically fixed algorithms, the variance of measurement can be large across a wide range of blood pressures. Significantly, however, MP is determined by the lowest cuff pressure of maximum oscillations and has been strongly supported by many clinical validations.

Self-Measurement

From the growing number of publications on the topic in recent years, it is evident that the interest in self-measurement of blood pressure has increased dramatically. There is also evidence that the management of patients with high blood pressure can be improved if clinic measurements are supplemented by home or ambulatory monitoring. Research has shown that blood pressure readings taken in the clinic can be elevated by as much as 75 mmHg in SP and 40 mmHg in DP when taken by a physician. The tendency for blood pressure to increase in specific individuals in the presence of a physician due to stress response is generally known as "white-coat" hypertension. When reasonably priced and easy to use, oscillometric devices became commonly available in the early 1970s, public interest in the self-measurement of blood pressure increased, and this has made it possible for greater patient involvement in the detection and management of hypertension [8, 9]. Health care costs may also be reduced by home monitoring. Indeed, a recent study found that costs were almost 30% lower for patients who measured their blood pressure than for those who did not. Measurements taken at the patient's home are more highly correlated to 24-h blood pressure levels than clinic readings are. It has also been shown that most patients can monitor their blood pressure and may be more relaxed as well as assured by doing so, particularly when experiencing symptoms.

Ambulatory Monitoring

There is great significance for ambulatory monitoring of blood pressure. Throughout 24 h, blood pressure is subject to numerous situational and periodic fluctuations. The pressure readings have a pronounced diurnal rhythm in an individual, with a decrease of 10–20 mmHg during sleep and a prompt increase on getting up and walking in the morning. Readings tend to be higher during working hours and lower at home, and they depend on the pattern of activity. After a bout of vigorous exercise or strenuous work, blood pressure may be reduced for several hours. The readings may be raised if the patient is talking during the measurement period. Smoking a cigarette and drinking coffee, especially if they are combined, may both raise the pressure. When assessing the efficacy of antihypertensive drugs, ambulatory blood pressure monitoring can provide considerable information and validation of the drug treatment.

Although the technique of noninvasive ambulatory blood pressure monitoring was first described more than three decades ago, it has only recently become accepted as a clinically useful procedure for evaluation of patients with abnormal regulation of blood pressure. It gives the best evaluation for patients who have white-coat hypertension. Technical advances in microelectronics and computer technology have led to the introduction of ambulatory monitors with improved accuracy and reliability, small size, quiet operation, and reasonably low price [10]. They can take and store several hundred readings throughout 24 h while patients may not be compromised with their everyday activities, thus becoming usable for purposes of clinical diagnosis. Theoretically, ambulatory monitoring can provide information about the level and variability covering the full range of blood pressure experienced during day-to-day activities. It is now recognized to be a beneficial procedure in clinical practice since blood pressure varies significantly during 24 h, especially useful in detecting white-coat hypertension. However, many studies have found that the accuracy of monitoring using current ambulatory monitors is acceptable only when patients are at rest, but not during physical activity or under truly ambulatory conditions. The report of error codes during operation in the latter situations is much higher [11].

Cuff Size

Both the length and width of an occluding cuff are important for accurate and reliable measurement of blood pressure by indirect methods. A too-short or too-narrow cuff results in false high blood pressure readings. Several studies have shown that a cuff of inappropriate size about the patient's arm circumference can cause considerable error in blood pressure measurement. The cuff should also fit around the arm firmly and comfortably. Some manufacturers have designed cuffs with a fastener spaced so that a cuff of appropriate width only fits an arm of appropriate diameter. With this design, the cuff will not stay on the arm during inflation unless it fits accordingly.

7.1.3 Direct Blood Pressure Measurement

Direct measurement is also called invasive measurement because bodily entry is made. For direct arterial blood pressure measurement, an artery is cannulated. The equipment and procedure require proper setup, calibration, operation, and maintenance. Such a system yields blood pressures dependent upon the location of the catheter tip in the vascular system. It is beneficial for the continuous determination of pressure changes at any instant in dynamic circumstances. When massive blood loss is anticipated, powerful cardiovascular medications are suddenly administered, or a patient is induced to general anesthesia, continuous monitoring of blood pressures becomes vital.

Most commonly used sites to make continuous observations are the brachial and radial arteries. The femoral or other sites may be used as points of entry to sample pressures at different locations inside the arterial tree, or even the left ventricle of the heart. Entry through the venous side of the circulation allows checks of pressures in the central veins close to the heart, the right atrium, the right ventricle, and the pulmonary artery [2, 12]. A catheter with a balloon tip carried by blood flow into smaller branches of the pulmonary artery can occlude flow in the artery from the right ventricle so that the tip of the catheter reads the pressure of the left atrium, just downstream. These procedures are very complex, and there is always a concern about the risk of harm as opposed to the benefit.

Invasive access to a systemic artery involves considerable handling of a patient. The longer a catheter stays in a vessel, the more likely an associated thrombus will form. The Allen's test can be performed by pressing on one of the two main arteries at the wrist when the fist is clenched, then opening the hand to see if blanching indicates inadequate perfusion by the other artery. However, it has proved an equivocal predictor of possible ischemia. In the newborn, when the arterial catheter is inserted through an umbilical artery, there is a particular hazard of infection and thrombosis, since thrombosis from the catheter tip in the aorta can occlude the arterial supply to vital abdominal organs. Some of the recognized contraindications and complications include poor collateral flow, severe hemorrhage diathesis, occlusive arterial disease, arterial spasm, and hematoma formation.

Despite well-studied potential problems, direct blood pressure measurement is generally accepted as the gold standard of arterial pressure recording. It presents the only satisfactory alternative when conventional cuff techniques are not successful. This also confers the benefit of continuous access to the artery for monitoring gas tension and blood sampling for biochemical tests. It also has the advantage of assessing cyclic variations and beat-to-beat changes of pressure continuously, and permits assessment of short-term variations.

Catheter-Tubing-Sensor System

A large variety of vascular catheters exists. Catheter materials have undergone testing to ensure that they have a minimal tendency to form blood clots on their surface. The catheter chosen may be inserted percutaneously over a hollow stylet into the blood vessel. Guide wires can be helpful to facilitate longer or larger-diameter catheters into vessels, after the guide wires have been placed through a smaller catheter or needle. Less often, entry to a vessel requires a "cutdown," a direct exposure of the vessel after a skin incision. Ultrasonic devices may assist in locating the vessels that are not readily apparent at the skin surface.

Although pressure sensors can be located at the catheter tip, this presents a problem for calibration if left in place and a clot forms near the tip of the catheter, damping the pressure signal. Instead, most catheters connect to an external pressure sensor via fluid-filled low-compliance tubing. The signal from the sensor then transforms into a display or recording. The sensor may take one of several forms, from a variable resistance diaphragm to a silicon microchip. A basic system can consist of an intravascular catheter connected to a rigid fluid-filled catheter and tubing that communicates the pressure to an elastic diaphragm, the deflection of which is detected electrically [13]. There is a direct relationship between the deflection of the diaphragm and the voltage. The higher the voltage, the greater the pressure. Continuous low-rate infusion of heparinized saline is carried out to keep the catheter patent or free from coagulation. The advent of disposable sensor kits has dramatically simplified the clinical use of intravascular monitoring. With the cost continually being lowered with the development of the semiconductor industry, disposable sensors become more and more cost-effective.

Although direct recording is considered the most accurate method, its accuracy may be limited by variations in the kinetic energy of the fluid in the catheter or the dynamic frequency response of the measurement system. The hydraulic link between the patient and the sensor is the primary source of potential errors and hazards for the monitoring. Damping and degrading the system's natural frequency, caused by trapped air bubbles, small catheters, various narrow connections, compliant and too long tubing, and too many components connected, are the two characteristic problems with a pressure sensor system. Extreme care should be exercised to eliminate all air bubbles from the fluid to provide adequate dynamic response. The sensor should be zeroed at the level of the heart to eliminate hydrostatic error. A fast flush test is easy to use for inspecting the dynamic response of the entire catheter, tubing, and sensor system. It can also help direct adjustments for the system to minimize dynamic artifacts.

7.1.4 Recommendations, Standards, and Validation Requirements

The AHA recommendations provide a systematic step-by-step procedure for measuring blood pressure, including equipment, observer, subject, and technique. It extends considerations of blood pressure recording in special populations such as infants and children, the elderly, pregnant, and obese subjects. It also provides recommendations for self-measurement or home measurement of blood pressure, as well as ambulatory blood pressure measurement.

The Association for the Advancement of Medical Instrumentation (AAMI) and American National Standards Institute (ANSI) published and revised a national standard for evaluating electronic or automated sphygmomanometers. This standard establishes labeling requirements, safety and performance requirements, and reference test methods for electronic or automated sphygmomanometers used in the indirect measurement of blood pressure. Specific requirements for ambulatory blood pressure monitors were also included. Recently, AAMI/ANSI amended this SP10 standard to include neonatal devices as well. Some of the specific requirements, procedures, and limits were modified to fit neonatal applications, such as the maximum cuff pressure, ranges of age and weight, reference standards for validation, minimum sample size of data, etc. The overall system efficacy for both neonatal and adult devices requires that for systolic and diastolic pressures treated separately, the mean difference between the paired measurements of the test system and the reference standard shall be five mmHg or less, with a standard deviation of 8 mmHg or less.

For manual or nonautomated indirect blood pressure measuring devices, the ANSI/AAMI SP9 standard applies. The British Hypertension Society (BHS) also published and revised a protocol for assessing the accuracy and reliability of blood pressure measurement using automatic and semiautomatic devices. Many automatic and semiautomatic devices, including ambulatory devices, have been evaluated according to the BHS protocol. Such evaluation provided a quality-control mechanism for manufacturers and an objective comparison for customers [14]. However, there are many more devices available on the market that have not been evaluated accordingly. Different from the AAMI SP10 standard, in which either indirect or direct blood pressure may be used as a reference standard, the BHS protocol relies exclusively on references of sphygmomanometric blood pressure measurement. It does not recommend comparison with intra-arterial blood pressure values. This could make accurate validation of ambulatory devices difficult because sphygmomanometric measurements during exercise and under ambulatory conditions are not accurate [15].

Manufacturer, Product, Price, Efficacy, and Technology

The annual publication of the Medical Device Register is a comprehensive reference work that provides a wealth of detailed information on U.S. and international medical devices, medical device companies, OEM suppliers, and the key personnel in the industry. Blood pressure devices are listed in the sphygmomanometer directory. Price information of specific models for some providers is also published. For example, A&D Engineering, Inc. listed prices from $51.95 (model UA701) to $179.95 (model UA-751) for a whole line of sphygmomanometers in the 1997 Medical Device Register. Since technology and the market can change rapidly, models, features, specifications, and prices may change accordingly. More specific and updated information may be available by contacting the manufacturers or distributors directly.

Many of the listed blood pressure devices have multiple evaluation studies, and only a few study results are presented here. Given reference standards for comparison, although direct and indirect methods yield similar measurements, they are rarely identical because the direct method measures pressure, and the indirect method is more indicative of flow. They found that the average offsets of all tested devices amounted to –11.7 mmHg for systolic and 1.6 mmHg for diastolic blood pressure, which were close to those of the mercury sphygmomanometer (–14.2 mmHg for systolic and – 0.1 mmHg for diastolic pressure), indicating a significant difference between the two assessment standards. When selecting a blood pressure device for a specific application, the evaluation using the reference that is of a common practice in the intended population or environment may be practically more informative, since that reference has been the everyday basis for decision making in blood pressure diagnosis and treatment [16, 17].

Different evaluation results for the same brand product can also be due to different versions of a model used for validation, where a later version may have performance improvement over the earlier one. Another source of discrepancies can come from utilizing different study protocols or only partially following the same protocols. It is recommended that the original clinical evaluation report be carefully examined in determining the desired efficacy that may meet the users' requirements. If the devices were FDA-approved for marketing in the U.S., one may request a copy of their clinical validation report directly from the manufacturer.

Advancement of Indirect Blood Pressure Measurement

Since the introduction of Dinamap™, an automated blood pressure monitor based on the oscillometric principle, many variants of oscillometric algorithms have been developed [17]. However, the fixed or variable fractions of the maximum oscillations are still the fundamental algorithms of the oscillometry. Typically, mean blood pressure was determined by the lowest cuff pressure with the most significant average oscillation. Systolic and diastolic blood pressure were determined by the cuff pressure, with the amplitude of oscillation being a fixed fraction of the maximum. Performance of the algorithms may be improved by introducing a greater level of complexity or variables into the considerations. The Dinamap™ 1846SX (Critikon, Tampa, FL) oscillometric device offered two measurement modes. The normal mode uses two matching pulse amplitudes at each cuff pressure step to establish an oscillometric envelope or curve. Therefore, measurement time is heart rate dependent. The second mode, which the manufacturer refers to as "stat mode," is capable of faster determination by disabling the dual pulse-matching algorithm, which was designed for artifact rejection [1]. The stat mode does not appear to compromise accuracy in anesthetized patients, in which rapid measurement of blood pressure is often more desirable, particularly during induction and management of anesthesia.

Protocol Systems developed another variant of the oscillometric algorithm. In addition to using pulse amplitude for primary artifact rejection, it further calculates the impulse value, the principal area of the pulse waveform, in constructing an oscillometric curve. This curve is smoothed by employing a Kalman filter that also provides an expected mean and acceptable upper and lower bounds of prediction for the principal area of the subsequent pulse waveform. Smoothing of the oscillometric curve is accomplished by using the difference between the predicted and calculated area data of the pulse waveform for each cuff pressure step. Blood pressures are derived from the final smoothed oscillometric curve.

In more recent study, oscillometric algorithms using an artificial neural network have been reported to produce better estimates of reference blood pressures than the standard oscillometric algorithm [3]. By using neural network training and processing, subtle features and nonlinear relationships of the oscillometric envelope have been modeled. Empiricism of the oscillometric fixed fraction criteria is overcome, and variances of measurements are significantly reduced [11]. Because of its low risk and cost, noninvasive continuous blood pressure monitoring represents another need in critical-care monitoring to supplement invasive arterial catheterization [12, 15]. A significant development in this field is the arterial counterpulsation principle, proposed by Penaz and further developed by two major groups of people. Finapres™, a continuous finger arterial blood pressure monitor, was engineered and developed by Ohmeda, Denver, CO. Many clinical evaluation reports of these devices have been published since then.

Recently, several other continuous blood pressure monitors have been made commercially available. Examples of these are Cortronic APM770, which monitors pulsation of the brachial artery with a slightly pressurized arm cuff and calibrates it to a continuous pressure waveform; Sentinel ARTRAC 7000, which monitors pulse transit time and correlates that to pressure change; Colin CBM-3000 and JENTOW (Colin Electronics, Komaki, Japan) and Nellcor NCAT N-500 (Nellcor, Hayward, CA), which are tonometric devices monitoring the radial artery pulse waveform by a matrix pressure sensor. All of these monitors require a frequent calibration reference. Except for a few favorable reports with the tonometric method and devices, many reports so far have been unfavorable [1, 18]. Nevertheless, noninvasive continuous monitoring represents an important and growing field of biomedical sensor and instrumentation research and development. Continuous monitors, which maintain cuff pressure, must periodically relieve pressure to prevent the risk of venous congestion, edema, swelling, and tissue damage.

Reproducibility, Accuracy, and Reliability Issues and Recommendations for Corrective Measures

For each blood pressure assay technique, there is an issue of reproducibility of measurements under approximately similar conditions. Reproducibility quantifies the internal uncertainty of an individual method and instrument, whereas accuracy quantifies the external uncertainty when compared with a reference. When dealing with blood pressure measurement, it is important to bear in mind that even for standard methods, there is a certain amount of non-repeatable random error. Consequently, taking the average of repeated measurements or multiple readings is always advised before any serious recommendation or management is made.

7.1.5 Blood Pressure Impact, Challenge, and Future

Hypertension is one of the most common and important risk factors for health in industrialized countries. It is the leading cause of death in the U.S. It is treatable by a variety of effective medications. It can cause severe damage to the heart and arteries, leading to cardiac infarction, stroke, or renal failure. Significant sudden changes in blood pressure may also precede a major physiological catastrophe such as cardiac arrest. There is now almost universal acceptance that basic physiological parameters such as blood pressure should constantly be monitored in the clinical setting.

There has been increasing interest in automatic blood pressure monitoring devices in recent years, and some clinicians are now advising patients to record their blood pressure at home over up to 3 months before starting antihypertensive medication. Self-monitoring of blood pressure has become very common with the development of microchip technology and oscillometric monitors. The patients no longer have to learn how to listen for Korotkoff sounds. This has also removed bias and observer errors, allowing more accurate measurement than by conventional techniques using a stethoscope and a mercury sphygmomanometer.

Special populations have unique blood pressure assessment requirements. Newborns require miniaturized equipment. The act of taking a blood pressure in a newborn may stimulate a series of movements, causing motion artifact. The very obese may find it hard to fit correctly with a cuff at the upper arm if the upper arm is too conical rather than cylindrical. In pregnancy, auscultatory and oscillometric methods, although useful to follow trends, may correspond poorly with central pressures, and even the proper Korotkoff sound (IV or V) to designate as diastolic pressure is uncertain.

Observing blood pressure has limitations. It may suggest what is happening with blood volume, but sometimes does not reveal that blood volume has become inadequate until circulatory collapse has occurred. Venous and left atrial pressures are often used in an attempt to clarify blood volume problems, but with uncertain results. Similarly, a satisfactory blood pressure does not always indicate adequate tissue perfusion. Some medications that increase blood pressure can do so at the expense of general perfusion [18]. Since blood pressure is measured at specific sites in the arterial tree, if circulation has become nonhomogeneous, such as can happen in arteriosclerosis, the region distal to the arteriosclerosis can be compromised without warning from blood pressure readings sampled at another site. Even mean blood pressure, so useful otherwise, can fail in these circumstances.

Despite inherent problems, observation of blood pressure through both old and new technologies retains more than enough usefulness to have remained an essential aspect of patient care. The promise of improved technology to solve problems such as those of motion artifacts, noninvasive continuous monitoring, long-distance telemetry, rapid analysis of accumulated or concurrent data, and assessment of new inaccessible regions of blood flow represents continued challenges for future biomedical research and development.

Recently, exciting research has revealed that comparing pressures taken at the arm and the ankle results in a handy but straightforward index for assessment of lower extremity vascular disease, with implications for general cardiovascular risk factors. The possibility of obtaining noninvasive blood pressure readings from arteries in the forehead by stick-on oscillometric patches has also been proven. At least in anesthetized patients, the forehead noninvasive blood pressure corresponded reasonably well with central arterial pressures. Finger blood pressure monitors have found some applications in continuous ambulatory and sleep blood pressure assessments. A technology that is capable of continuously monitoring brachial or even central blood pressure continues to be a clinical demand and a future challenge.

References

1. Panula T, Sirkia JP, Wong D, Kaisti M (2022) Advances in non-invasive blood pressure measurement techniques. IEEE Rev Biomed Eng 15:230–243

2. Badnjević A (2023) Evidence-based maintenance of medical devices: current shortage and pathway towards solution. Technol Health Care 31(1):293–305
3. Li S, Zhang C, Xu Z, Liang L, Tian Y, Li L et al (2022) Cuffless blood pressure monitoring: academic insights and perspectives analysis. Micromachines. 13(8):1327
4. Carey RM, Wright JT, Taler SJ, Whelton PK (2021) Guideline-driven management of hypertension: an evidence-based update. Circ Res 128(7):827–846
5. Bećirović F, Spahić L, Merdović N, Gurbeta Pokvić L, Badnjević A (2025) Advancement of post-market surveillance of medical devices leveraging artificial intelligence: patient monitors case study. Technol Health Care 33(2):974–980
6. Giorgini P, Weder AB, Jackson EA, Brook RD (2014) A review of blood pressure measurement protocols among hypertension trials: implications for evidence-based clinical practice. J Am Soc Hypertens 8(12):870–875
7. Hodgkinson JA, Sheppard JP, Heneghan C, Martin U, Mant J, Roberts N, McManus RJ (2013) Accuracy of ambulatory blood pressure monitors: a systematic review of validation studies. J Hypertens 31(2):239–250
8. Hundur M, Spahić L, Bećirović F, Gurbeta Pokvić L, Badnjević A (2025) Advancement of post-market surveillance of medical devices leveraging artificial intelligence: ECG devices case study. Technol Health Care
9. Ayan M, Kadavath S, Campbell PT (2018) The role of clinic blood pressure for the diagnosis of hypertension. Curr Opin Cardiol 33(4):372–377
10. Martinović M, Kosović M, Spahić L, Softić A, Pokvić LG, Badnjević A (2025) Machine learning for improved medical device management: a focus on dialysis machines. Technol Health Care. Epub ahead of print
11. Papaioannou TG, Vavuranakis M, Stefanadis C, Tousoulis D (2015) New technological and clinical trends in blood pressure theranostics: is it time to consider a spatio-temporal approach? Curr Pharm Des 21(6):723–732
12. Stergiou GS, Parati G, Vlachopoulos C, Achimastos A, Andreadis E, Asmar R et al (2016) Methodology and technology for peripheral and central blood pressure and blood pressure variability measurement: current status and future directions. J Hypertens 34(9):1665–1677
13. Spahić L, Pokvić LG, Badnjević A (2024) Inspection and testing of infusion pumps. In: Badnjević A, Cifrek M, Magjarević R, Džemić Z (eds) Inspection of medical devices. Springer, Cham, pp 253–268
14. Pokvić LG, Deumić A, Softić A, Badnjević A (2024) Application of artificial intelligence for management of maintenance and prediction of performances. In: Badnjević A, Cifrek M, Magjarević R, Džemić Z (eds) Inspection of medical devices. Springer, Cham, pp 123–134
15. Yamakoshi K (2009) In the spotlight: bioinstrumentation. IEEE Rev Biomed Eng 2:92–102
16. Badnjević A, Pokvić LG, Deumić A, Bećirović LS (2022) Post-market surveillance of medical devices: a review. Technol Health Care 30(6):1315–1329
17. Dieterle T (2012) Blood pressure measurement—an overview. Swiss Med Wkly 142:w13517
18. Filipovský J (2022) Recommendations of the European society of hypertension for measuring blood pressure in the medical office and outside the medical facility. Vnitr Lek 68(2):107–111

Chapter 8
Review and Practice Problems

8.1 Summary of Key Concepts

10–20 System—International standard for EEG electrode placement.
2D, 3D, and 4D Ultrasound

- **2D**: Real-time cross-sectional images
- **3D**: Volumetric imaging using multiple 2D slices
- **4D**: Real-time 3D imaging with time as the fourth dimension

Abdominal Ultrasound—A sonographic exam used to evaluate solid and hollow abdominal organs including the liver, gallbladder, pancreas, kidneys, and bladder.
Acoustic Impedance (Z)—Product of tissue density and sound speed; determines wave reflection at tissue interfaces.
Acoustic Window—Area or approach providing optimal transmission and reception of ultrasound waves (e.g., subcostal for heart).
Action Potential—Rapid electrical impulse triggered by depolarization, allowing communication along neurons.
Active Contour (Snakes)—A method where a deformable curve evolves to outline object boundaries based on image features and smoothness constraints.
Adaptive Filtering—A dynamic filter that adjusts parameters in real-time based on input signal conditions.
Adaptive filters (LMS, RLS)—Real-time filters that adjust using reference data (e.g. accelerometer signals) for artifact cancellation.
Adhesive Gel Electrode—A surface electrode with a conductive gel that improves signal transfer and reduces impedance.
AI-Based Feature Extraction—Use of ML or deep learning (e.g., CNN, LSTM) to identify patterns and extract meaningful ECG features.
AI-Enabled Telemonitoring—Remote analysis of respiratory patterns using artificial intelligence for predictive health management.

© The Author(s), under exclusive license to Springer Nature Switzerland AG 2026
A. Badnjević and L. Spahić, *Biomedical Signals and Systems*, Series in BioEngineering,
https://doi.org/10.1007/978-3-032-10982-8_8

AI-Enhanced ECG—ECG systems using machine learning for real-time interpretation, risk prediction, and feature extraction.

Airway Resistance—Opposition to airflow in the respiratory tract, often elevated in obstructive diseases like asthma and COPD.

Aliasing—A distortion that occurs when a signal is sampled below the Nyquist rate, causing overlapping of frequency components.

Alpha Asymmetry—Difference in alpha band power between brain hemispheres; linked to emotional and motivational states.

Ambulatory Blood Pressure Monitoring (ABPM)—24-h monitoring of BP during normal daily activities, capturing variability and detecting masked hypertension.

Amplification—Increasing a weak biosignal's amplitude to the ADC's input range without distortion, preserving waveform integrity.

Amplitude—Vertical measurement (in mV) from baseline to peak of a wave; reflects signal strength and physiological status.

Analog Filter—A filter circuit (passive or active) implemented before digitization to suppress noise and prevent aliasing.

Analog-to-Digital Conversion (ADC)—The process of converting continuous (analog) signals into discrete (digital) data points.

Anti-Aliasing—Filtering high-frequency components before digitization to avoid aliasing artifacts.

Area-Based Features—Measures like QRS area or ST area that integrate amplitude and time, helpful in dyssynchrony and infarction assessment.

Arrhythmia—Abnormal heart rhythm, detectable via ECG through changes in RR intervals, P waves, or QRS morphology.

Artifact—Any non-neural signal contaminating EEG (e.g., eye blinks, muscle activity, power line noise).

ASR (Artifact Subspace Reconstruction)—Automated method that detects and reconstructs contaminated EEG segments in real-time.

Astrocytes—Glial cells that maintain the blood–brain barrier and modulate synaptic transmission.

Atrioventricular (AV) Node—A conduction delay node that ensures atrial contraction precedes ventricular contraction.

Atrium (Right and Left)—The two upper heart chambers that receive blood returning to the heart.

Attenuation—Loss of ultrasound energy as it propagates due to absorption, scattering, and reflection.

Attenuation Coefficient (α)—Rate of intensity loss per unit distance; depends on frequency and tissue type.

Auscultatory Method—Traditional non-invasive method using a stethoscope and sphygmomanometer to detect Korotkoff sounds for SP and DP.

Autonomy—A patient's right to make informed decisions about their healthcare without coercion.

Ballistocardiography (BCG)—A method that measures the body's recoil as blood is ejected from the heart.

Band-Pass Filter (BPF)—Allows signals within a specific frequency range to pass while attenuating frequencies outside this band.

Bandwidth—Frequency range over which amplifier maintains consistent gain; must cover the signal's spectral content.

Baseline Drift—A slow-changing fluctuation in the signal baseline that can obscure important features.

Baseline Wander—Low-frequency noise (0.05–0.5 Hz) typically caused by respiration or movement that shifts the ECG baseline.

BCI Illiteracy—Inability of some individuals to produce reliable EEG signals for BCI use, often due to cortical or cognitive factors.

Beamforming—Signal processing technique used to focus and steer ultrasound beams, enhancing resolution and image quality.

Beneficence—The ethical obligation to act in the best interest of the patient.

Bilateral Filter—A filter that uses both spatial and intensity differences to smooth an image while preserving edges.

Biomedical Signal—A measurable physiological signal generated by the body's organs and systems (e.g., ECG, EEG, EMG).

Biomedical Signal Acquisition—The process of detecting physiological activity and converting it into analyzable electrical or optical signals.

Biomedical Signal Processing—The analysis and transformation of physiological signals to extract clinically meaningful data.

Biophysical Profile (BPP)—A third-trimester assessment combining ultrasound and non-stress testing to evaluate fetal well-being using parameters like movement, tone, and amniotic fluid volume.

Bispectral Index (BIS)—Quantitative EEG-derived metric used to assess depth of anesthesia.

Blood–Brain Barrier (BBB)—Selective filter protecting brain tissue from toxins and pathogens.

B-Mode Imaging—Brightness mode; the standard 2D grayscale ultrasound imaging mode.

Body Plethysmography—Technique to assess absolute lung volumes (e.g., RV, TLC) using pressure–volume relationships in a sealed chamber.

Boyle's Law—Pressure and volume are inversely related; governs airflow during breathing.

Brainstem—Controls autonomic functions like respiration and heart rate; includes midbrain, pons, and medulla.

Bundle of His—Conduction fibers transmitting signals from the AV node to the ventricles.

Canny Edge Detector—A multi-step, advanced method for precise edge detection including noise reduction and edge linking.

Capnography—Measurement of exhaled carbon dioxide ($EtCO_2$); used to assess ventilation efficiency and airway status.

Cerebellum—Coordinates motor activity and balance; also implicated in cognitive and emotional processing.

Channel Selection—Choosing the most relevant electrodes for specific EEG analyses or applications.
Clinical Decision Support System (CDSS)—Software tools that use processed biomedical signals to support diagnosis and treatment decisions.
Color Doppler—Ultrasound mode overlaying flow direction and velocity information in color on grayscale anatomical images.
Common-Mode Rejection Ratio (CMRR)—A measure of an amplifier's ability to reject interference common to both input electrodes (e.g., powerline noise).
Confidentiality—The obligation of healthcare providers to keep patient information private unless given consent to share it.
Conformity Testing—Regulatory process that ensures ECG devices meet performance standards in detecting clinically significant patterns.
Continuous EEG (cEEG)—Ongoing EEG recording in ICUs to detect nonconvulsive seizures and monitor sedation.
Continuous Wavelet Transform (CWT)—A form of wavelet analysis that uses continuously scaled and translated versions of a wavelet for detailed time–frequency decomposition.
Continuous-Wave Doppler—Uses separate transmit and receive elements to measure high-velocity flow without depth specificity.
Contrast Echocardiography—Echocardiographic enhancement technique using microbubble contrast agents to improve visualization of endocardial borders and detect intracardiac shunts or masses.
Convex Transducer—Emits low-frequency, sector-shaped beams allowing deep tissue penetration, suitable for abdominal and obstetric imaging.
Convolutional Neural Networks (CNNs)—Deep learning models used to learn and segment complex features in ultrasound images automatically.
CPAP Therapy—Continuous Positive Airway Pressure treatment for OSA, monitored through airflow and pressure signal trends.
CRT (Cardiac Resynchronization Therapy)—A therapy for heart failure patients with prolonged QRS (>120 ms), often evaluated using ECG.
Cuff Size—The dimensions of the blood pressure cuff; incorrect sizing can lead to measurement errors (e.g., too small = falsely high readings).
Cutoff Frequency (fc)—The threshold frequency at which a filter starts attenuating a signal.
Data Integrity—The accuracy, consistency, and reliability of biomedical signals throughout acquisition and processing.
Data Ownership—Legal and ethical questions regarding who controls biomedical data (patients, researchers, institutions, or companies).
Data Protection—Legal and technical measures (e.g., encryption, access control) to secure biomedical signal data from unauthorized access or breaches.
Deep Learning (DL)—A subfield of ML that uses neural networks (e.g., CNNs) for feature learning and classification.
Diastolic Pressure (DP)—The lowest arterial pressure during heart relaxation (diastole), representing pressure in arteries between heartbeats.

Digital Filter—A mathematical algorithm that processes discrete-time signals to enhance or remove specific frequency components.
Digital Twin—A real-time, AI-driven model of an individual's physiological state used to simulate and predict treatment outcomes.
Direct (Invasive) BP Measurement—Measurement of arterial pressure via catheterization, providing real-time, beat-to-beat pressure readings—considered the gold standard.
Discrete Fourier Transform (DFT)—A sampled version of the Fourier Transform used for digital signals.
Doppler Effect—A change in frequency of the echo signal due to motion, used to measure blood flow velocity and direction.
Doppler Ultrasound—A technique that measures blood flow velocity using the Doppler effect, commonly used in cardiovascular and fetal monitoring.
Dry Electrode—A type of electrode that does not require conductive gel, used in wearable devices but with higher impedance.
Duration—Time (in ms) a wave or interval spans (e.g., PR interval, QRS duration, QT interval).
Electrocardiogram (ECG)—A test that measures the electrical activity of the heart.
ECG Preprocessing—The initial stage in ECG analysis where noise and artifacts are removed to enhance signal quality and diagnostic accuracy.
Echo—Reflected ultrasound wave from tissue boundaries; forms the basis of image generation.
Edge Detection—A method to identify significant intensity changes representing structural boundaries in an image.
EEG Bands—Distinct frequency bands (delta, theta, alpha, beta, gamma) in brain activity associated with different mental states.
EEG Brain Death Confirmation—Use of isoelectric EEG over a defined time to support clinical diagnosis of brain death.
EEG Feature Extraction—The process of transforming raw EEG signals into descriptive attributes (features) for interpretation or machine learning.
EEGLAB—MATLAB-based open-source toolbox widely used for EEG preprocessing and analysis.
Einthoven's Triangle—A geometric model illustrating the three standard bipolar leads (I, II, III) formed between the limb electrodes.
Electrical Axis—The average direction of ventricular depolarization, used to detect deviations such as left or right axis deviation.
Electrical Isolation—Galvanic separation (optical, capacitive, magnetic) between the patient and electronics, essential for safety.
Electrode—A sensor placed on the body to detect electrical signals from the heart for ECG recording.
Electrode impedance noise—Signal instability from fluctuations in electrode–skin interface resistance.
Electrode Motion Artifacts—Distortions due to changes in electrode–skin contact impedance, mimicking arrhythmias or masking waveforms.

Electroencephalogram (EEG)—A device that records the brain's electrical activity via scalp electrodes.
Electroencephalography—Non-invasive method to record brain electrical activity via scalp electrodes, widely used in clinical and research settings.
Electromyography (EMG)—A technique that records the electrical activity of muscles, used to diagnose neuromuscular disorders.
Electromyogram—A test that captures electrical signals produced by muscle contractions.
EMG Artifact—High-frequency contamination (>20 Hz) caused by muscle activity, common during speech or movement.
EMG noise—Muscle-generated interference overlapping ECG (30–100 Hz) and EEG (20–300 Hz) frequency bands.
End-Tidal CO_2 ($EtCO_2$)—The CO_2 concentration at the end of exhalation; a surrogate marker of ventilation status.
Ensemble averaging—Improves SNR by averaging repetitive events, assuming noise is random.
Entropy—A measure of signal complexity or irregularity; low entropy can suggest seizures or unconsciousness.
Epileptiform Activity—Abnormal EEG patterns such as spikes, sharp waves, and seizures, commonly associated with epilepsy.
Event-Based Segmentation—Aligning feature windows with physiological markers (e.g., R-peaks in ECG).
Event-Related Potentials (ERP)—Time-locked EEG responses to stimuli (e.g., P300, N170, N400).
Event-Related Spectral Perturbation (ERSP)—Time–frequency map showing stimulus-locked changes in spectral power.
Expiration—Normally passive process during rest; becomes active during exertion via abdominal and internal intercostal muscles.
Fast Fourier Transform (FFT)—An efficient algorithm to compute the DFT, commonly used in real-time applications.
Feature Extraction—Identifying critical parameters (e.g., respiratory rate, FEV_1, tidal volume) from raw signals to assess respiratory function.
Fetal Ultrasound—Sonographic imaging used in pregnancy to monitor fetal development, assess growth and well-being, and detect congenital anomalies.
Filtering—A process to remove unwanted components (noise) from a signal while preserving clinically important data.
Finite Impulse Response (FIR) Filter—A filter whose output depends only on current and past input values and has a finite response duration.
Fixed-Window Segmentation—Dividing signals into equal-length segments for consistent feature extraction.
Fourier Transform (FT)—A mathematical method to transform time-domain signals into their frequency components.
fQRS (Fragmented QRS)—Additional R' notches or slurs in QRS, indicative of myocardial scarring or conduction delay.

Fractal Dimension (FD)—A nonlinear metric reflecting waveform self-similarity and complexity.
Frequency-Domain Analysis—A method of analyzing signals based on their frequency components rather than their variation over time.
Frequency-Domain Features—Attributes computed from EEG spectral content, often via Fourier analysis (e.g., band power, spectral entropy).
Gain—The ratio of output voltage to input voltage; indicates how much amplification is applied.
Gain–Bandwidth Product (GBP)—A trade-off relationship: higher gain reduces bandwidth unless multiple stages are used.
Gaussian Filtering—A linear smoothing filter using a Gaussian kernel to reduce high-frequency noise while preserving edges.
General Data Protection Regulation (GDPR)—A European Union regulation that governs data privacy and security.
Genitourinary Ultrasound—Sonographic assessment of the kidneys, bladder, uterus, ovaries, prostate, and testes for structural abnormalities, masses, and pathology.
Goldberger's Leads—Augmented unipolar leads (aVR, aVL, aVF) providing additional perspectives from the frontal plane.
Half-Cell Potential—The voltage difference between an electrode and the surrounding biological tissue or electrolyte.
Heart—A muscular organ responsible for pumping blood through the circulatory system via rhythmic contractions.
Heart Rate Variability (HRV)—The variation in the time interval between heartbeats, influenced by autonomic nervous system regulation.
High-Density EEG (hdEEG)—EEG with 64 + channels for spatially resolved recordings.
High-Pass Filter (HPF)—Allows signals above a cutoff frequency to pass and attenuates lower frequencies.
Hilbert-Huang Transform (HHT)—Adaptive method using empirical mode decomposition for analyzing non-linear, non-stationary signals.
Health Insurance Portability and Accountability Act (HIPAA)—U.S. legislation that protects sensitive patient health information.
Hjorth Parameters—Three statistical descriptors of EEG: activity (power), mobility (mean frequency), and complexity (waveform shape).
Hybrid methods—Combine traditional filtering (e.g. wavelet) with AI/ML for enhanced denoising.
Hypothalamus—Regulates homeostasis, hormone secretion, circadian rhythms, and autonomic functions.
Independent Component Analysis (ICA)—Blind source separation technique for isolating and removing artifacts from EEG.
Impedance—Opposition to AC current at the electrode-amplifier interface. High input impedance prevents signal loss; low output impedance ensures efficient signal transfer.

Impedance Pneumography—Measures chest impedance to derive respiratory rate and rhythm using skin electrodes.
Infinite Impulse Response (IIR) Filter—A filter that uses both past input and past output values, with a theoretically infinite response.
Informed Consent—The ethical and legal process of informing a patient about a procedure, including its risks, benefits, and alternatives, to ensure voluntary agreement.
Infrared Thermography—A method for visualizing heat emissions from the body to detect temperature-related pathologies.
Inspiration—Active phase of breathing, driven primarily by diaphragm and external intercostal muscle contraction.
Instrumentation Amplifier—A precision amplifier architecture offering high input impedance, adjustable gain, and superior CMRR.
Instrumentation Safety Standards (IEC 60,601–1, IEC 60,601–2-47, ISO 14971)—Regulations governing safety, performance, and risk in medical devices.
Intellectual Property (IP)—Legal rights protecting inventions, including algorithms or processing methods developed from biomedical signals.
Inter-individual Variability—Differences in biomedical signals between different individuals due to genetics, age, lifestyle, etc.
Intra-individual Variability—Variations in biomedical signals within the same person over time.
Intrapleural Pressure (Pip)—Slightly negative pressure within the pleural cavity that keeps the lungs inflated.
Invasive Sensor—A sensor that requires insertion into the body (e.g., catheter-based pressure sensor) for direct signal acquisition.
Inverse Fourier Transform—Used to reconstruct the original time-domain signal from its frequency representation.
J Point—The junction between the QRS complex and the ST segment.
Jitter—Variation in sampling interval timing; affects temporal accuracy.
Korotkoff Sounds—Distinct sounds heard over an artery during cuff deflation in auscultatory BP measurement, indicating SP (Phase I) and DP (Phase V).
k-space—The frequency domain in MRI where raw imaging data is stored before inverse transformation into an image.
Kurtosis—Measures the "peakedness" of the signal distribution; high kurtosis can indicate spikes or outliers.
Lead—A specific arrangement of electrodes used to measure electrical activity from different angles.
Leakage Current—Unwanted current through a device in contact with the patient; must stay below safety limits ($< 10\ \mu A$).
Lewis Lead—A modified lead configuration used to better visualize atrial activity.
Limb Leads—Electrodes placed on the limbs to form Leads I, II, and III (bipolar) and aVR, aVL, aVF (augmented unipolar).
Linear Transducer—Produces high-frequency, high-resolution images of superficial structures using parallel beams in a rectangular field of view.

Long-Term Potentiation (LTP)—A form of synaptic strengthening involved in learning and memory.

Low-Pass Filter (LPF)—Allows signals with frequencies below a cutoff frequency to pass while attenuating higher frequencies.

Lung Ultrasound—Point-of-care imaging modality to assess for pneumothorax, pleural effusion, pulmonary edema, and pneumonia.

Lyapunov Exponent—A measure of signal sensitivity to initial conditions; positive values indicate chaos.

Machine Learning (ML)—Algorithms like SVMs, CNNs, and LSTMs trained to classify breathing patterns and detect abnormalities automatically.

Mean—The average value of the signal over a time segment; useful for detecting baseline drift.

Mean Arterial Pressure (MAP)—The average arterial pressure throughout one cardiac cycle; a critical indicator of tissue perfusion.

Median Filtering—A non-linear noise reduction technique that replaces a pixel's value with the median value from its local neighborhood.

Microelectrode—A fine-tipped electrode used to measure signals from individual neurons or cells.

Montage—A specialized electrode configuration to enhance the detection of specific cardiac conditions.

Mother Wavelet—The base function used in wavelet transforms from which all scaled and shifted wavelets are derived.

Motion Artifact—Noise or distortion in a biomedical signal caused by patient movement.

Multirate Processing—Using different sampling rates during preprocessing to optimize resolution and computational efficiency.

Multiresolution Analysis—A key property of wavelet transforms where low frequencies are analyzed with better frequency resolution and high frequencies with better time resolution.

Multi-sensor fusion—Uses data from multiple sensors (e.g., accelerometers, gyros) to improve artifact detection and removal.

Muscle Artifacts (EMG Noise)—High-frequency noise (20–200 Hz) from skeletal muscle contractions, common during stress tests.

Musculoskeletal Ultrasound—High-resolution ultrasound imaging of soft tissues, joints, tendons, and muscles, useful for both diagnosis and image-guided procedures.

Myelin—Insulating sheath around axons that increases conduction speed via saltatory conduction.

Near-Infrared Spectroscopy (NIRS)—A technique to measure oxygenation and blood flow in tissues using near-infrared light.

Needle Electrode—An invasive electrode inserted into tissue to record signals like intramuscular EMG.

Neuron—Fundamental signaling cell of the nervous system, composed of dendrites, a soma, and an axon.

Neuroplasticity—The brain's ability to reorganize synaptic connections, especially after injury or learning.
Noise—Unwanted components in a signal, including motion artifacts, muscle activity, and powerline interference.
Noise Filtering—The use of techniques such as low-pass, high-pass, band-pass, and adaptive filters to remove artifacts and improve signal clarity.
Noninvasive Continuous Monitoring—Techniques (e.g., Finapres™, tonometry) that continuously estimate arterial pressure without direct arterial access.
Nonlinear Features—Complexity-based measures capturing chaos, irregularity, or self-similarity in EEG (e.g., entropy, fractal dimension).
Nonlinearity—A property of a system where the output is not directly proportional to the input, common in biological signal behavior.
Non-Local Means (NLM) Filtering—A noise reduction technique that averages pixel values based on patch similarity rather than spatial proximity.
Non-maleficence—An ethical principle requiring healthcare professionals to avoid causing harm to patients.
Non-Overlapping Windows—Disjoint segments; computationally efficient for embedded systems.
Non-stationary Signal—A signal whose frequency components vary over time, making time–frequency analysis especially useful.
Notch Filter—A filter used to remove specific frequency components, typically 50/60 Hz powerline interference.
Nyquist Criterion—Sampling rate must be $\geq 2 \times$ the highest frequency of interest to avoid aliasing.
Obstructive Sleep Apnea (OSA)—Repeated episodes of airway collapse during sleep despite respiratory effort, detected via airflow and effort signals.
Ocular Artifact (EOG)—Contamination from eye blinks/movements affecting frontal electrodes, often in delta/theta bands.
Offline Preprocessing—Signal enhancement applied after data acquisition, often used for diagnostics or research.
Oscillometric Method—Automated, non-invasive method that uses pressure oscillations in a cuff to calculate SP, DP, and MAP using empirical algorithms.
Overlapping Windows—Segments share data points to capture dynamic changes; improves temporal sensitivity.
Oversampling & Decimation—Sampling faster than Nyquist and averaging to improve resolution and reduce noise.
P Wave—Represents atrial depolarization.
P300—Event-related potential linked to attention and cognitive processing; used in both psychiatry and BCIs.
Paradoxical Breathing—Abnormal breathing pattern where chest and abdomen move out of sync; indicates diaphragmatic dysfunction.
Passband—The frequency range that is allowed to pass through the filter with minimal attenuation.
Principal Component Analysis (PCA)—A method to reduce feature dimensionality while preserving most relevant variance.

Pericardium—The double-walled sac that encloses the heart and provides protection and lubrication.
Phonocardiography (PCG)—The graphical representation of heart sounds, useful for detecting murmurs or valvular defects.
Photoplethysmography (PPG)—An optical technique for measuring blood volume changes in tissue, often used for heart rate and SpO_2.
Piezoelectric Effect—The property of certain crystals to generate an electric signal in response to mechanical pressure (and vice versa).
Polysomnography (PSG)—Multichannel sleep study that includes respiratory signals to diagnose sleep disorders.
Population Screening—Use of ECG in asymptomatic groups to identify early signs of heart disease (e.g., athletes, workers).
Posterior Leads—Leads V7–V9 placed on the back to detect posterior wall myocardial infarction.
Post-Market Surveillance (PMS)—Continuous performance monitoring of ECG devices after deployment, with a focus on feature accuracy.
Postoperative Respiratory Monitoring—Use of respiratory signals in recovery to detect hypoventilation or apnea, especially due to anesthesia or opioids.
Power Doppler—Displays the strength of the Doppler signal to highlight low-velocity blood flow, useful in small vessels.
Power Spectral Density (PSD)—Distribution of signal power across frequency bands, often used in sleep or cognitive state analysis.
Powerline Interference—Sinusoidal noise (50 or 60 Hz) from electrical devices that contaminates the ECG signal.
PR Interval—Time from atrial depolarization to ventricular depolarization.
Precordial Leads (V1–V6)—Chest leads that capture the heart's electrical activity in the horizontal plane.
PREP Pipeline—Standardized EEG preprocessing framework that includes referencing and artifact removal.
Preprocessing—The initial stage of image processing aimed at reducing noise and enhancing the visibility of key structures.
Prewitt Operator—A basic edge detector similar to Sobel but using simpler convolution kernels.
Privacy—The right of individuals to control access to their personal health data, including biomedical signals.
Pulmonary Ventilation—The physical movement of air into and out of the lungs during inspiration and expiration.
Pulse Oximetry (SpO_2)—Non-invasive method to measure arterial oxygen saturation via light absorption.
Pulse Pressure (PP)—The difference between systolic and diastolic pressures; PP = SP—DP.
Pulse Transit Time—A noninvasive method for estimating BP by measuring the time it takes a pulse wave to travel between two arterial sites.
Pulsed-Wave Doppler—Measures flow at specific depths using short ultrasound pulses; enables localization of flow.

Pulse-Echo Principle—Basic imaging principle where transmitted pulses reflect off tissue boundaries and are received as echoes to form images.
Purkinje Fibers—Fibers distributing the electrical impulse through ventricular myocardium for coordinated contraction.
Pyramidal Neurons—Cortical neurons generating synchronized postsynaptic potentials contributing to scalp EEG signals.
QRS Complex—Represents ventricular depolarization.
QRS Detection—The identification of the R-peak, crucial for rhythm classification, heart rate computation, and waveform segmentation.
QT Interval—Time from Q wave to end of T wave; prolonged QT can lead to dangerous arrhythmias like torsades de pointes.
Quality Factor (Q)—A measure of filter selectivity; higher Q yields a narrower bandwidth.
Quantization—The process of mapping sampled analog values into discrete numerical levels during ADC.
Quantization Noise—Error introduced by assigning continuous values to the nearest discrete level.
Rayleigh Scattering—Type of scattering occurring when ultrasound encounters small structures like red blood cells.
Real-Time Preprocessing—Immediate signal processing, crucial in ICU, telemetry, or wearable systems.
Referencing—Method for calculating EEG voltages (e.g., average reference, linked ears, Laplacian).
Reflection Coefficient—Ratio of reflected to incident ultrasound energy at an interface between two media.
Region Growing—A segmentation method that adds neighboring pixels to a region if they meet similarity criteria.
Regulatory Compliance—Adherence to legal and industry standards governing medical devices and data (e.g., FDA, MDR, IVDR).
Residual Volume (RV)—The volume of air remaining in the lungs after maximal exhalation; not measurable by spirometry.
Resolution (Bits)—Number of bits per sample; determines the number of quantization levels (2^n).
Resolution vs. Penetration Trade-off—High frequencies provide better resolution but lower tissue penetration, and vice versa.
Respiratory Inductance Plethysmography (RIP)—Non-invasive method using chest/abdominal bands to monitor breathing patterns and respiratory effort.
Respiratory Rate Variability (RRV)—Metric of respiratory rhythm that reflects changes in autonomic regulation and clinical status.
Respiratory Signal Analysis—Interpretation of signals like airflow, tidal volume, and $EtCO_2$ to assess respiratory and systemic health.
Respiratory Signal Processing—The transformation and analysis of raw respiratory data (e.g., airflow, volume, muscle activity) into clinically useful information.
Respiratory Signals—Physiological data derived from breathing activity, including mechanical, airflow, gas exchange, and neural/electrical signals.

Resting Membrane Potential—Electrical charge difference across the neuronal membrane (~ −70 mV at rest).
Right-Sided Leads—ECG leads placed on the right side of the chest to assess right ventricular involvement.
Root Mean Square (RMS)—A measure of signal power; widely used in EMG to assess muscle activation.
Sampling—Measuring an analog signal at discrete time intervals. Governed by Nyquist–Shannon theorem to prevent aliasing.
Sampling Rate—How often the signal is sampled per second, measured in Hz; must exceed twice the highest signal frequency.
Scalogram—A time–frequency plot similar to a spectrogram, but generated using wavelet transform coefficients.
Segmentation—Dividing an image into meaningful regions (e.g., organs or tumors) based on pixel similarities.
Sensor—A device that detects a specific type of physiological signal (e.g., electrical, optical, acoustic) and converts it into an input signal for analysis.
Sensorimotor Rhythms (SMRs)—Mu and beta oscillations modulated during motor imagery, critical for motor BCIs.
Short-Time Fourier Transform (STFT)—A time–frequency analysis method that applies the Fourier Transform over successive time windows.
Signal Acquisition—The process of capturing reflected ultrasound waves and converting them into electrical signals for imaging.
Signal Conditioning—The sequence of operations (amplification, filtering, impedance matching, isolation) that prepare a raw biomedical signal for reliable digitization and analysis.
Signal Energy—Total power in the signal; calculated as the sum of squared amplitude values.
Signal Quality Index (SQI)—A measure of ECG signal clarity, often used to trigger re-filtering or alert clinicians.
Signal-to-Noise Ratio (SNR)—Ratio of EEG signal strength to background noise; higher is better.
Sinoatrial (SA) Node—The heart's natural pacemaker, initiating electrical impulses.
Skewness—Describes asymmetry in the signal amplitude distribution; abnormal skew may reflect pathology.
Slope—Rate of change of a waveform (steepness), often used in ischemia and conduction analysis.
Slope Sign Changes (SSC)—The number of changes in signal slope; indicates sharp transitions such as seizures or arrhythmias.
Smart ADC—ADC with adaptive sampling, self-calibration, AI integration for power optimization and accuracy.
Smart noise removal—AI-driven, context-aware, real-time artifact correction, often embedded in wearables.
Snell's Law—Governs refraction of ultrasound waves between media with different sound speeds.

Sobel Operator—A gradient-based method for detecting vertical and horizontal edges using convolution kernels.
Speckle Noise—A granular noise pattern in ultrasound caused by interference of returning echoes, which reduces image clarity.
Spectral Entropy—A frequency-domain metric quantifying the unpredictability of spectral content.
Spectrogram—A visual representation of the signal's frequency content over time.
Spirometry—A clinical test used to measure lung volumes and capacities.
ST Segment—Period when ventricles are fully depolarized.
Standard Deviation (SD)—Quantifies the dispersion of signal values; lower SD may reflect signal stability.
STEMI/NSTEMI—Types of myocardial infarction characterized by ST elevation or depression, respectively.
Stopband—The frequency range that is heavily attenuated by the filter.
Stress Echocardiography—Ultrasound performed under exercise or pharmacologic stress to detect myocardial ischemia by assessing wall motion abnormalities.
Surface Electrode—A non-invasive electrode placed on the skin to detect signals like ECG or EEG.
Synapse—The junction between neurons where neurotransmitters are released to propagate signals.
Systemic noise—Electronic internal noise (thermal, shot, flicker) and ADC quantization noise.
Systolic Pressure (SP)—The maximum arterial pressure during heart contraction (systole), representing the force of blood against artery walls when the heart pumps.
T Wave—Represents ventricular repolarization.
T Wave Alternans—Beat-to-beat variability in T wave amplitude; a marker of ventricular instability.
Telemetry—Continuous ECG monitoring system, especially in hospital settings.
Template subtraction—Removes predictable artifacts (e.g. eye blinks) by subtracting a known waveform.
Thalamus—Relay station for most sensory inputs (except smell); contributes to consciousness and alertness.
Theta/Beta Ratio—Frequency-domain index commonly used in ADHD diagnostics and neurofeedback protocols.
Thresholding—A basic segmentation technique where pixels are classified based on whether their intensity is above or below a certain threshold.
Thyroid Ultrasound—Imaging of the thyroid gland and neck structures to evaluate nodules, guide biopsies, and monitor disease recurrence.
Tidal Volume (TV)—The volume of air moved in or out of the lungs during a normal breath (~500 mL in adults).
Time-Domain Analysis—Analysis of signal variations over time to measure breath cycles, rate, and volume.

Time-Domain Features—Signal characteristics derived directly from EEG waveforms (e.g., mean, variance, zero-crossings).
Time-of-Flight—The time taken for ultrasound waves to travel to a tissue boundary and return; used to determine depth.
Total Harmonic Distortion (THD)—Harmonic noise introduced by conversion, impacting frequency analysis.
Total Lung Capacity (TLC)—Sum of all lung volumes (TV + IRV + ERV + RV).
Transducer—A device containing piezoelectric crystals that emit and receive ultrasound waves by converting mechanical energy to electrical energy and vice versa.
Transducer—Device containing piezoelectric crystals that both emits and receives ultrasound waves.
Transduction—The conversion of a biological or physical phenomenon into an electrical signal.
Transesophageal Echocardiography (TEE)—An invasive cardiac ultrasound technique involving a probe inserted into the esophagus for clearer imaging of posterior cardiac structures and detailed evaluation of valves and atrial abnormalities.
Transfer Function—Mathematical representation of a filter's behavior in the frequency domain.
Transform-Based Features—Derived from frequency, time–frequency, or nonlinear analyses (e.g., Fourier, wavelet, entropy measures).
Transthoracic Echocardiography (TTE)—A non-invasive ultrasound of the heart performed with a transducer on the chest wall, used to assess cardiac structure, function, ejection fraction, and valvular disease.
Triphasic Waves—EEG pattern typically associated with metabolic encephalopathies like hepatic encephalopathy.
U Wave—May represent repolarization of the Purkinje fibers; often seen in bradycardia or hypokalemia.
Ultrasound Image Processing—A set of techniques to improve ultrasound image quality by reducing noise, enhancing contrast, and extracting relevant anatomical features for clinical interpretation.
Ultrasound Imaging—A mechanical signal modality that uses sound waves to visualize internal body structures.
Valves (Tricuspid, Pulmonary, Mitral, Aortic)—Structures ensuring unidirectional blood flow through the heart.
Variance—Measures signal variability from the mean; high variance in EMG can indicate fatigue.
Vascular Ultrasound—Non-invasive ultrasound imaging of arteries and veins using B-mode, color Doppler, and spectral Doppler to assess for stenosis, thrombosis, or insufficiency.
Vector—A directional representation of electrical flow in the heart, used to interpret ECG signals.

Ventricle (Right and Left)—The two lower chambers that pump blood out of the heart to the lungs (right) and body (left).
Video-EEG Monitoring—Long-term EEG recording synchronized with video to correlate clinical events with neural activity.
Vital Capacity (VC)—Maximum amount of air that can be exhaled after maximum inhalation; key for assessing lung function.
Wave Delineation—Determining the onset, peak, and offset of ECG components such as the P wave, QRS complex, and T wave.
Waveform Length (WL)—Cumulative length of signal changes; higher WL implies increased complexity or bursts.
Wavelet Denoising—A method of reducing noise by transforming an image into different frequency bands and applying thresholding to high-frequency components.
Wavelet Filtering—Time–frequency decomposition used to suppress transient noise/artifacts while preserving signal features.
Wavelet Transform (WT)—Decomposes signals across scales; useful for non-stationary signal denoising and artifact detection.
Wearable Devices—Portable systems embedded with sensors to monitor physiological parameters continuously and remotely.
White-Coat Hypertension—Elevated blood pressure in a clinical setting due to anxiety, which may not reflect the patient's usual BP levels.
Wiener Filter—An adaptive filter that balances smoothing and detail preservation based on local variance.
Window Function—A mathematical function (e.g., Hamming, Gaussian) applied to a signal segment in STFT to localize analysis.
Windowing—A method of breaking a signal into smaller segments before analysis to handle non-stationary behavior.
Zero-Crossing Rate (ZCR)—Counts the number of times a signal crosses zero; high ZCR may suggest tremor.
Z-score Normalization—A method of scaling signal data based on its mean and standard deviation.

8.2 Practice Problems

8.2.1 Section 1 Practice Problems

Fill in the Blanks

(a) The ________ is a biomedical device used to record the electrical activity of the brain.
(b) Biomedical signals are often __________, meaning they don't follow a constant pattern over time.

(c) ECG waveforms consist of multiple parts, including the P wave, ______ complex, and T wave.
(d) Artifacts in EEG signals can be caused by ________ movement or electrical interference.
(e) The non-invasive measurement of oxygen saturation is known as ________ _ _______.
(f) The process of turning a biological event into an electrical signal is known as __________.
(g) Before sampling, a signal is passed through a(n) __________ to remove unwanted high-frequency components.
(h) Sampling below the Nyquist rate causes a phenomenon known as _________ _.
(i) EEG signals are extremely weak and require __________ to be useful for analysis.
(j) In pulse oximetry, an optical signal is detected using a __________ sensor.
(k) __________ transform is widely used to analyze non-stationary biomedical signals like EEG and ECG.
(l) A __________ filter is used to eliminate powerline interference from biomedical signals.
(m) The process of converting physiological activity into electrical signals is called __________.
(n) __________ is used to reduce the effect of individual differences by centering and scaling signals.
(o) The __________ band in EEG corresponds to deep sleep and brain injuries.
(p) __________ is a legal framework in the EU that regulates data privacy and protection.
(q) The principle of __________ requires healthcare providers to avoid causing harm to patients.
(r) Before acquiring biomedical signals, __________ must be obtained from the patient.
(s) A(n) __________ filter might improve signal clarity but must not distort diagnostic features.

Multiple Choice Questions

Q1 What type of signal is recorded by an electroencephalogram (EEG)?

(A) Mechanical
(B) Thermal
(C) Electrical
(D) Acoustic

Q2 Which of the following is NOT a property of biomedical signals?

(A) Stationarity
(B) Variability
(C) Nonlinearity
(D) Multi-component nature

Q3 What is the typical frequency range for EEG signals?

(A) kHz range
(B) GHz range
(C) Hz range
(D) MHz range

Q4 Which of the following is used to measure lung mechanical activity?

(A) EMG
(B) PCG
(C) Spirometry
(D) NIRS

Q5 What signal analysis method is used to remove noise from biomedical signals?

(A) Filtering
(B) Modulation
(C) Sampling
(D) Amplification

Q6 Which of the following is the correct order of biomedical signal acquisition steps?

(A) Amplification → Detection → Transduction → Digitization
(B) Sensing → Transduction → Signal Conditioning → Digitization
(C) Digitization → Filtering → Sensing → Transduction
(D) Transduction → Sensing → Amplification → Sampling

Q7 What is the purpose of an anti-aliasing filter?

(A) To increase signal strength
(B) To convert analog signals to digital
(C) To block high frequencies that can cause aliasing
(D) To reduce quantization noise

Q8 A sampling rate of 1000 Hz can capture signals with frequency components up to:

(A) 250 Hz
(B) 500 Hz
(C) 1000 Hz
(D) 2000 Hz

Q9 Which sensor is used to capture respiratory airflow?

(A) Optical sensor
(B) Pressure sensor
(C) Magnetic sensor
(D) Acoustic sensor

Q10 Quantization error can be reduced by:

(A) Using a low sampling rate
(B) Using fewer ADC bits
(C) Increasing bit resolution
(D) Amplifying noise

Q11 Which of the following signal types is primarily used to monitor brain activity?

(A) ECG
(B) EMG
(C) EEG
(D) PPG

Q12 Which domain analysis is most effective for identifying frequency bands like alpha and beta in EEG signals?

(A) Time-domain
(B) Spatial-domain
(C) Frequency-domain
(D) Thermal-domain

Q13 What is the function of a Clinical Decision Support System (CDSS)?

(A) To regulate heart rhythm
(B) To preprocess ECG signals
(C) To assist clinicians in diagnosis and treatment decisions
(D) To acquire biosignals from sensors

Q14 Which transform allows localized frequency analysis of signals using windowing?

(A) PCA
(B) Fourier Transform
(C) STFT
(D) Notch Filter

Q15 Which deep learning model is most commonly used for image or signal classification tasks in biomedical applications?

(A) Decision Tree
(B) Support Vector Machine
(C) Convolutional Neural Network
(D) Linear Regression

Q16 What does informed consent primarily protect?

(A) Signal processing algorithms
(B) Device certification
(C) Patient autonomy
(D) Institutional funding

Q17 Which law governs the protection of health information in the United States?

(A) MDR
(B) GDPR
(C) IVDR
(D) HIPAA

Q18 What is a potential risk of over-filtering biomedical signals?

(A) Increased security
(B) Enhanced resolution
(C) Loss of critical clinical information
(D) Reduced processing time

Q19 Who is responsible for regulatory oversight of medical devices in the U.S.?

(A) WHO
(B) ISO
(C) FDA
(D) IEEE

Q20 In the context of data ownership, which party must consent to the use of their biomedical data for commercial purposes?

(A) The manufacturer
(B) The healthcare provider
(C) The patient
(D) The government

Short Answer

(1) Why are biomedical signals considered non-stationary and nonlinear? Give one example to support your answer.
(2) Describe how EEG and ECG differ in terms of signal origin, frequency, and application.
(3) What are common sources of noise in biomedical signal acquisition, and how can they be reduced?
(4) Explain the importance of the Nyquist rate in biomedical signal acquisition. Provide an example from ECG or EEG.
(5) Why is filtering critical in biomedical signal processing, and how does it improve signal quality?
(6) Describe how a wearable ECG device acquires and processes heart signals from a user.
(7) Differentiate between invasive and non-invasive sensors with examples.
(8) Explain the difference between time-domain and frequency-domain analysis in biomedical signal processing.
(9) Why is wavelet transform preferred over Fourier transform for EEG signal analysis?

(10) Describe the role of AI in wearable biomedical devices for early disease detection.
(11) How does a notch filter improve signal quality in ECG recordings?
(12) Why is informed consent important in biomedical signal acquisition and processing?
(13) Describe how signal processing might unintentionally harm a patient if not applied properly.
(14) What are the key differences between privacy, confidentiality, and data protection in healthcare?
(15) How does regulatory compliance ensure the safety and reliability of biomedical signal processing devices?

8.2.2 Section 2 Practice Problems

Fill in the Blanks

(a) An __________ electrode is used non-invasively to record signals like EEG and ECG.
(b) __________ is a key concern in wearable sensors because patient motion may distort the signal.
(c) A __________ electrode is inserted into tissue for precise signal capture from deep sources like muscles.
(d) The __________ equation governs the half-cell potential at the electrode-tissue interface.
(e) __________ is a technique that uses light to monitor blood volume changes, such as in pulse oximeters.
(f) __________ ensures a weak biosignal is scaled to the ADC's input voltage range.
(g) __________ rejection ratio helps suppress interference picked up equally by both input leads.
(h) A __________ filter removes slow-frequency baseline drift from ECG signals.
(i) A __________ filter is designed to eliminate 50 or 60 Hz powerline interference.
(j) High input impedance in biomedical amplifiers helps prevent __________ effects at the electrode interface.
(k) __________ is the minimum sampling frequency to accurately reconstruct a signal.
(l) A 12-bit ADC provides __________ quantization levels.
(m) __________ noise arises from assigning input values to discrete levels.
(n) The real-world resolution of an ADC, accounting for noise, is called __________.
(o) A __________ ADC uses oversampling and noise shaping for high resolution.
(p) __________ removes 50/60 Hz hum from biomedical recordings.
(q) Low-frequency baseline drift (~0.15 Hz) is commonly caused by __________.

(r) __________ filters adjust in real time using reference signals like accelerometer data.
(s) __________ decomposes signals into scale-based components for targeted denoising.
(t) In EEG, __________ isolates independent signal sources to remove eye blink artifacts.
(u) The __________ of a signal measures how often it crosses the zero line.
(v) High RMS values in EMG are typically associated with __________.
(w) A significant drop in EEG __________ may suggest seizure activity.
(x) __________ segmentation aligns windows to physiological events such as heartbeats.

Multiple Choice Questions

Q1 Which electrode type is best for long-term non-invasive ECG monitoring?

(A) Needle electrode
(B) Dry electrode
(C) Adhesive gel electrode
(D) Microelectrode

Q2 What is a common drawback of dry electrodes compared to gel-based electrodes?

(A) Invasiveness
(B) Higher impedance
(C) Short lifespan
(D) Lack of real-time capability

Q3 Which of the following measures brain activity?

(A) ECG
(B) EMG
(C) EEG
(D) PPG

Q4 Photoplethysmography (PPG) primarily measures:

(A) Electrical muscle activity
(B) Brainwave frequency
(C) Blood volume changes
(D) Respiratory rate

Q5 What is the role of impedance in electrode signal acquisition?

(A) It stabilizes the heart rate
(B) It blocks unwanted brain activity
(C) It affects signal quality and strength
(D) It eliminates signal artifacts

Q6 Why is amplifier bandwidth important in biosignal acquisition?

(A) Affects signal amplitude only
(B) Determines power consumption
(C) Ensures full signal spectrum is captured
(D) Blocks interference entirely

Q7 Which amplifier type offers adjustable gain and high CMRR?

(A) Simple op-amp
(B) Instrumentation amplifier
(C) FET buffer
(D) Passive RC circuit

Q8 What is the main purpose of an anti-aliasing filter?

(A) To amplify signals near Nyquist rate
(B) To prevent high-frequency noise from folding into signal band
(C) To balance electrode impedance
(D) To increase gain

Q9 Which noise type do notch filters target?

(A) Baseline drift
(B) High-frequency EMG
(C) Powerline interference
(D) Thermal noise

Q10 What is considered safe leakage current for patient-applied parts?

(A) < 1 mA
(B) < 100 μA
(C) < 10 μA
(D) < 1 μA

Q11 What happens if sampling rate is below Nyquist rate?

(A) Higher SNR
(B) Higher resolution
(C) Aliasing
(D) Reduced jitter

Q12 Which ADC is best suited for microvolt EEG signals?

(A) SAR
(B) Delta-Sigma
(C) Flash
(D) Pipeline

Q13 ENOB stands for:

(A) Effective Number of Bits
(B) Encoded Noise of Bits

(C) Energy Number of Bits
(D) Electronic Number of Bits

Q14 What does jitter affect most?

(A) Amplitude resolution
(B) Signal power
(C) Timing accuracy
(D) Encoding scheme

Q15 Which ADC type offers sub-microWatt power consumption and moderate speed?

(A) Flash
(B) Pipeline
(C) Integrating
(D) SAR

Q16 Which type of artifact is due to electrode movement?

(A) Thermal noise
(B) Motion artifact
(C) Power-line interference
(D) Quantization noise

Q17 What technique is best for removing eye-blink interference in EEG?

(A) Notch filter
(B) Ensemble averaging
(C) ICA
(D) High-pass filter

Q18 Which filter is used to remove baseline wander in ECG?

(A) Low-pass
(B) High-pass
(C) Notch
(D) Band-stop

Q19 Muscle-generated noise overlapping ECG $\geq$ 30 Hz is called:

(A) Baseline wander
(B) Electrode impedance noise
(C) EMG noise
(D) Power-line hum

Q20 Adaptive filters often rely on data from:

(A) EEG channels only
(B) Accelerometers
(C) Temperature sensors
(D) Optical sensors

Q21 Which of the following best represents signal energy?

(A) Sum of amplitudes
(B) Square root of variance
(C) Sum of squared amplitudes
(D) Number of zero crossings

Q22 In EMG, which feature is commonly used to estimate muscle activation intensity?

(A) Entropy
(B) RMS
(C) Kurtosis
(D) Median

Q23 Which time-domain feature describes the asymmetry of a signal's distribution?

(A) Variance
(B) Kurtosis
(C) Entropy
(D) Skewness

Q24 What is the primary advantage of overlapping windows in real-time signal monitoring?

(A) Lower computational load
(B) Simpler statistics
(C) Better detection of dynamic changes
(D) Automatic artifact rejection

Q25 Time-domain features are especially useful in:

(A) High-resolution spectral analysis
(B) Real-time wearable systems
(C) Deep frequency filtering
(D) Color imagin

Q26 Which of the following statements about the Fast Fourier Transform (FFT) is correct?

(A) It computes inverse transform only
(B) It is slower than DFT
(C) It uses a divide-and-conquer strategy
(D) It only works on analog signals

Q27 Power Spectral Density is used to:

(A) Measure signal energy in the time domain
(B) Convert signals to the spatial domain
(C) Identify dominant frequencies and their power
(D) Reconstruct k-space images

Q28 Spectrograms are especially useful when:

(A) The signal is constant over time
(B) The signal is periodic
(C) The signal changes over time
(D) The signal is in image format

Q29 Which frequency-domain method enables simultaneous time and frequency analysis?

(A) Inverse DFT
(B) PSD
(C) STFT
(D) Wavelet Transform

Q30 Which biomedical application most directly relies on frequency-domain analysis for image reconstruction?

(A) Pulse oximetry
(B) ECG classification
(C) MRI
(D) EEG seizure detection

Q31 Which filter type has an infinite impulse response and can use feedback from past outputs?

(A) FIR Filter
(B) Low-Pass Filter
(C) IIR Filter
(D) Notch Filter

Q32 A filter designed to suppress 50 Hz power line noise in an ECG signal is:

(A) High-pass filter
(B) Band-pass filter
(C) Notch filter
(D) FIR filter

Q33 What is the main advantage of FIR filters over IIR filters?

(A) Lower computational cost
(B) Infinite duration
(C) Guaranteed stability
(D) Sharper cutoff

Q34 In a high-pass filter, frequencies __________ the cutoff are attenuated.

(A) Equal to
(B) Below
(C) Above
(D) None of the above

Q35 Which filter type would be most suitable for isolating alpha waves (8–12 Hz) in EEG analysis?

(A) Low-pass filter
(B) Band-pass filter
(C) High-pass filter
(D) Notch filter

Q36 Which transform uses scalable and translatable functions for time–frequency analysis?

(A) DFT
(B) Inverse DFT
(C) Wavelet Transform
(D) Convolution

Q37 What is the main drawback of STFT?

(A) Cannot process digital signals
(B) Fixed time–frequency resolution
(C) Requires high sampling rates
(D) Only works on images

Q38 Which signal is most likely to require time–frequency analysis for meaningful interpretation?

(A) A steady sine wave
(B) A static image
(C) An ECG signal with arrhythmia
(D) A barcode

Q39 Multiresolution analysis in wavelets means:

(A) Only one resolution is used
(B) Equal time and frequency resolution
(C) Variable resolution for different frequency bands
(D) Analysis is only in the time domain

Q40 In EEG analysis, which method offers better detection of transient events like seizures?

(A) Fourier Transform
(B) Wavelet Transform
(C) Histogram Analysis
(D) PCA

Short Answer

(1) Explain the difference between surface and needle electrodes in terms of invasiveness and signal quality.

(2) Why might motion artifacts present a major challenge for wearable biosignal monitors?
(3) Describe how the half-cell potential influences the electrical behavior of an electrode.
(4) How does the selection of electrode type impact clinical applications such as EEG, ECG, or EMG?
(5) Explain why high input impedance is crucial in the amplifier stage of ECG or EEG systems.
(6) Describe how gain and GBP are related, and why multi-stage amplification is beneficial.
(7) What are the advantages and disadvantages of analog vs. digital filtering?
(8) How does electrical isolation protect patient safety in biomedical devices?
(9) Why is anti-aliasing filtering essential before sampling a biosignal?
(10) Describe quantization noise and how increased bit-depth minimizes it.
(11) Explain the benefits of Delta-Sigma ADCs in EEG applications.
(12) Discuss trade-offs between sampling rate, resolution, and power in wearable devices.
(13) Why is motion artifact particularly problematic in wearable PPG devices?
(14) Explain why wavelet transform is effective for non-stationary noise removal.
(15) How does ensemble averaging improve SNR in ECG noise reduction?
(16) Describe the benefits of combining analog and digital noise removal techniques.
(17) Describe how time-domain features such as mean and variance can be used in HRV analysis.
(18) Explain the difference between fixed-window and event-based segmentation in ECG.
(19) What does a high zero-crossing rate in an EMG signal indicate?
(20) How do entropy and kurtosis support seizure detection in EEG signals?
(21) Why might frequency-domain analysis be preferred over time-domain analysis for detecting arrhythmias?
(22) How is Power Spectral Density used to analyze EEG signals?
(23) What are the advantages of using the Fast Fourier Transform in wearable devices?
(24) Describe the trade-off between time and frequency resolution in frequency-domain analysis.
(25) How does the Inverse Fourier Transform contribute to medical imaging?
(26) Why is time–frequency analysis particularly valuable for EEG and ECG signals?
(27) Compare the strengths of STFT and WT for analyzing non-stationary biomedical signals.
(28) How does the choice of window size affect time–frequency resolution in STFT?
(29) Describe how the Morlet wavelet is used in EEG analysis and why it's preferred.
(30) What are the trade-offs involved in using wavelet transform for real-time signal processing in wearable devices?

(31) What are the trade-offs between FIR and IIR filters in biomedical signal processing?
(32) Describe a real-world scenario where a notch filter would be necessary in EEG or ECG.
(33) How do low-pass filters help in smoothing and denoising biomedical signals?
(34) Explain the impact of filter order on the sharpness of a filter's frequency response.
(35) Why is filter stability a concern in IIR filter design?

8.2.3 *Section 3 Practice Problems*

Fill in the Blanks

(a) The __________ is known as the heart's natural pacemaker.
(b) The __________ valves prevent blood from flowing back into the atria.
(c) The __________ wave represents ventricular repolarization on an ECG.
(d) The __________ node introduces a delay between atrial and ventricular contractions.
(e) The __________ complex represents ventricular depolarization.
(f) The __________ triangle is formed by Leads I, II, and III.
(g) Lead __________ is commonly used for continuous rhythm monitoring.
(h) The __________ leads provide a horizontal view of the heart.
(i) A negative deflection in a lead indicates electrical activity moving _________ the positive electrode.
(j) The augmented lead __________ focuses on the electrical activity from the right arm.
(k) _______ wander can distort the ST segment and mimic ischemia.
(l) _______ filters adjust their parameters dynamically to suppress time-varying noise.
(m) Powerline interference is typically centered around _______ Hz.
(n) _______ processing changes the sampling rate for better efficiency or resolution.
(o) The _______ transform is often used for wavelet-based ECG denoising.
(p) The _______ complex represents ventricular depolarization.
(q) _______ interval prolongation can indicate risk for torsades de pointes.
(r) Wave delineation is essential for determining _______ features like QRS duration.
(s) _______ filters help emphasize large slope values in QRS detection.
(t) _______ analysis transforms the ECG into time–frequency space for noise-resilient feature extraction.
(u) ST-segment elevation in contiguous leads is characteristic of _______.
(v) A prolonged _______ interval increases the risk for torsades de pointes.
(w) _______ is a hallmark ECG feature of left bundle branch block.

Multiple Choice Questions

Q1 Which part of the heart pumps oxygenated blood into the systemic circulation?

(A) Right ventricle
(B) Left atrium
(C) Left ventricle
(D) Right atrium

Q2 The P wave in an ECG represents:

(A) Ventricular repolarization
(B) Atrial depolarization
(C) Atrial repolarization
(D) Ventricular depolarization

Q3 What is the primary function of the AV node?

(A) Initiate ventricular depolarization
(B) Contract the atria
(C) Delay impulse transmission to the ventricles
(D) Pump blood into the pulmonary artery

Q4 Which wave typically follows the QRS complex in an ECG?

(A) U wave
(B) P wave
(C) T wave
(D) PR interval

Q5 What does a prolonged QT interval on an ECG suggest?

(A) Atrial flutter
(B) Ventricular hypertrophy
(C) Risk of torsades de pointes
(D) Ischemia in the atria

Q6 Which lead configuration best assesses the inferior wall of the heart?

(A) Lead I
(B) Lead aVL
(C) Lead II
(D) Lead V5

Q7 Which lead is most useful for visualizing atrial activity and SA node function?

(A) Lead II
(B) aVF
(C) V1
(D) aVR

Q8 Which augmented lead is directed toward the left leg?

(A) aVR
(B) aVL
(C) aVF
(D) V6

Q9 Which of the following is NOT a precordial (chest) lead?

(A) V1
(B) V6
(C) aVR
(D) V4

Q10 What does the term "electrical axis" refer to in ECG interpretation?

(A) Heart rate variability
(B) Direction of atrial contraction
(C) Average direction of ventricular depolarization
(D) Number of ECG leads

Q11 Which filtering method is best for preserving waveform shape while removing baseline drift?

(A) Low-pass filter
(B) Median filter
(C) FIR high-pass filter
(D) Notch filter

Q12 Which of the following is a major source of high-frequency noise in ECG signals?

(A) Baseline wander
(B) Respiration
(C) EMG (muscle) activity
(D) Powerline interference

Q13 What is the advantage of wavelet filtering in ECG preprocessing?

(A) Removes all noise regardless of source
(B) Preserves both high and low frequency features
(C) Reduces the need for electrodes
(D) Only works offline

Q14 What is typically used as a reference signal in adaptive LMS filtering for motion artifact removal?

(A) Blood pressure
(B) Heart rate
(C) Accelerometer or impedance signal
(D) Pulse oximeter

Q15 What does an SQI (Signal Quality Index) help detect?

(A) Heart rate variability
(B) Blood oxygen saturation
(C) ECG signal noise and artifact contamination
(D) Patient stress levels

Q16 What does the PR interval represent?

(A) Time between T waves
(B) AV nodal conduction time
(C) Atrial repolarization
(D) Bundle branch conduction

Q17 Which of the following is most commonly used to detect the R-peak?

(A) PCA
(B) Pan-Tompkins algorithm
(C) FFT
(D) STFT

Q18 A prolonged QRS duration suggests:

(A) Bradycardia
(B) Ischemia
(C) Bundle branch block
(D) Atrial flutter

Q19 The area under the QRS complex provides information about:

(A) Atrial repolarization
(B) Ventricular dyssynchrony
(C) Blood pressure
(D) ST segment slope

Q20 Which transform is most suitable for analyzing changing rhythms over time?

(A) Fourier Transform
(B) PCA
(C) Wavelet Transform
(D) Linear Regression

Q21 Which ECG finding is most commonly associated with atrial fibrillation?

(A) Wide QRS complex
(B) Regular RR intervals
(C) Absence of P waves
(D) T wave inversion

Q22 Which condition is suggested by peaked T waves and widened QRS?

(A) Hypokalemia
(B) Hyperkalemia

(C) Hypocalcemia
(D) Hypercalcemia

Q23 What ECG finding is commonly used to assess CRT eligibility?

(A) QTc > 500 ms
(B) ST depression
(C) QRS $\geq$ 120 ms with LBBB
(D) T wave flattening

Q24 In myocardial infarction, a new Q wave on ECG indicates:

(A) Ischemia
(B) Ventricular hypertrophy
(C) Transmural infarction
(D) Pericarditis

Short Answer

(1) Describe the blood flow pathway through the heart, starting from the vena cava.
(2) Explain how the conduction system coordinates atrial and ventricular contractions.
(3) Why does the left ventricle have a thicker myocardium than the right ventricle?
(4) How can an ECG help diagnose myocardial infarction?
(5) What clinical conditions can alter the appearance of the T or U wave on an ECG?
(6) Explain how Einthoven's Triangle helps visualize the heart's electrical activity.
(7) Why are precordial leads essential in diagnosing anterior myocardial infarctions?
(8) Describe how Goldberger's augmented leads enhance the ECG's diagnostic utility.
(9) What does right axis deviation indicate in an ECG?
(10) When would a clinician choose to use posterior or right-sided leads?
(11) Explain why aggressive filtering might be harmful to diagnostic accuracy in ECG interpretation.
(12) Describe how baseline wander can lead to false clinical diagnoses.
(13) What preprocessing techniques are most effective in wearable ECG devices for mobile patients?
(14) Contrast real-time and offline preprocessing in terms of their strengths and use cases.
(15) How does adaptive filtering improve ECG signal clarity in dynamic environments?
(16) Why is accurate wave delineation critical in QT interval measurement?
(17) Describe how HRV is derived from ECG signals and its clinical significance.
(18) How do wearable ECG devices handle feature extraction in the presence of motion artifacts?

(19) Give two examples of nonlinear ECG features and their clinical relevance.
(20) What validation metrics are used in regulatory conformity assessments of ECG features?
(21) Explain how real-time T wave monitoring can help in early ischemia detection.
(22) What role does ECG play in managing patients with heart failure?
(23) Describe how wearable ECG systems contribute to post-discharge monitoring.
(24) How do AI models improve population-based screening using ECG data?
(25) List two ECG indicators that suggest risk for sudden cardiac death.

8.2.4 Section 4 Practice Problems

Fill in the Blanks

(a) The _______ is the main muscle responsible for inspiration.
(b) _______ is the amount of air left in the lungs after a forced expiration.
(c) According to _______ Law, increasing volume decreases pressure in a closed system.
(d) _______ is the device used to measure lung volumes like tidal volume and vital capacity.
(e) In healthy adults, tidal volume averages about _______ mL per breath.
(f) _______ is a non-invasive method for measuring chest and abdominal movements during breathing.
(g) The _______ volume remains in the lungs after maximum exhalation.
(h) _______ is a method that uses a sealed chamber to determine total lung capacity.
(i) _______ measures the concentration of carbon dioxide in exhaled breath.
(j) The most commonly used test to assess airflow limitation is _______.
(k) _______ is used to isolate specific frequency bands of respiratory signals for cleaner data interpretation.
(l) _______ analysis helps reveal dominant frequencies in breathing cycles.
(m) _______ are used to detect abnormal patterns such as apnea using capnographic or EMG data.
(n) The transformation of noisy raw respiratory data into readable, structured form is called _______.
(o) Chest impedance changes during breathing are measured using _______.
(p) _______ is the most widely used respiratory signal to evaluate CO_2 elimination.
(q) In sleep medicine, _______ is used to detect thoracoabdominal effort during apneic events.
(r) The gold standard test for diagnosing OSA is _______.
(s) AI-based telemonitoring systems often rely on trends in _______ and tidal volume.
(t) Pediatric monitoring systems must address challenges like motion artifacts and _______ breathing amplitude.

Multiple Choice Questions

Q1 Which lung volume cannot be measured directly by spirometry?

(A) Tidal Volume
(B) Inspiratory Reserve Volume
(C) Expiratory Reserve Volume
(D) Residual Volume

Q2 During quiet breathing, expiration is:

(A) Active and requires abdominal muscle contraction
(B) Passive, relying on lung recoil
(C) Caused by diaphragm contraction
(D) Driven by Boyle's law alone

Q3 A punctured pleural cavity that leads to lung collapse is called:

(A) Emphysema
(B) Pneumothorax
(C) Asthma
(D) COPD

Q4 Which of the following muscles is NOT involved in forced expiration?

(A) Abdominal muscles
(B) Internal intercostals
(C) Sternocleidomastoid
(D) Rectus abdominis

Q5 Which pressure must remain negative to prevent lung collapse?

(A) Intrapulmonary pressure
(B) Atmospheric pressure
(C) Intrapleural pressure
(D) Residual pressure

Q6 Which of the following signals reflects gas exchange efficiency?

(A) Tidal volume
(B) Pulse oximetry
(C) Airway resistance
(D) Lung compliance

Q7 What is the main limitation of spirometry?

(A) Cannot measure airflow
(B) Only usable during sleep
(C) Cannot assess residual volume
(D) Requires invasive access

Q8 Respiratory Inductance Plethysmography (RIP) is best for:

(A) Estimating alveolar CO_2
(B) Measuring intrapleural pressure
(C) Monitoring real-time breathing effort
(D) Assessing blood oxygen directly

Q9 Paradoxical breathing is typically caused by:

(A) Hyperventilation
(B) Diaphragm paralysis
(C) COPD
(D) Increased lung compliance

Q10 Which technique estimates absolute lung volumes using Boyle's Law?

(A) Spirometry
(B) RIP
(C) Capnography
(D) Body Plethysmography

Q11 Which technique is best for analyzing non-stationary respiratory signals like sleep apnea?

(A) Fourier Transform
(B) Kalman Filter
(C) Wavelet Transform
(D) PCA

Q12 What does a sudden drop in capnographic CO_2 levels most likely indicate?

(A) Hypercapnia
(B) Hypoventilation
(C) Apnea
(D) Bradypnea

Q13 Which ML model is ideal for classifying sequential respiratory data?

(A) Support Vector Machine
(B) Decision Tree
(C) Long Short-Term Memory (LSTM)
(D) K-Nearest Neighbors

Q14 A high-pass filter is useful for removing:

(A) High-frequency tremors
(B) Baseline drift
(C) Respiratory peaks
(D) Cardiac oscillations

Q15 What respiratory monitoring method uses electrodes to measure electrical impedance across the chest?

(A) Plethysmography
(B) Capnography
(C) Impedance Pneumography
(D) RIP

Q16 What is a common clinical use of $EtCO_2$ monitoring during surgery?

(A) Diagnosing COPD
(B) Detecting airway obstruction
(C) Measuring oxygen saturation
(D) Evaluating cardiac rhythm

Q17 Which signal is most useful in diagnosing central sleep apnea?

(A) Nasal airflow
(B) SpO_2
(C) Thoracoabdominal movement
(D) Respiratory effort (RIP)

Q18 Respiratory signal analysis during PACU monitoring helps detect:

(A) Sleep apnea
(B) Anemia
(C) Opioid-induced hypoventilation
(D) Deep vein thrombosis

Q19 What technology enables continuous respiratory monitoring in home care?

(A) Spirometry
(B) Capnography
(C) AI-enhanced wearable sensors
(D) Invasive ventilators

Q20 Which of the following is NOT typically analyzed in PSG for OSA diagnosis?

(A) Capnography
(B) Nasal airflow
(C) ECG
(D) Chest X-ray

Short Answer

(1) Describe how Boyle's Law explains airflow during the breathing cycle.
(2) A patient has difficulty exhaling forcefully. Which lung volume is likely affected?
(3) How does the respiratory system respond to increased CO_2 levels in the blood?
(4) Explain the role of spirometry in diagnosing obstructive vs. restrictive lung diseases.

(5) Why does residual volume prevent lung collapse?
(6) Describe the types of respiratory signals and what each reveals about lung function.
(7) Why is body plethysmography superior to spirometry in measuring residual volume?
(8) A patient shows abnormal chest-abdomen movement patterns. What condition could be suspected using RIP?
(9) How does capnography help in detecting airway obstruction or hypoventilation?
(10) A COPD patient has elevated airway resistance. How can this be quantified using plethysmography?
(11) Explain how filtering improves the quality of respiratory signals in wearable sensors.
(12) What are the differences between time-domain and frequency-domain analysis in signal processing?
(13) How can machine learning assist in early detection of respiratory distress from breathing patterns?
(14) Describe the use of capnography in identifying respiratory abnormalities like apnea.
(15) What is the role of adaptive filtering in respiratory signal processing during motion?
(16) How does capnography support early detection of respiratory failure in critical care?
(17) Describe how wearable sensors contribute to OSA monitoring at home.
(18) A post-op patient shows a sudden drop in $EtCO_2$. What might this indicate?
(19) How do respiratory signals help personalize CPAP therapy for OSA?
(20) What are the benefits of using AI in respiratory signal analysis for chronic disease management?

8.2.5 Section 5 Practice Problems

Fill in the Blanks

(a) The _______ lobe is primarily responsible for decision-making and voluntary motor control.
(b) Saltatory conduction occurs in _______ axons and increases conduction velocity.
(c) The main inhibitory neurotransmitter in the CNS is _______.
(d) The _______ regulates hunger, thirst, and temperature through hormonal control.
(e) Action potentials are initiated at the _______ of the neuron.
(f) EEG signals primarily reflect synchronized activity of __________ neurons.
(g) The _________ frequency band is typically observed during deep sleep.
(h) _________ is a standardized system used for EEG electrode placement.

(i) Event-related potentials (ERPs) are extracted by _________ EEG epochs across trials.
(j) In EEG systems, high ________ ensures diagnostically relevant waveforms are preserved.
(k) EEG preprocessing enhances the ____________ ratio to improve signal clarity.
(l) Eye movements introduce __________ artifacts, especially over frontal channels.
(m) High-frequency muscle noise is known as ________ contamination.
(n) __________ is a blind source separation technique often used to isolate EEG artifacts.
(o) Adaptive filters can dynamically respond to changes in __________ conditions.
(p) ___________ features capture the amplitude dynamics of EEG without spectral transformation.
(q) Power spectral density is used to compute _________ features.
(r) The __________ transform provides a variable time–frequency resolution using scalable wavelets.
(s) _________ entropy estimates EEG signal unpredictability in the frequency domain.
(t) A high _________ exponent indicates chaotic EEG dynamics.
(u) EEG is the primary tool for detecting _________ discharges in epilepsy.
(v) The _________ index is used during anesthesia to monitor sedation depth.
(w) In ADHD, a common EEG finding is an elevated _________ ratio.
(x) P300-based BCIs rely on the detection of a time-locked brainwave following a ________ stimulus.

Multiple Choice Questions

Q1 Which structure relays nearly all sensory information to the cortex?

(A) Medulla
(B) Hippocampus
(C) Thalamus
(D) Amygdala

Q2 Which glial cell is primarily responsible for myelinating axons in the CNS?

(A) Schwann cell
(B) Microglia
(C) Astrocyte
(D) Oligodendrocyte

Q3 The hippocampus is most associated with:

(A) Vision
(B) Motor coordination
(C) Memory formation
(D) Autonomic regulation

Q4 In an action potential, what ion is primarily responsible for depolarization?

(A) K^+
(B) Cl^-
(C) Na^+
(D) Ca^{2+}

Q5 Which structure produces cerebrospinal fluid (CSF)?

(A) Meninges
(B) Hypothalamus
(C) Choroid plexus
(D) Corpus callosum

Q6 What EEG feature is most associated with cognitive processing and attention?

(A) Alpha waves
(B) Gamma waves
(C) Delta waves
(D) Theta waves

Q7 Which type of electrode requires no conductive gel?

(A) Wet Ag/AgCl
(B) Dry electrode
(C) Invasive electrode
(D) ECG electrode

Q8 Which of the following is NOT a physiological artifact?

(A) Eye blink
(B) Power line interference
(C) Jaw movement
(D) Heartbeat

Q9 Which ERP component is often used in brain-computer interfaces for detecting target stimuli?

(A) N400
(B) N170
(C) P300
(D) K-complex

Q10 What is the minimum sampling rate for analyzing signals up to 100 Hz?

(A) 50 Hz
(B) 100 Hz
(C) 200 Hz
(D) 1000 Hz

Q11 What EEG preprocessing method is commonly used to remove blink and muscle artifacts?

(A) PCA
(B) ICA
(C) FFT
(D) ERP

Q12 What type of artifact does a 60 Hz notch filter typically remove?

(A) EMG
(B) Eye blinks
(C) Power line interference
(D) Cardiac signal

Q13 Which technique decomposes EEG signals into time–frequency components?

(A) ASR
(B) PCA
(C) Wavelet transform
(D) Regression

Q14 What is a benefit of Artifact Subspace Reconstruction (ASR)?

(A) High-density EEG layout
(B) Hardware-based signal acquisition
(C) Real-time noise removal without signal loss
(D) Detection of seizure onset zones

Q15 In EEG referencing, which method enhances spatial localization by subtracting nearby signals?

(A) Linked ears
(B) Average reference
(C) Laplacian
(D) Ground reference

Q16 Which feature type is best for detecting transient changes in EEG oscillations?

(A) Time-domain
(B) Frequency-domain
(C) Time–frequency
(D) Nonlinear

Q17 Which metric is typically used in EEG-based ADHD detection?

(A) Alpha peak frequency
(B) Theta/Beta ratio
(C) Spectral slope
(D) Kurtosis

Q18 What does a positive Lyapunov exponent indicate in EEG signals?

(A) Sleep state
(B) Regular periodicity
(C) Chaotic dynamics
(D) Muscle artifacts

Q19 Hjorth's "mobility" parameter estimates which EEG characteristic?

(A) Entropy
(B) Mean frequency
(C) Peak amplitude
(D) Fractal dimension

Q20 Which transform adapts frequency resolution based on scale and is ideal for EEG?

(A) FFT
(B) STFT
(C) DCT
(D) Wavelet Transform

Q21 Which EEG pattern is most characteristic of Alzheimer's disease?

(A) Sleep spindles
(B) Triphasic waves
(C) Posterior alpha attenuation
(D) P300 burst

Q22 What EEG-based index is typically used during general anesthesia?

(A) SMR
(B) BIS
(C) P300
(D) SSVEP

Q23 Which of the following disorders often shows left frontal alpha asymmetry?

(A) Epilepsy
(B) Depression
(C) Narcolepsy
(D) ADHD

Q24 What does SSVEP stand for in the context of BCIs?

(A) Short-range Somatosensory Visual Encoding Potential
(B) Steady-State Visual Evoked Potential
(C) Slow Stimulus Visual Evoked Pattern
(D) Signal Stimulus Visual Event Potential

Q25 In sleep studies, REM onset is identified in EEG by:

(A) K-complexes
(B) Delta bursts
(C) Low-voltage mixed-frequency activity
(D) Triphasic waves

Short Answer

(1) Explain how neurotransmitter release occurs at the synapse and name two major types.
(2) A patient with Parkinson's disease has reduced dopamine in the substantia nigra. What motor symptoms might occur and why?
(3) What would happen to action potential propagation in a demyelinated neuron?
(4) Describe the role of astrocytes in maintaining the extracellular environment and BBB.
(5) How does long-term potentiation contribute to learning and memory?
(6) Why is differential amplification critical in EEG acquisition?
(7) Describe one clinical use of resting-state EEG and one of event-related EEG.
(8) A patient's EEG shows elevated theta activity during wakefulness. What might this indicate?
(9) How does dry electrode impedance compare to wet electrodes, and what are the implications for signal quality?
(10) Explain how motion artifacts can affect EEG and list two strategies to mitigate them.
(11) Why is artifact removal essential before applying machine learning models to EEG data?
(12) Describe the advantage of combining ICA with ASR in artifact removal pipelines.
(13) A user moves frequently during wearable EEG recording. What artifacts might arise, and how would you address them?
(14) How can referencing strategy impact the interpretation of ERP components?
(15) What are the risks of over-filtering EEG data, and how can they be avoided?
(16) Describe one advantage and one limitation of using FFT for EEG feature extraction.
(17) In what situations might you prefer nonlinear features like entropy or fractal dimension over spectral features?
(18) A BCI system is being designed for real-time control. What feature types would you prioritize and why?
(19) How does spectral entropy differ from power spectral density?
(20) Explain how alpha asymmetry might relate to mood disorders such as depression.
(21) Explain how EEG can help differentiate between drug-induced coma and true brain death.

(22) Describe a clinical use of continuous EEG in the ICU.
(23) A child shows high theta and low beta activity in frontal regions. What condition might be suspected?
(24) What are the benefits of using video-EEG monitoring in epilepsy diagnosis?
(25) How do EEG-guided neurofeedback protocols aid stroke rehabilitation?

8.2.6 *Section 6 Practice Problems*

Fill in the Blanks

(a) The core component of an ultrasound transducer is the __________ crystal.
(b) High-frequency ultrasound provides better __________ but poorer penetration.
(c) The __________ effect explains how crystals generate and detect sound waves.
(d) Doppler ultrasound assesses __________ using frequency shifts.
(e) Attenuation is higher in __________ frequency waves.
(f) The __________ effect allows transducers to generate and receive ultrasound signals.
(g) __________ transducers are ideal for musculoskeletal and thyroid imaging.
(h) The __________-shaped beam from convex transducers offers wide coverage and deep penetration.
(i) The __________ principle is the foundation of diagnostic ultrasound imaging.
(j) The __________ effect is used in Doppler ultrasound to measure blood flow velocity.
(k) Ultrasound images are typically affected by __________ noise, which gives them a grainy appearance.
(l) The __________ operator uses gradient values to detect edges in ultrasound images.
(m) __________ segmentation methods classify pixels by comparing their intensities to a fixed value.
(n) __________ filtering considers pixel similarity across the entire image rather than just neighboring pixels.
(o) The __________ edge detector includes steps like Gaussian smoothing, non-maximum suppression, and hysteresis thresholding.
(p) __________ echocardiography provides high-resolution heart imaging by placing a probe in the esophagus.
(q) In the __________ trimester, ultrasound is used to perform a detailed anatomical scan of the fetus.
(r) __________ ultrasound is the primary tool for diagnosing gallstones and gallbladder inflammation.
(s) A __________ profile uses ultrasound and non-stress testing to evaluate fetal health.
(t) __________ echocardiography uses microbubbles to enhance image quality in difficult-to-visualize patients.

Multiple Choice Questions

Q1 Which property most directly affects the amount of ultrasound reflection at a tissue boundary?

(A) Tissue color
(B) Acoustic impedance difference
(C) Water content
(D) Patient age

Q2 What does the 4th dimension represent in 4D ultrasound?

(A) Depth
(B) Frequency
(C) Time
(D) Contrast

Q3 Which type of tissue causes the strongest ultrasound reflection?

(A) Fat
(B) Muscle
(C) Bone
(D) Liver

Q4 What determines the speed of ultrasound in a medium?

(A) Shape and size of organ
(B) Electric conductivity
(C) Density and elasticity of the medium
(D) Patient posture

Q5 Which imaging mode is used to visualize blood flow?

(A) M-mode
(B) 3D mode
(C) Doppler mode
(D) Elastography

Q6 Which transducer is best suited for imaging deep abdominal organs?

(A) Linear
(B) Phased array
(C) Convex
(D) Pencil

Q7 What does the Doppler shift measure in clinical ultrasound?

(A) Tissue density
(B) Electrical impedance
(C) Blood flow velocity
(D) Transducer frequency

Q8 What is a main limitation of linear transducers?

(A) High noise sensitivity
(B) Poor image resolution
(C) Shallow penetration depth
(D) Lack of real-time imaging

Q9 Which Doppler mode is best for detecting low-velocity blood flow?

(A) Color Doppler
(B) Spectral Doppler
(C) Power Doppler
(D) Continuous-wave Doppler

Q10 What component is essential for emitting and receiving ultrasound in a transducer?

(A) Diode
(B) Magnet
(C) Piezoelectric crystal
(D) Capacitor

Q11 What is the primary cause of speckle noise in ultrasound?

(A) Power fluctuations in the transducer
(B) Beam misalignment
(C) Interference from reflected wavefronts
(D) Electromagnetic interference

Q12 Which filtering method best preserves edges while reducing random noise?

(A) Median filter
(B) Bilateral filter
(C) Mean filter
(D) Sobel operator

Q13 What advantage does the Canny edge detector have over Sobel and Prewitt?

(A) Lower computational complexity
(B) Better detection in homogeneous regions
(C) Multi-step processing for improved precision
(D) Works only on high-resolution images

Q14 Which of the following segmentation techniques uses both gradient and smoothness constraints to evolve a contour?

(A) Thresholding
(B) Region Growing
(C) Active Contours (Snakes)
(D) Sobel Filter

Q15 CNN-based segmentation is best described as:

(A) Rule-based segmentation
(B) Data-driven learning of features
(C) Edge-only segmentation
(D) Manual region drawing

Q16 Which type of ultrasound is best for evaluating valvular vegetations in suspected endocarditis?

(A) Transthoracic echocardiography
(B) Transesophageal echocardiography
(C) Stress echocardiography
(D) Contrast echocardiography

Q17 What fetal measurement is used in the first trimester to determine gestational age?

(A) Biparietal diameter
(B) Abdominal circumference
(C) Crown-rump length
(D) Femur length

Q18 Which condition is most accurately diagnosed using vascular ultrasound of the lower extremities?

(A) Aortic aneurysm
(B) Deep vein thrombosis (DVT)
(C) Peripheral neuropathy
(D) Stroke

Q19 In lung ultrasound, the presence of multiple B-lines is most indicative of:

(A) Pneumothorax
(B) Pulmonary embolism
(C) Pulmonary edema
(D) Normal lung

Q20 What is the primary use of musculoskeletal ultrasound?

(A) Imaging brain structures
(B) Evaluating liver fibrosis
(C) Visualizing tendons and joints
(D) Assessing fetal heart tones

Short Answer

(1) Explain how the principle of acoustic impedance mismatch is used to generate contrast in ultrasound images.
(2) A radiologist switches to a lower frequency probe. What trade-off are they likely making, and why?

(3) Describe how Doppler ultrasound measures blood flow direction and velocity.
(4) Why is ultrasound gel used during imaging procedures?
(5) A fetus's facial movement is being studied in real time. What type of ultrasound is most appropriate and why?
(6) Compare linear and convex transducers in terms of frequency, resolution, and application.
(7) Why is beamforming critical for achieving high-resolution ultrasound images?
(8) How does the angle between the ultrasound beam and blood flow direction affect Doppler accuracy?
(9) A physician wants to image the fetal heart and umbilical artery. Which transducer and Doppler mode should they use?
(10) Explain why low-frequency ultrasound waves are better for deep imaging but offer less detail.
(11) Explain why ultrasound image preprocessing must strike a balance between noise reduction and edge preservation.
(12) Compare the performance of Gaussian filtering and wavelet-based denoising for speckle noise reduction.
(13) A radiologist is unable to clearly delineate the boundary between a liver cyst and surrounding tissue in a grayscale image. What edge detection method would you recommend and why?
(14) How would you segment a tumor with poorly defined edges in a speckled ultrasound image?
(15) Describe how U-Net architecture is used in deep learning for ultrasound segmentation.
(16) A patient with obesity and poor transthoracic imaging windows needs valve evaluation. What ultrasound modality would you recommend and why?
(17) Describe two key differences between first and second-trimester fetal ultrasounds in terms of clinical goals.
(18) A critically ill patient presents with respiratory distress. What ultrasound findings might suggest pulmonary edema?
(19) Explain how contrast echocardiography helps in identifying a left ventricular thrombus.
(20) A young athlete complains of shoulder pain. How can musculoskeletal ultrasound assist in diagnosis and management?

8.2.7 Section 7 Practice Problems

Fill in the Blanks

(A) The __________ pressure is the maximum pressure during ventricular contraction.
(B) Mean arterial pressure is essential for ensuring adequate __________ to organs.
(C) Korotkoff sounds are used in the __________ method of blood pressure measurement.

(D) The __________ method uses oscillations in the cuff to determine blood pressure values automatically.
(E) __________ blood pressure monitoring is used to detect variations throughout the day, especially during normal activity.

Multiple Choice Questions

Q1 What is the formula commonly used to estimate MAP using systolic and diastolic pressures?

(A) MAP = (SP + DP)/2
(B) MAP = SP – DP
(C) MAP = DP + 1/3(SP – DP)
(D) MAP = SP + DP – PP

Q2 Which of the following is not an advantage of oscillometric devices?

(A) Operator-independent readings
(B) Continuous real-time waveform
(C) Easy self-monitoring at home
(D) Reduced observer bias

Q3 In the auscultatory method, which Korotkoff phase is typically used to determine diastolic pressure?

(A) Phase I
(B) Phase II
(C) Phase III
(D) Phase V

Q4 What is a major drawback of ambulatory BP monitoring during physical activity?

(A) It underestimates SP
(B) It overestimates DP
(C) It often gives error codes
(D) It requires arterial catheterization

Q5 What is the gold standard for measuring arterial blood pressure?

(A) Oscillometric measurement
(B) Auscultatory method
(C) Direct intra-arterial measurement
(D) Palpation method

Short Answer

(1) Explain why cuff size is important in indirect BP measurement.
(2) A patient's clinic BP readings are consistently high, but their home readings are normal. What condition might they have, and how can it be confirmed?

(3) Describe one advantage and one limitation of using the oscillometric method for blood pressure measurement.
(4) When is invasive blood pressure monitoring preferred over noninvasive methods?
(5) How does white-coat hypertension differ from masked hypertension, and what tools help distinguish between them?

MIX
Papier aus verantwortungsvollen Quellen
Paper from responsible sources
FSC® C105338

If you have any concerns about our products,
you can contact us on
ProductSafety@springernature.com

In case Publisher is established outside the EU,
the EU authorized representative is:
Springer Nature Customer Service Center GmbH
Europaplatz 3, 69115 Heidelberg, Germany

Printed by Libri Plureos GmbH
in Hamburg, Germany